(2.6) $(x + y)^3 = x^3 + 3x^2y + 3xy^2 + y^3$

(2.7) $(x - y)^3 = x^3 - 3x^2y + 3xy^2 - y^3$

(2.8) $(x + y)(x^2 - xy + y^2) = x^3 + y^3$

(2.9) $(x - y)(x^2 + xy + y^2) = x^3 - y^3$

(2.10) $(x + y + z)^2 = x^2 + y^2 + z^2 + 2xy + 2xz + 2yz$

(4.1) If $f(x)$, $g(x)$, and $h(x)$ are expressions, then $f(x) = g(x)$ and $f(x) + h(x) = g(x) + h(x)$ are equivalent equations

(4.2) If k is a nonzero constant, then $f(x) = g(x)$ and $k \cdot f(x) = k \cdot g(x)$ are equivalent equations

(4.3) If $k(x)$ is a nonzero expression, then $f(x) = g(x)$ and $k(x) \cdot f(x) = k(x) \cdot g(x)$ are equivalent equations

(5.7) $(a/b)^n = a^n/b^n;\ b \neq 0$

(5.8) $a^{-t} = 1/a^t$

(5.9) $a^{1/k} = \sqrt[k]{a}$

(5.10) $b^{j/k} = (b^j)^{1/k} = \sqrt[k]{b^j} = (b^{1/k})^j = (\sqrt[k]{b})^j;\ \sqrt[k]{b}$ real

(5.11) $\sqrt[k]{ab} = \sqrt[k]{a}\,\sqrt[k]{b};\ a \geq 0,\ b \geq 0$

(5.12) $\sqrt[k]{a/b} = \sqrt[k]{a}/\sqrt[k]{b};\ a \geq 0,\ b > 0$

(5.13) $\sqrt[k]{\sqrt[j]{b}} = \sqrt[kj]{b}$

(6.4) Roots of quadratic equation: $(-b \pm \sqrt{b^2 - 4ac})/2a$

(6.5) Sum of roots: $r + s = -b/a$

(6.6) Product of roots: $rs = c/a$

(6.7) $ax^2 + bx + c = a(x - r)(x - s)$

(8.1) Linear equations are independent if and only if $A/a \neq B/b$

(8.2) Linear equations are inconsistent if and only if $A/a = B/b \neq C/c$

(8.3) Linear equations are dependent if and only if $A/a = B/b = C/c$

(8.4) Cramer's rule: $x = \dfrac{D_x}{D},\ y = \dfrac{D_y}{D},\ z = \dfrac{D_z}{D}$

(continued on back endpaper)

intermediate
algebra

Intermediate Algebra

FIFTH EDITION

PAUL K. REES
Louisiana State University

FRED W. SPARKS
Texas Technological University

CHARLES SPARKS REES
The University of New Orleans

McGraw-Hill Book Company
New York   St. Louis   San Francisco   Auckland
Bogotá   Düsseldorf   Johannesburg   London   Madrid
Mexico   Montreal   New Delhi   Panama   Paris
São Paulo   Singapore   Sydney   Tokyo   Toronto

INTERMEDIATE ALGEBRA

1 2 3 4 5 6 7 8 9 0 K P K P 7 8 3 2 1 0 9 8

This book was set in Caledonia by Black Dot, Inc.
The editors were A. Anthony Arthur and Shelly Levine Langman;
the designer was Elliot Epstein;
the production supervisor was Dominick Petrellese.
The drawings were done by J & R Services, Inc.
Kingsport Press, Inc., was printer and binder.

Library of Congress Cataloging in Publication Data

Rees, Paul Klein, date
    Intermediate algebra.

    1. Algebra. I. Sparks, Fred Winchell, date joint author. II. Rees, Charles Sparks, joint author. III. Title.
QA152.R34   1978   512'.9'042   77-22409
ISBN 0-07-051731-2

contents

3
RATIONAL EXPRESSIONS 61

4
LINEAR AND FRACTIONAL EQUATIONS 78

5
EXPONENTS, ROOTS, AND RADICALS 102

6
QUADRATIC EQUATIONS   123

7
RELATIONS, FUNCTIONS, AND GRAPHS   149

14
INEQUALITIES AND SYSTEMS OF INEQUALITIES   283

preface

In writing the manuscript for the fifth edition of *Intermediate Algebra*, we have carefully considered which topics should be treated and which should be deleted from earlier editions. We have also given considerable thought to how to present the topics.

The axioms for the real number system are now all in Chapter 1. We decided to omit the chapter on polynomial equations, since that is a topic ordinarily treated in college algebra. We also decided to put all the work on inequalities in one chapter, whereas it has been treated in parts of several chapters in earlier editions. We did this so as to have an uninterrupted treatment of equalities, and to make it easier to concentrate on inequalities.

The entire book has been gone over carefully, with readability constantly in mind. We have rewritten some sections, added illustrative examples, and given more explanatory notes and more problems, including simple ones. The topics that have received a major overhaul include the set of real numbers (Sec. 1.3); axioms for the real number system (Sec. 1.4); fundamental operations on fractions (Sec. 1.6); algebraic expressions (Sec. 2.1); factors of a quadratic trinomial (Sec. 2.11); linear equations (Sec. 4.4); complex numbers (Sec. 6.3); solution by completing the square (Sec. 6.4); addition of radicals (Sec. 6.7); relations (Sec. 7.1); functions (Sec. 7.2); the inverse (Sec. 7.8); variation (Sec. 10.3) to include the change from the British to the metric system; arithmetic series (Sec. 11.1); geometric series (Sec. 11.2); the binomial formula (Sec. 12.1); scientific notation and approximations (Sec. 13.1); logarithms (Secs. 13.7 and 13.8) to include the use of calculators; the graph of $y = \log_b x$ and of $y = b^x$ (Sec. 13.12); and all of Chapter 14 on inequalities and linear programming.

We have increased the number of problems by nearly 20 percent to about 3400. There are a summary and a review exercise at the end of each chapter that has more than two exercises. The exercises are still a normal lesson apart, and the problems are still in groups of four similar ones except for the review exercises.

We are grateful to those users who have sent comments to us. The book is a better one than it would have been without the suggestions contained in those comments.

The original authors want to take this opportunity to welcome Charles Sparks Rees as a coauthor. He is the son of one of us and the namesake of the other.

Paul K. Rees
Fred W. Sparks
Charles Sparks Rees

1
the real
number system

If you open a technical journal or turn to the science section of a popular magazine, you will probably find a statement, a formula, or an equation expressed in numerals, letters, and other mathematical symbols. Some knowledge of algebra is necessary for understanding such statements. In this text we present topics of algebra necessary for further progress in mathematics and in fields that use mathematics.

The basis for our discussion consists of (1) a few undefined terms, (2) definitions of other terms and operations, and (3) a set of axioms. The axioms are the rules that we agree to follow. From these we develop the properties of numbers and operations and the rules of procedure.

The real number system is used in the first few chapters of this text, and the concept and terminology of sets is used throughout the book. We discuss these in this chapter.

1.1 SETS

One of the basic and useful concepts of mathematics is denoted by the word *set*. This word is used every day in such phrases as "a set of dishes," "a croquet set," "a set of drawing instruments," and in other expressions referring to a *collection* of objects. We assume that the reader is familiar with the word *collection* and define a set as follows:

Set A *set* is a collection of well-defined objects called *elements*.

By "well defined" we mean that there exists a criterion that enables us to make one of the following decisions about an object, or an element, that we shall designate by a:

1 a belongs to the set.
2 a does not belong to the set.

If a is an element of the set S, we say that a *belongs to* S and express the statement by the notation $a \in S$. The notation $a \notin S$ means that a *does not* belong to S. Thus $a \in \{a, b\}$, $a \in \{a\}$, but $a \notin \{b, c\}$.

We now list three sets of well-defined objects. We designate each set by the letter S and state the criterion that defines an element of S.

- S is the football squad of Trinity College. The criterion that determines the membership of the squad is the list of names selected by the coach.
- S is the herd of sheep in the south pasture of the Bar X ranch. If an animal is a sheep and is in the specified pasture, it is an element of the set. A sheep not in the specified pasture is not an element of the set, and an animal that is not a sheep is not in the set.
- S is the set of counting numbers less than 7 that are divisible by 2. The description of the set establishes the criterion. The numbers 2, 4, and 6 are the elements of S since each is divisible by 2 and is less than 7.

Listing Method

Descriptive Method

As implied above, we frequently use a capital letter to stand for a set. A set is described in two ways. In one, we list the elements of the set, such as letters, numerals, or the names of objects, and enclose the list in braces { }. In the other, we enclose a descriptive phrase in braces and understand that the elements of the set are those, and only those, that satisfy the description. For example, if W is the set of the names of the days of a week, then we may designate W by the listing method as

$W = \{$Sunday, Monday, Tuesday, Wednesday, Thursday, Friday, Saturday$\}$

or we may use the descriptive method by writing

$W = \{x \mid x$ is the name of a day of a week$\}$

The vertical line $\mid$ is read "such that."

Equality of Sets

Two sets are equal if every element of each set is an element of the other set.

The relation of equality does not require that the elements of the sets be arranged in the same order. For example,

$$\{s, t, a, r\} = \{r, a, t, s\} = \{t, a, r, s\}$$

In the sets $A = \{a, b, c, d, e\}$ and $B = \{a, c, e\}$, each element of B is an element of A. This illustrates the following definition:

Subset

Proper Subset

If each element of set B is an element of a set A, then B is a *subset* of A. Furthermore, if each element of B is an element of A but there are elements in A that are not elements of B, then B is a *proper subset* of A.

We use the notation $B \subseteq A$ to indicate that B is a subset of A, and $B \subset A$ to denote that B is a proper subset of A.

Example 1 If $A = \{1, 2, 3, 4, 5\}$, $B = \{1, 2, 3, 5\}$, and $C = \{3, 5, 1\}$, then $B \subseteq A$ and $B \subset A$; also $C \subseteq B$.

Example 2 If $T = \{x \mid x$ is a member of a football squad$\}$ and $S = \{x \mid x$ is a member of the squad who plays end$\}$, then $S \subset T$.

Now if $B \subseteq A$ and $A \subseteq B$, it follows that each element of B belongs to A and each element of A belongs to B. Hence, $A = B$. Therefore the definition of equality of A and B can be stated in this way:

If $B \subseteq A$ and $A \subseteq B$, then $A = B$ (1.1)

It may happen that a subset of A is also a subset of B. For example, if $A = \{1, 2, 3, 4, 5, 6\}$, and $B = \{2, 4, 6, 8, 10\}$, then the set $\{2, 4, 6\}$ is a subset of A and of B. This set is called the *intersection of A and B* and illustrates the following definition.

Intersection of Two Sets

The *intersection* of the sets A and B is designated by $A \cap B$ and consists of all elements of A that also belong to B. The notation $A \cap B$ is read "A cap B," or "the intersection of A and B."

This definition may also be stated as follows:

$$A \cap B = \{x \mid x \in A \text{ and } x \in B\}$$

Example 3 If $A = \{a, c, e, g\}$ and $B = \{c, a, f, e\}$, then $A \cap B = \{a, c, e\}$.

Example 4 If $A = \{x \mid x$ is an alderman of Alton$\}$ and $B = \{x \mid x$ is a member of the Alton Kiwanis Club}, then $A \cap B = \{x \mid x$ is an alderman and a Kiwanian in Alton}.

If in Example 4, no alderman is a Kiwanian, then $A \cap B$ contains no elements and illustrates the following definition:

Empty or Null Set The *empty* or *null* set is designated by $\emptyset$, and is the set that contains no elements.

Other examples of the null set are

1 $\{x \mid x$ is a woman who has been president of the United States}
2 $\{x \mid x$ is a two-digit positive integer less than 10}
3 $\{x \mid x$ is a former governor of California} $\cap \{x \mid x$ is a former governor of Texas}

Disjoint Sets If $S \cap T = \emptyset$, then the sets S and T are *disjoint* sets. (1.2)

The totality of elements that are involved in any specific situation is called the *universal set*, and is designated by the capital letter U. For example, the states in the United States are frequently classified into sets, such as the New England states, the Midwestern states, the Southern states, and in several other ways. Each of these sets is a subset of the universal set, which, in this example, is composed of all the states in the United States.

Another concept associated with the theory of sets is the complement of one set with respect to another. As an example, if $A = \{x \mid x$ is a student of a given college} and $B = \{x \mid x$ is on the football squad of the college}, then the complement of B with respect to A is $C = \{x \mid x$ is a student of the college who is not on the football squad}. This illustrates the following definition:

Complement of a Set The *complement* of the set B with respect to A is designated by $A - B$, and $A - B = \{x \mid x \in A$ and $x \notin B\}$.

The complement of the set B with respect to the universal set U is $B' = U - B$. Accordingly, we may write $A - B$ as $A \cap B'$.

As a second example, we consider the sets $T = \{x \mid x$ is a female student in college $C\}$ and $S = \{x \mid x$ is a member of the senior class of $C\}$, then $T - S = \{x \mid x$ is a female student not classified as a senior}. Note that, in this case, S is not a subset of T.

A method for picturing sets and certain relations between them was devised by the Englishman John Venn (1834-1923), who

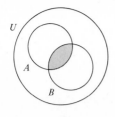 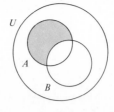 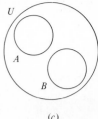

(a)   (b)   (c)

FIGURE 1.1   The shaded area is $A \cap B$ The shaded area is $A - B$ $A \cap B = \phi$

used a simple plane figure to represent a set. We shall illustrate the method by the use of circles and shall define the universal set U as all points within and on the circumference of a circle C. The various subsets of U will be represented by circles wholly within the circle C. In Fig. 1.1 we show the *Venn diagrams* for $A \cap B$, $A - B$, and the situation in which $A \cap B = \varnothing$.

Venn Diagrams

If $S = \{1, 2, 3, 4, 5, 6\}$ and $T = \{2, 4, 6, 8, 10\}$, the elements 1, 3, and 5 belong to S but not to T; the elements 8 and 10 belong to T but not to S; and the elements 2, 4, and 6 belong to both S and T. Hence the elements of $V = \{1, 2, 3, 4, 5, 6, 8, 10\}$ are in S or in T or are in both S and T. The set V is called the *union of the sets S and T* and illustrates the following definition:

Union of
Two Sets

The *union* of the sets S and T is designated by $S \cup T$ and is the set of all elements x such that $x \in S$ or $x \in T$ or $x \in$ both S and T. The notation $S \cup T$ is read "S cup T" or "S union T."

Figure 1.2 shows the Venn diagram for the union of two sets. The following examples are illustrations of the union and the intersection of two sets and of the complement of one set relative to another.

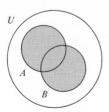

FIGURE 1.2   The shaded area is $A \cup B$

Example 5   If $A = \{m, r, t\}$ and $B = \{r, t, s\}$, then

$$A \cup B = \{m, r, t, s\}$$
$$A \cap B = \{r, t\}$$
$$A - B = \{m\}$$
$$B - A = \{s\}$$

Example 6   If $A = \{1, 3, 5\}$, $B = \{2, 4, 6, 7\}$, and $C = \{3, 4, 5\}$, then

$$A \cup B = \{1, 2, 3, 4, 5, 6, 7\}$$
$$A \cap B = \varnothing$$
$$A - B = A$$
$$A \cup C = \{1, 3, 4, 5\}$$
$$A - C = \{1\}$$

Example 7   Using a Venn diagram, prove that

$$T \cup (S - T) = S \cup T$$

Solution   We represent S and T by the circles in Fig. 1.3.   Now $S - T$ is the set of all points in the shaded region bounded by the two circles. Furthermore, all points in the shaded region together with all points in T constitute the set of all points in the region bounded by the two circles.   Therefore, $T \cup (S - T) = S \cup T$.

FIGURE 1.3

EXERCISE 1.1   Operations on Sets

1 If $A = \{2, 3, 4, 7, 8\}$ and $B = \{2, 4, 8\}$, find $A \cup B$, $A \cap B$, and $A - B$.

2 If $A = \varnothing$ and $B = \{1, 3, 5, 7, 9\}$, find $A \cup B$, $A \cap B$, and $A - B$.

3 If $A = \{x \,|\, x$ is a member of the marching band$\}$ and $B = \{x \,|\, x$ plays a trumpet in the marching band$\}$, find $A \cup B$, $A \cap B$, and $A - B$.

4 If $W = \{a, c, e, g, h\}$ and $P = \{b, c, d, e, f\}$, find $W \cup P$, $W \cap P$, and $W - P$.

5 If $C = \{x \,|\, x$ is in a French class$\}$ and $D = \{x \,|\, x$ is in an algebra class$\}$, find $C \cup D$, $C \cap D$, and $C - D$.

6 If $M = \{x \,|\, x$ has black hair$\}$ and $N = \{x \,|\, x$ likes strawberry ice cream$\}$, find $M \cup N$, $M \cap N$, and $M - N$.

7 If $A = \{x \,|\, x$ is a college student who has red hair$\}$ and $B = \{x \,|\, x$ is a coed who has red hair$\}$, find $A \cup B$, $A \cap B$, and $A - B$.

8 If $A = \{x \,|\, x$ is a senator who weighs 170 lb or more$\}$ and $\{B = x \,|\, x$ is a senator who weighs less than 170 lb$\}$, find $A \cup B$, $A \cap B$, and $A - B$.

The following sets are used in Probs 9 to 12:

$A = \{x \,|\, x$ is a counting number less than, or equal to, 11$\}$

$B = \{x \,|\, x \in A$ and x is not divisible by 2$\}$

$C = \{x \,|\, x \in A$ and x is divisible by 3$\}$

$D = \{x \,|\, x \in A$ and x is not divisible by 3$\}$

$E = \{3, 5, 7, 9\}$

9 Find $A \cup C \cup D \cup E$, $A \cap B \cap D$, and $A \cap D \cap E$.

10 Find $(A - C) \cup (D - E)$ and $(A \cap C) - (B \cup D)$.

11 Find $(A \cup B) - (B \cap C)$.

12 Prove that $C \cup (D \cap E) = (C \cup D) \cap (C \cup E)$.

Using a Venn diagram, show that each of the following statements is true.

13 $(S - T) \subseteq S$

14 $T \cup (T - S) = T$

15 $T \cap (T - S) = T - S$

16 $T \cap (T \cup S) = T$

17 $A \cap (B \cup C) = (A \cap B) \cup (A \cap C)$

18 $D \cup (E \cap F) = (D \cup E) \cap (D \cup F)$

19 $A \cap (B - C) = (A \cap B) - C$

20 $(A \cup B)' = A' \cap B'$

1.2   CONSTANTS AND VARIABLES

In the remainder of this text we shall use letters and other symbols to stand for numbers. Symbols used in this way are called *variables* and are defined more precisely below.

Variable and Replacement Set   A letter or a symbol that stands for a number that is an element of a given set is a *variable*, and the given set is the *replacement set*.

Constant   If the replacement set for a given letter or symbol contains only one element, then that letter or symbol is a *constant*.

For example, the Greek letter π stands for the ratio of the circumference of a circle to the diameter, and is approximately equal to 3.1416. Hence, π is a constant since there is only one number in the replacement set. Also the symbol for each real number such as 2, -3, $\frac{3}{4}$, or $\sqrt{3}$ is a constant. Furthermore, if v stands for the total number of votes cast in the 1968 United States presidential election, then v is a constant. However, if v stands for the total number of votes cast in a United States presidential election, then v is a variable. If d stands for the distance in kilometers between any two unspecified cities in the world, then d is also a variable.

1.3   THE SET OF REAL NUMBERS

In this book, we shall be doing many things with real numbers—adding them, multiplying them, comparing their sizes, etc. In order to make sure that everyone knows what can and what cannot be done with them, we shall in the remainder of this chapter write down some basic assumptions about real numbers, and some rules that follow from these assumptions.

Counting Numbers We begin with the *counting numbers* 1, 2, 3, 4, 5, ... The three dots mean that there are many other counting numbers that fall in the same pattern as the first five, but there are so many that we cannot write them all. The counting numbers are also *called the natural numbers* or the *positive integers.*

Natural Numbers Positive Integers

Just as $5 + 8 = 13$ and $8 + 13 = 21$, the sum of any two positive integers is also a positive integer. This is not true for differences. Although $15 - 9$ is the positive integer 6, the difference $12 - 19$ is not a positive integer. We thus expand our set to include the *negative integers* $-1, -2, -3, -4, -5, ...$ These numbers are as important as positive integers, for example in working with a checkbook, in measuring temperatures above and below freezing, in reporting weight gain or loss, and in reporting movements in stock prices.

Negative Integers

Zero We must also include *zero* since, for instance, $21 - 21 = 0$.

Integers The set of *integers* is defined to be

$$\{. . . , -5, -4, -3, -2, -1, 0, 1, 2, 3, 4, 5, . . .\} =$$
$$\{. . . , -5, -4, -3, -2, -1\} \cup \{0\} \cup \{1, 2, 3, 4, 5, . . .\}$$

and so every integer is by definition either a positive integer, a negative integer, or zero.

In addition to adding and subtracting integers, we may multiply and divide them. For example

$$2 \times 3 = 6 \qquad 4 \times 7 = 28 \qquad 8 \div 4 = 2 \qquad 91 \div 7 = 13$$

Sometimes the division does not come out even, and we write things like

$$\tfrac{7}{2} = 3.5 \qquad \tfrac{11}{5} = 2.2 \qquad \tfrac{1}{3} = .33333 \cdots$$

We also have, by long division, that $\tfrac{2}{7} = .285714285714 \cdots$, where the dots indicate that the sequence of digits 285714 keeps repeating forever. In fact,

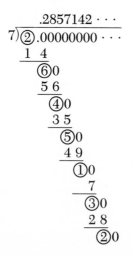

The circled numbers determine the pattern in the quotient, and since we are dividing by 7 there can be at most 6 nonzero numbers in the circles. As soon as we repeat a circled number, say 2, we begin repeating the pattern in the quotient.

In finding $\frac{2}{11}$, there will be at most 10 different circled numbers, and

$$
\begin{array}{r}
.181\cdots \\
11)\,②.00000\cdots \\
\underline{1\ 1} \\
⑨0 \\
\underline{8\ 8} \\
②0
\end{array}
$$

In this case there are in fact just two different circled numbers, but we knew before we began that there were at most 10. If we used long division to find $\frac{16}{27}$, we would find that the quotient repeats after at most 26 digits, and in $\frac{19}{255}$ repetition would begin after at most 254 digits.

Rational Number   If m and n are integers and $n \neq 0$, then m/n is called a *rational number*.

Rational numbers are called rational because they are *ratios* of integers.

We saw in the paragraphs before this definition that every rational number may be expressed as a decimal whose digits form a repeating pattern. This includes rationals such as $\frac{1}{3} = .333\cdots$ and $\frac{3}{11} = .2727\cdots$, where the repetition is obvious, and all terminating decimals such as $\frac{1}{4} = .25 = .25000\cdots$, where the repetition comes from all the final zeros.

The opposite situation is also true: every decimal whose digits form a repeating pattern represents a rational number. This will be illustrated in Examples 1 and 2 of Sec. 11.3, where it is shown that $.351351351\cdots = \frac{13}{37}$.

We have thus seen that *a decimal represents a rational number if and only if the sequence of digits forms a repeating pattern*.

Since rationals are the repeating decimals, we now consider nonrepeating decimals. The numbers $2.03003000300003\cdots$ and $.1234567891011121314\cdots$ are nonrepeating. It may be shown that π and $\sqrt{3}$ are also represented by nonrepeating decimals.

Irrational Number   An *irrational number* is a decimal whose digits do not form a repeating pattern.

We have now progressed from positive integers all the way to irrational numbers, and we may make the following definition.

Real Numbers   The set of *real numbers* is $Q \cup I$, where Q = the set of rationals and I = the set of irrationals.

It follows from the definitions that *a real number is irrational if and only if it is not a quotient of integers.* The set of real numbers is the set of all decimals.

The relationships between the various subsets of the real numbers are given in Fig. 1.4.

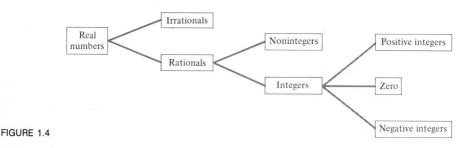

FIGURE 1.4

EXERCISE 1.2   Real Numbers

Perform the indicated operations in Probs. 1 to 32.

1	$3 + 11$	**2**	$14 + 35$	**3**	$32 + 51$	**4**	$106 + 213$
5	$18 + 37$	**6**	$48 + 25$	**7**	$473 + 768$	**8**	$547 + 666$
9	$9 - 5$	**10**	$-5 + 9$	**11**	$5 - 9$	**12**	$-9 + 5$
13	$87 - 44$	**14**	$373 - 212$	**15**	$717 - 632$	**16**	$555 - 468$
17	9×4	**18**	12×3	**19**	18×20	**20**	60×60
21	18×21	**22**	18×22	**23**	46×31	**24**	63×36
25	$38 \div 2$	**26**	$38 \div 4$	**27**	$7 \div 3$	**28**	$3 \div 7$
29	$\frac{4}{5}$	**30**	$\frac{5}{9}$	**31**	$\frac{14}{9}$	**32**	$\frac{6}{11}$

Which name best describes the numbers in each of Probs. 33 to 36?

33 $\frac{2}{3}, -5, 1.4070707 \cdots, \frac{1}{2} + \frac{1}{2}$

34 $\frac{3}{5}, 17, .252225222225 \cdots, .252525 \cdots$

35 $\sqrt{2}, \sqrt{3}, -4.1727374757 \cdots$

36 $4, -2, 4 -2, 17.000 \cdots$

Classify each statement in Probs. 37 to 44 as true or false (R = reals, I = irrationals, Q = rationals, J = integers, P = positive integers, N = negative integers).

37 $N \cap P = \varnothing$ **38** $0 \in Q$ **39** $J = P \cup N$

40 $P \subset Q - J$ **41** $1.4141414 \cdots \in I$ **42** $\sqrt{3} \in R - J$

43 $5 \times \frac{3}{5} \in Q - J$ **44** $1.232232223 \cdots + 1.323323332 \cdots \in Q$

A magic square is a square array of numbers in which every row and every column and each main diagonal has the same sum. Show that each square in Probs. 45 to 48 is a magic square.

45

8	1	6
3	5	7
4	9	2

46

67	1	43
13	37	61
31	73	7

47

16	2	13	3
5	11	8	10
4	14	1	15
9	7	12	6

48

23	6	19	2	15
10	18	1	14	22
17	5	13	21	9
4	12	25	8	16
11	24	7	20	3

1.4 AXIOMS FOR THE REAL NUMBER SYSTEM

The real number system can be determined by three words: complete ordered field. We shall go into the axioms that define these three words in this section. Basically, however, these axioms merely formalize the properties of real numbers. Undoubtedly, you are familiar with many of them already.

Fields   A *field* is a set S with two operations + and · which allow any two elements in S to be added or multiplied. The laws which govern addition and multiplication are as follows, where a, b, and c are any three real numbers:

Closure $a + b$ and $a \cdot b$ are real numbers

Associativity $(a + b) + c = a + (b + c)$ and $(a \cdot b) \cdot c = a \cdot (b \cdot c)$

Identity   There are unique elements 0 and 1 such that

$a + 0 = a$ for all a
and
$b \cdot 1 = b$ for all b

Inverse   Given any a, there is a unique number $-a$ such that $a + (-a) = 0$; and given any $b \neq 0$, there is a unique number b^{-1} such that

$b \cdot (b^{-1}) = 1$

Commutativity $a + b = b + a$ and $a \cdot b = b \cdot a$

| *Distributivity* | $a \cdot (b + c) = (a \cdot b) + (a \cdot c)$ |

We sometimes write $a \cdot b$ as $a \times b$ or ab or $(a)(b)$.

Example 1   The real numbers themselves, with ordinary addition and multiplication, are our prime example of a field. It is possible to construct fields whose elements are not real numbers, and operations other than the usual addition and multiplication.

Example 2   The rational numbers with ordinary addition and multiplication are also a field. We shall go into the addition and multiplication and field properties of rationals more fully in Sec. 1.5.

Example 3   The set $\{a + b\sqrt{2} \mid a \text{ and } b \text{ are rational}\}$ with ordinary addition and multiplication is a field. We shall see this field again in Chap. 5.

Example 4   It is possible to define addition and multiplication of points in the plane so that the set of all points in the plane forms a field. For this purpose, we take $(a, b) + (c, d) = (a + c, b + d)$ and $(a, b) \cdot (c, d) = (ac - bd, ad + bc)$.

There are many uses of the field properties. In multiplying 14 and 32, we normally write

$$
\begin{array}{r}
14 \\
32 \\
\hline
28 \\
42 \\
\hline
448
\end{array}
$$

The reason the 42 is "moved over" to the left is that when we use the distributive law, we get $14 \cdot 32 = 14 \cdot (30 + 2) = 14 \cdot 30 + 14 \cdot 2 = 420 + 28$, so the normal procedure could as well be written

$$
\begin{array}{r}
14 \\
32 \\
\hline
28 \\
420 \\
\hline
448
\end{array}
$$

Other examples are $4 \cdot 7 = 7 \cdot 4$ and $6 + (8 + 3) = (6 + 8) + 3$ (that is, $6 + 11 = 14 + 3$).

Notice that a field requires only addition and multiplication. We define *subtraction* by means of the inverse for addition:

Subtraction

$$a - b \text{ means } a + (-b) \tag{1.3}$$

Division and *division* is defined (for $b \neq 0$) by means of the inverse for multiplication:

$$a \div b \text{ or } \frac{a}{b} \text{ means } a \cdot (b^{-1}) \qquad (1.4)$$

Suppose that $a + b = a + c$. Then $(-a) + (a + b) = (-a) + (a + c)$; hence $(-a + a) + b = (-a + a) + c$, and it follows that *Cancellation* $0 + b = 0 + c$, or $b = c$. We have just proved the *cancel- Law of Addition* lation law of addition*:

If $a + b = a + c$ then $b = c$

This allows us to handle negative signs [see (1.5) and (1.6)].
We propose now to prove

$$-(-a) = a \qquad (1.5)$$

Since $0 = a + (-a)$, and also $0 = (-a) + [-(-a)] = [-(-a)] + (-a)$, it follows that $a + (-a) = [-(-a)] + (-a)$, and by the cancellation law, $a = -(-a)$.
It is also true that

$$-(a + b) = -a - b \qquad (1.6)$$

In fact $(a + b) + [-(a + b)] = 0$, and $a + b + (-a - b) = (a + b - a) - b = (a - a + b) - b = (0 + b) - b = b - b = 0$, so $a + b + [-(a + b)] = a + b + (-a - b)$, and again by the cancellation law, $-(a + b) = -a - b$.
For example, $5 = -(-5)$; $-(3 + 8) = -3 - 8$; and $-(2 + 3 + 4) = -2 - 3 - 4 = -9$.
In order to calculate $-(a - b)$, we write

$$-(a - b) = -[a + (-b)] \qquad \text{by Eq. (1.3)}$$
$$= -a - (-b) \qquad \text{by Eq. (1.6)}$$
$$= -a + [-(-b)] \qquad \text{by Eq. (1.3)}$$
$$= -a + b \qquad \text{by Eq. (1.5)}$$

More than two terms may be combined. Thus,

$$a - (b + c - d) = a - b - c + d$$
$$5 - (3 - 4) = 5 - 3 + 4 = 6$$
$$-4 - (-2 + 5) = -4 + 2 - 5 = -7$$
$$-(-3 - 5 + 6 - 1) = 3 + 5 - 6 + 1 = 3$$

We shall present some more of the consequences of the field properties in Sec. 1.5.
Ordered Field A field S is *ordered* if there is a subset P (the positive elements) of S which satisfies the following three conditons:

If a and b are in P, then $a + b \in P$ (1.7)

If a and b are in P, then $a \cdot b \in P$ (1.8)

If $c \in S$, then exactly one of the following is true:

$$c \in P \quad \text{or} \quad -c \in P \quad \text{or} \quad c = 0 \qquad (1.9)$$

Trichotomy   The property (1.9) is called *trichotomy* because one of three possibilities must hold.

The set P is the set of positive elements in the field.

It is an important fact that the reals can be ordered in only one way—that is, the positive numbers are the only set P which satisfies all three conditions above. The rationals (Example 2) may also be ordered in just one way. The field of Example 3 may be ordered in two ways since there are two sets, each of which satisfies (1.7), (1.8), and (1.9). The field of Example 4 may not be ordered at all. See Probs. 48 and 49 in the Review Exercise.

The most important consequence of a field being ordered is that it allows us to say that one element is larger than another element.

$a < b$ If a and b are real numbers (or elements in any ordered field), then $a < b$ means $b - a > 0$ (or $b - a \in P$).

If $a < b$, we may also write $b > a$. Thus $5 < 10$ and $10 > 5$; $-15\frac{1}{2} < -6$ and $-6 > -15\frac{1}{2}$; and $18 > 17.01020304\cdots$.

In addition to comparing the size of two numbers, we may compare the numbers without their signs. This is done by use of absolute values.

Absolute   If a is any real number, then the *absolute value* of a is written
Value $|a|$ and is defined by

$$|a| = \begin{cases} a & \text{if } a > 0 \\ -a & \text{if } a < 0 \\ 0 & \text{if } a = 0 \end{cases} \qquad (1.10)$$

Accordingly, $|7| = 7$ since $7 > 0$; $|-4| = -(-4) = 4$ since $-4 < 0$; and $|0| = 0$. Thus the absolute value of any real number is positive or zero.

$a \leq b$ If either $a < b$ or $a = b$, we write $a \leq b$. It follows that $4 \leq 9$ and $4 \leq 4$.

From the definition of $|a|$, we see that $|a| \geq 0$ for all real numbers a. Also, $-|a| \leq a \leq |a|$. For example, $|-3| = 3 > 0$, and $-|-3| \leq -3 \leq |3|$, that is, $-3 \leq -3 \leq 3$.

We shall now consider the absolute value of a product and of a

FIGURE 1.5

sum.   Considering the four different cases ($a > 0$ and $b > 0$; $a > 0$ and $b < 0$; $a < 0$ and $b > 0$; and $a < 0$ and $b < 0$) we may show

$$|ab| = |a| \cdot |b| \tag{1.11}$$

$$|a + b| \le |a| + |b| \tag{1.12}$$

The latter is known as the *triangle inequality*.

The proofs are left to the exercises.

Completeness   There are several ways of defining an ordered field to be *complete*. For our purposes, it is sufficient to say that *completeness means there is a one-to-one correspondence between the set of real numbers and the set of points on a line* (extended infinitely far to both the left and right). In Fig. 1.5 we have indicated such a correspondence. This simply means that to each real number we may associate exactly one point on the line, and vice versa.

The rational numbers do not have this property (completeness)—there are more points on a line than there are rational numbers. To put it another way, if each rational number were assigned a point, we could use up all the rationals and still have some "holes" left in the line. One hole occurs at $\sqrt{7}$.

The real numbers are the only complete ordered field.

EXERCISE 1.3   Axioms for the Real Number System

Which field axiom is used in each of Probs. 1 to 8?

1   $3 \cdot 17 = 17 \cdot 3$

2   $(14 + 3) + 20 = 14 + (3 + 20)$

3   $4.17 + 2.22222 \cdots = 6.39222 \cdots$

4   $2 \cdot (4 + 3) = (2 \cdot 4) + (2 \cdot 3)$

5   $1 \times 234{,}567 = 234{,}567$

6   $5 + (-5) = 0$

7   $0 + 97{,}531 = 97{,}531$

8   $3.4 \times 5.67 = 19.278$

Classify each statement in Probs. 9 to 16 as true or false.

9   The associative law of multiplication is true for rational numbers.

10   The negative numbers are closed under addition.

11   $|a| = |-a|$ for all real numbers a.

12   The sum of any two irrational numbers is irrational.

13   If a is negative, then $-(-a)$ is negative.

14   The absolute value of $|a|$ is $|a|$.

15   $-(-a + b) = a + b$.

16   If $a < b$, then $-a < b$.

The number $|a - b|$ *represents the distance from a to b. Verify this for the numbers given in Probs. 17 to 20.*

17 $a = 4, b = 1$ **18** $a = 2, b = 7$

19 $a = -5, b = -16$ **20** $a = -18, b = -4$

Calculate the numbers in Probs. 21 to 24.

21 $4 - (3 - 2)$ **22** $6 + 2[-4 - (-3)]$

23 $-\{2 + 3[-1 - (6 + 2)] + 3\}$ **24** $4\{-[-2 + 3(-1 + 2)] + 6\}$

25   Show that if $a < b$ and $b < c$, then $a < c$.

26   Show that $|ab| = |a| \cdot |b|$.

27   Show that $|a + b| \le |a| + |b|$.

28   Verify each of Probs. 25 to 27 for $a = -3$ and $b = 5$, and for $a = 4$ and $b = 7$.

1.5   THEOREMS ABOUT THE REAL NUMBER SYSTEM

Recall that $a \times b$ is also written as ab or $a \cdot b$ or $(a)(b)$, for convenience. Furthermore we defined $a \div b$ or a/b as $a(b^{-1})$; hence, by definition,

$$\frac{a}{b} = a(b^{-1}) = a\left(\frac{1}{b}\right)$$

By definition of the multiplicative identity,

$$a(1) = a \tag{1.13}$$

for every real number a. We shall now show that, for the additive identity zero,

Multiplication by Zero
$$a(0) = 0 \tag{1.14}$$

for every real number a.

$$0 = 0 + 0 \qquad \text{**by definition of 0**}$$
$$a(0) = a(0 + 0) \qquad \text{**multiplication by a**}$$
$$a(0) + 0 = a(0) + a(0) \qquad \text{**definition of 0 and distributive law**}$$
$$0 = a(0) \qquad \text{**cancellation law for addition**}$$

In words, this becomes *zero multiplied by any real number is zero.*

Example 1 $5(1) = 5$; $6(0) = 0$; and $0 \times 31 = 0$.

We shall now consider zero in division. In $a \div b$, we shall consider what happens when $b = 0$ and when $a = 0$. First, if $b = 0$, then to write $a \div 0$ would be the same as $a \times 0^{-1}$. However in the field axioms, an inverse is provided for every number except

0, so 0^{-1} is not defined, and we are forced to say (see Prob. 44 in the Review Exercises)

Division by Zero $\dfrac{a}{0}$ **is not defined for any real number** a $\qquad\qquad$ (1.15)

Now consider $a \div b$, where $b \neq 0$ and $a = 0$. Then $a \div b = 0 \times b^{-1} = 0$, since zero times any number is zero. Thus,

$$\frac{0}{b} = 0 \quad\text{if}\quad b \neq 0 \qquad\qquad (1.16)$$

We shall prove that for any real numbers a and b,

$$a(-b) = -(ab) \qquad\qquad (1.17)$$

It is also true that

$$(-a)(-b) = ab \qquad\qquad (1.18)$$

The proof of (1.18) will be left for the exercises. It is similar to the following proof of (1.17).

$$
\begin{aligned}
a(-b) &= a(-b) + (ab - ab) && \text{since } ab - ab = 0 \\
&= [a(-b) + ab] - (ab) && \text{associative law for addition} \\
&= [a(-b + b)] - (ab) && \text{distributive law} \\
&= a(0) - (ab) && -b + b = 0 \\
&= 0 - (ab) && a(0) = 0 \\
&= -(ab)
\end{aligned}
$$

As for (1.18), notice that according to the pattern $(-8)(3) = -24$, $(-8)(2) = -16$, $(-8)(1) = -8$, $(-8)(0) = 0$, we should have $(-8)(-1) = 8$, and $(-8)(-2) = 16$.

Example 2 $(3)(-4) = -(3)(4) = -12$
$(-2)(7) = -(2)(7) = -14$
$(-3)(-6) = 3(6) = 18$

Law of Signs for Multiplication   We may state the law of signs for multiplication by saying that *the product of two nonzero numbers is positive if they have the same sign, and negative if they have opposite signs.*
$\qquad$ Suppose $a/b = c$. Then $a \times b^{-1} = c$, so $a \times (b^{-1} \times b) = c \times b$ and $a = b \times c$. The argument goes in reverse also, so if $b \neq 0$,

$$\frac{a}{b} = c \quad\text{if and only if}\quad a = bc \qquad\qquad (1.19)$$

Law of Signs for Division   From this and the law of signs for multiplication, we get a similar law of signs for division, namely, *the quotient of two nonzero numbers is positive if the numbers have the same sign, and negative if they have opposite signs.*

Example 3
$$\frac{-14}{2} = -\frac{14}{2} = -7$$

$$\frac{18}{-3} = -\frac{18}{3} = -6$$

$$\frac{-9}{-3} = \frac{9}{3} = 3$$

We define a^n, where n is a positive integer, as follows:

$$a^n = a \cdot a \cdot a \cdots a \qquad n \text{ factors} \tag{1.20}$$

It follows that

$$a^m a^n = \underbrace{(a \cdot a \cdot a \cdots a)}_{m \text{ factors}}\underbrace{(a \cdot a \cdot a \cdots a)}_{n \text{ factors}}$$

Hence, since the expression on the right is the product of $m + n$ factors, each of which is a, we have

Product of Powers $\quad a^m a^n = a^{m+n} \qquad$ **m and n are positive integers** $\tag{1.21}$

Example 4 Find the product of $4a^3$ and $6a^5$.

Solution
$$\begin{aligned} 4a^3 \cdot 6a^5 &= 4 \cdot 6 \cdot a^3 \cdot a^5 && \text{by commutative axiom} \\ &= 24a^{3+5} && \text{by law of exponents (1.21)} \\ &= 24a^8 \end{aligned}$$

We use (1.21) for finding the power of a power. For example,

$$(x^2)^3 = x^2 \cdot x^2 \cdot x^2 = x^{2+2+2} = x^6$$

In general,

$$(x^m)^n = \underbrace{x^m \cdot x^m \cdot x^m \cdots x^m}_{n \text{ factors}}$$

$$= x^{m+m+m+\cdots+m}, n \text{ terms}$$

$$= x^{nm}$$

Consequently, we have

Power of a Power $\quad (x^m)^n = x^{mn} \qquad$ **x real, m and n positive integers** $\tag{1.22}$

Furthermore, since

$$(ab)^n = \underbrace{ab \cdot ab \cdot ab \cdots ab}_{n \text{ factors}}$$

$$= \underbrace{(a \cdot a \cdot a \cdots a)}_{n \text{ factors}}\underbrace{(b \cdot b \cdot b \cdots b)}_{n \text{ factors}} \qquad \text{by associative and commutative axioms}$$

$$= a^n b^n$$

we have the following law for the power of a product:

Power of a Product $(ab)^n = a^n b^n$ **a and b real, n is a positive integer**   **(1.23)**

The theorem that we now prove is called the *law of exponents for division*. If m and n are positive integers with $m > n$, we have, by (1.19),

$$\frac{a^m}{a^n} = x \quad \text{if and only if} \quad a^n x = a^m$$

and we shall determine x. Since $m > n$ and m and n are positive integers, $m - n$ is a positive integer and can be used as an exponent. Now if we replace x in $a^n x$ by a^{m-n}, we get

$$\begin{aligned} a^n a^{m-n} &= a^{n+m-n} &&\text{by law of exponents for multiplication} \\ &= a^{m+n-n} &&\text{by commutative axiom} \\ &= a^m &&\text{since } n - n = 0 \end{aligned}$$

Hence, if $x = a^{m-n}$, then $a^n x = a^m$. Therefore, we have the law of exponents for division:

Law of Exponents for Division $\dfrac{a^m}{a^n} = a^{m-n} \quad m > n$ **(1.24)**

We shall later prove that this law holds if $m < n$, and now we examine the situation if $m = n$. In this case, we have

$$\frac{a^n}{a^n} = a^{n-n} = a^0$$

Since our definition of a^n requires that n be a positive integer, a^0 has no meaning. However,

$$\frac{a^n}{a^n} = 1$$

and it is logical to define a^0 as the number 1. Hence, by this definition, we have:

Zero as an Exponent $a^0 = 1 \quad a \neq 0$ **(1.25)**

Example 5 $2^3 \cdot 2^8 \cdot 2^6 = 2^{3+8+6} = 2^{17}$

$$\frac{8^5}{4^3} = \frac{(2^3)^5}{(2^2)^3} = \frac{2^{15}}{2^6} = 2^9$$

$$12^4 = (2^2 \cdot 3)^4 = (2^2)^4 (3^4) = 2^8 \cdot 3^4$$

$$\frac{(3 \cdot 5)^3}{27} = \frac{3^3 \cdot 5^3}{3^3} = 3^0 \cdot 5^3 = 5^3$$

Example 6 $(-3x^2 y)(-4xy^3)(-5x^4 y^2) = (-3)(-4)(-5)x^2 \cdot x \cdot x^4 \cdot y \cdot y^3 \cdot y^2$

$$\begin{aligned} &= 12(-5)x^{2+1+4} y^{1+3+2} \\ &= -60x^7 y^6 \end{aligned}$$

Note that -2^4 means the negative of 2^4 (that is, -16), while $(-2)^4$ means $(-2)(-2)(-2)(-2) = 16$.

If a and b are positive integers, we represent their greatest common divisor by (a, b). It is the largest integer which divides both a and b. We find it by writing the product of all of the different factors of a and b which are common to both of them, each factor being raised to the lowest power to which it enters in either a or b. Now $240 = 2^4 \cdot 3 \cdot 5$ and $756 = 2^2 \cdot 3^3 \cdot 7$, and thus $(240, 756) = 2^2 \cdot 3 = 12$ since the only common factors of 240 and 756 are 2 and 3 with 2 occurring to the second power.

If a and b are positive integers, we let $[a, b]$ equal the least common multiple of a and b. It is the smallest integer which is a multiple of both a and b. We find it by writing the product of all of the different factors of a and b which occur in either a or b, each factor being raised to the highest power to which it enters in either a or b. Using the factors in the above paragraph gives $[240, 756] = 2^4 \cdot 3^3 \cdot 5 \cdot 7 = 15{,}120$.

Example 7 (a) $(24, 18) = 2 \cdot 3 = 6$, and $[24, 18] = 2^3 \cdot 3^2 = 72$ since
$$24 = 2^3 \cdot 3 \qquad \text{and} \qquad 18 = 2 \cdot 3^2$$

(b) $(36, 10) = 2$, and $[36, 10] = 2^2 \cdot 3^2 \cdot 5 = 180$ since
$$36 = 2^2 \cdot 3^2 \qquad \text{and} \qquad 10 = 2 \cdot 5$$

(c) $(12, 30) = 2 \cdot 3 = 6$, and $[12, 30] = 2^2 \cdot 3 \cdot 5 = 60$ since
$$12 = 2^2 \cdot 3 \qquad \text{and} \qquad 30 = 2 \cdot 3 \cdot 5$$

(d) $(44, 11) = 11$, and $[44, 11] = 2^2 \cdot 11 = 44$ since
$$44 = 2^2 \cdot 11 \qquad \text{and} \qquad 11 = 11$$

EXERCISE 1.4 Theorems about the Real Number System

Find each number in Probs. 1 to 16.

1 $(-4)(-8)$

2 $3(-15)$

3 $4(7)(-2)$

4 $5(-1)(-3)$

5 $2^2 \cdot 2^3$

6 $(-3) \cdot 3^2 \cdot 2$

7 $5^2 \cdot 2^2$

8 $(-2^3)(-3^2)$

9 $(2^2)^3$

10 $(3^3)^2$

11 $(2^4 \cdot 3)^2$

12 $(3^2 \cdot 5 \cdot 2)^2$

13 $\dfrac{3^2 \cdot 2^7}{(2^2)^3}$

14 $\dfrac{(2^5)(-5^2)}{(-4)(-5)}$

15 $\dfrac{(-3^2)^3 \cdot 5^2}{(3 \cdot 5)^2}$

16 $\dfrac{2^2 \cdot 3^4 \cdot 5^6}{(2 \cdot 5^2)^3}$

Express each number in Probs. 17 to 32 by using exponents.

17 $3^4 \cdot 3^5$

18 $3^5 \cdot (-3^4)$

19 $5^4 \cdot 5^{10}$

20 $(-2^5)(2^8)$

21 $\dfrac{5^{17}}{-5^{12}}$

22 $\dfrac{-11^{14}}{-11^{11}}$

23 $\dfrac{7^{10}}{-7^{10}}$

24 $\dfrac{5^{19}}{5^3}$

25 $(5^3)^8$

26 $(-3^5)^8$

27 $\dfrac{(5^4)^6}{5^{15}}$

28 $\dfrac{3^7(3^4)^3}{(3^3)^6}$

29 $\dfrac{(3^{21} \cdot 2^6)^2}{(-3^8)^5}$

30 $\dfrac{(5^2 \cdot 3^4)^3}{(5 \cdot 3^2)^4}$

31 $\dfrac{(2^2 \cdot 3^3 \cdot 5^5)^3}{(-3^4 \cdot 5^7)^2}$

32 $\dfrac{-(4^3 \cdot 3^4 \cdot 5^5)^2}{(2^3 \cdot 3^2 \cdot 5^2)^4}$

It is true that $(a, b)[a, b] = ab$ for all positive integers a and b? Verify this for Probs. 33 to 40.

33 $a = 3, b = 5$

34 $a = 12, b = 8$

35 $a = 15, b = 10$

36 $a = 4, b = 30$

37 $a = 48, b = 20$

38 $a = 28, b = 12$

39 $a = 18, b = 32$

40 $a = 14, b = 21$

41 Why are $(0)(5)$ and $\frac{0}{5}$ both zero, but neither $\frac{5}{0}$ nor $\frac{0}{0}$ is zero?

42 Show that $(x^m)^n = (x^n)^m$ for all positive integers m and n.

43 Prove that $(-a)(-b) = ab$.

44 Prove the cancellation law for multiplication: if $ab = ac$ and $a \neq 0$, then $b = c$.

1.6 FUNDAMENTAL OPERATIONS ON FRACTIONS

In this section we shall deal with the equality of two fractions, and with addition, subtraction, multiplication, and division of fractions.

If $a/b = c/d$, then by the definition of division, $ab^{-1} = cd^{-1}$. Multiplying by bd and using the commutative law gives $(ab^{-1})bd = (cd^{-1})db$ or $ad = bc$. Thus (for $b \neq 0$ and $d \neq 0$),

$$\frac{a}{b} = \frac{c}{d} \quad \text{if and only if} \quad ad = bc \quad\quad (1.26)$$

We shall now show (for any e and f not zero) that

$$\frac{a}{b} = \frac{ae}{be} = \frac{a/f}{b/f} \quad\quad (1.27)$$

We have the first equality in (1.27) if and only if $a(be) = b(ae)$. These, by the associative and commutative laws, are equal. That $a/b = (a/f)/(b/f)$ may be proved similarly.

Example 1

$$\frac{3}{6} = \frac{1}{2}$$
    since 3(2) = 1(6), using (1.26)

$$\frac{-4}{6} = \frac{2}{-3}$$
    since (−4)(−3) = (2)(6), using (1.26)

$$\frac{16}{24} = \frac{16/8}{24/8} = \frac{2}{3}$$
    by (1.27)

$$\frac{27}{45} = \frac{27/9}{45/9} = \frac{3}{5}$$
    by (1.27)

$$\frac{4}{5} = \frac{4 \cdot 2}{5 \cdot 2} = \frac{8}{10}$$
    by (1.27)

$$\frac{8}{11} = \frac{8 \cdot 4}{11 \cdot 4} = \frac{32}{44}$$
    by (1.27)

In order to add $\frac{4}{5}$ and $\frac{7}{5}$, we may write the sum as $4(\frac{1}{5}) + 7(\frac{1}{5})$.

This is just like adding 4 apples and 7 apples, except we are adding fifths instead of apples. The sum is $11(\frac{1}{5}) = \frac{11}{5}$.

We could also have used the distributive law:

$$\frac{4}{5} + \frac{7}{5} = 4(\frac{1}{5}) + 7(\frac{1}{5}) = (4 + 7)(\frac{1}{5}) = 11(\frac{1}{5}) = \frac{11}{5}$$

In general we may *add fractions with the same denominator* according to the following rule:

Addition of Fractions $\quad \dfrac{a}{b} + \dfrac{c}{b} = \dfrac{a + c}{b}$ $\hspace{4cm}$ (1.28)

Similarly, for subtraction we have

Subtraction of Fractions $\quad \dfrac{a}{b} - \dfrac{c}{b} = \dfrac{a - c}{b}$ $\hspace{4cm}$ (1.29)

Example 2 $\qquad \frac{3}{4} + \frac{5}{4} = \frac{8}{4} = 2$ $\qquad$ **by (1.28)**

$\qquad\qquad\quad \frac{6}{7} + \frac{5}{7} = \frac{11}{7}$ $\qquad$ **by (1.28)**

$\qquad\qquad\quad \frac{14}{9} - \frac{10}{9} = \frac{4}{9}$ $\qquad$ **by (1.29)**

$\qquad\qquad\quad \frac{13}{5} - \frac{6}{5} = \frac{7}{5}$ $\qquad$ **by (1.29)**

In order to add two fractions that do not have the same denominator, we begin by writing them as a set of equal fractions with the same denominator. Accordingly, to add $\frac{5}{6} + \frac{3}{4}$, we first write each fraction with the same denominator—the least common multiple of both denominators. Since $[6, 4] = 12$, we write

$$\frac{5}{6} + \frac{3}{4} = \frac{5 \cdot 2}{6 \cdot 2} + \frac{3 \cdot 3}{4 \cdot 3}$$
$$= \frac{10}{12} + \frac{9}{12}$$
$$= \frac{19}{12}$$

Subtraction is treated in a similar manner.

Example 3 $\qquad \dfrac{1}{2} + \dfrac{2}{3} = \dfrac{3}{6} + \dfrac{4}{6} = \dfrac{7}{6}$

$\qquad\qquad\quad \dfrac{8}{5} - \dfrac{9}{7} = \dfrac{56}{35} - \dfrac{45}{35} = \dfrac{11}{35}$

$\qquad\qquad\quad \dfrac{7}{9} - \dfrac{13}{15} = \dfrac{35}{45} - \dfrac{39}{45} = \dfrac{-4}{45}$

$\qquad\qquad\quad \dfrac{5}{24} + \dfrac{7}{36} = \dfrac{15}{72} + \dfrac{14}{72} = \dfrac{29}{72}$

In order to multiply two fractions, we write

$$\frac{a}{b} \cdot \frac{c}{d} = ab^{-1} \cdot cd^{-1} = ac \cdot b^{-1}d^{-1}$$

Now $b^{-1}d^{-1} = (bd)^{-1}$ so $ac \cdot b^{-1}d^{-1} = (ac)(bd)^{-1} = ac/bd$. Thus,

Multiplication of Fractions to multiply two fractions, we write a fraction whose numerator is the product of the numerators and whose denominator is the product of the denominators. In symbols,

$$\frac{a}{b} \cdot \frac{c}{d} = \frac{ac}{bd}$$ (1.30)

Example 4 $\dfrac{2}{3} \cdot \dfrac{5}{7} = \dfrac{2 \cdot 5}{3 \cdot 7} = \dfrac{10}{21}$

$\dfrac{3}{4} \cdot \dfrac{5}{14} = \dfrac{3 \cdot 5}{4 \cdot 14} = \dfrac{15}{56}$

$\dfrac{3}{25} \cdot \dfrac{16}{7} = \dfrac{3 \cdot 16}{25 \cdot 7} = \dfrac{48}{175}$

When there are common factors in the numerators and denominators, we may cancel these first and then multiply the resulting numerators and denominators.

Example 5 $\dfrac{3}{10} \cdot \dfrac{5}{6} = \dfrac{3 \cdot 5}{(5 \cdot 2)(3 \cdot 2)} = \dfrac{1}{2 \cdot 2} = \dfrac{1}{4}$

$\dfrac{8}{15} \cdot \dfrac{21}{20} \cdot \dfrac{25}{14} = \dfrac{2^3 \cdot (3 \cdot 7) \cdot (5^2)}{(5 \cdot 3) \cdot (5 \cdot 2^2) \cdot (2 \cdot 7)} = 1$

$\dfrac{6}{11} \cdot \dfrac{22}{21} = \dfrac{(3 \cdot 2) \cdot (11 \cdot 2)}{11 \cdot (3 \cdot 7)} = \dfrac{4}{7}$

To find the quotient of two fractions, we write

Division of Fractions $\dfrac{a/b}{c/d} = \dfrac{ab^{-1}}{cd^{-1}} \cdot \dfrac{bd}{bd} = \dfrac{ad}{bc} = \dfrac{a}{b} \cdot \dfrac{d}{c}$ (1.31)

That is, to divide one fraction by another, we invert the denominator (replace c/d by d/c) and multiply this by the numerator. In other words, multiply the numerator by the reciprocal of the denominator.

Example 6 $\dfrac{5/6}{3/10} = \dfrac{5}{6} \cdot \dfrac{10}{3} = \dfrac{5(5 \cdot 2)}{(2 \cdot 3)3} = \dfrac{5^2}{3^2} = \dfrac{25}{9}$

$\dfrac{8/7}{4/21} = \dfrac{8}{7} \cdot \dfrac{21}{4} = \dfrac{2^3 \cdot (3 \cdot 7)}{7 \cdot (2^2)} = 2 \cdot 3 = 6$

EXERCISE 1.5   Fundamental Operations

In Probs. 1 to 8, show that each equation is true.

1 $\frac{3}{5} = \frac{6}{10}$ **2** $\frac{5}{8} = \frac{15}{24}$ **3** $\frac{7}{4} = \frac{28}{16}$ **4** $\frac{3}{11} = \frac{21}{77}$

5 $\frac{9}{15} = \frac{15}{25}$ **6** $\frac{16}{14} = \frac{40}{35}$ **7** $\frac{20}{45} = \frac{32}{72}$ **8** $\frac{12}{26} = \frac{30}{65}$

Perform the indicated operations in Probs. 9 to 44.

9 $\frac{9}{7} - \frac{4}{7}$	**10** $\frac{8}{3} + \frac{5}{3}$	**11** $\frac{6}{5} + \frac{8}{5}$	**12** $\frac{13}{10} - \frac{4}{10}$
13 $\frac{14}{11} - \frac{16}{22}$	**14** $\frac{16}{14} + \frac{15}{21}$	**15** $\frac{14}{8} - \frac{6}{4}$	**16** $\frac{15}{18} + \frac{32}{24}$
17 $\frac{1}{2} + \frac{1}{3}$	**18** $\frac{1}{3} + \frac{1}{4}$	**19** $\frac{1}{4} + \frac{1}{7}$	**20** $\frac{1}{5} + \frac{1}{8}$
21 $\frac{1}{4} + \frac{2}{3}$	**22** $\frac{1}{5} + \frac{3}{8}$	**23** $\frac{1}{6} + \frac{5}{7}$	**24** $\frac{1}{8} + \frac{9}{10}$
25 $\frac{3}{4} - \frac{2}{3}$	**26** $\frac{6}{5} - \frac{3}{4}$	**27** $\frac{3}{2} - \frac{5}{3}$	**28** $\frac{4}{5} - \frac{7}{6}$
29 $\frac{16}{9} - \frac{8}{3}$	**30** $\frac{3}{8} + \frac{8}{3}$	**31** $\frac{4}{7} - \frac{17}{4}$	**32** $\frac{13}{3} + \frac{2}{11}$
33 $\frac{5}{8} \times \frac{4}{15}$	**34** $\frac{4}{7} \times \frac{21}{8}$	**35** $\frac{5}{8} \times \frac{24}{25}$	**36** $\frac{9}{8} \times \frac{4}{15}$
37 $\frac{14}{9} \times \frac{6}{7} \times \frac{3}{4}$	**38** $\frac{6}{5} \times \frac{4}{3} \times \frac{5}{16}$	**39** $\frac{1}{5} \times \frac{10}{3} \times \frac{6}{7}$	**40** $\frac{2}{9} \times \frac{6}{7} \times \frac{21}{4}$
41 $\frac{6}{5} \div \frac{27}{10}$	**42** $\frac{3}{5} \div \frac{9}{10}$	**43** $\frac{7}{8} \div \frac{21}{4}$	**44** $\frac{10}{9} \div \frac{20}{27}$

1.7   SUMMARY

Sets are discussed in Sec. 1.1, including union, intersection, complement, and relative complement. Sets are then used in defining the real number system, which is a complete, ordered field. There are eleven axioms for a field, three for an ordered field, and one more for a complete ordered field. The reals are also all decimals, both repeating (rationals) and nonrepeating (irrationals). The absolute value of a real number is also discussed.

Among the most important properties of real numbers are:

$$a(b + c) = ab + ac$$
$$-(-a) = a \qquad -(a + b) = -a - b$$
$$a(1) = a \quad \text{and} \quad a(0) = 0 \qquad \text{for all reals } a$$
$$a^m a^n = a^{m+n} \qquad a^m b^m = (ab)^m \qquad (a^m)^n = a^{mn}$$

Addition, subtraction, multiplication, and division of fractions are also discussed.

EXERCISE 1.6   Review

Let $A = \{1, 3, 5, 7, 9\}$, $B = \{2, 4, 6, 8, 10\}$, $C = \{1, 4, 7\}$, $D = \{4, 6, 8\}$, *and* $U = \{1, 2, 3, 4, 5, 6, 7, 8, 9, 10\}$ *in Probs. 1 to 7.*

1 Find $(B \cap C) \cup (B - D)$.   **2** Find $(D - C) \cup A$.

3 Find $D - (C \cup A)$.   **4** Find $(B \cup C)'$.

5 Show that $A \cap (B \cup C) = (A \cap B) \cup (A \cap C)$.

6 Show that $(A \cap C)' = A' \cup C'$.

7 Show that $D' - B = (D \cup B)'$.

Show that the number in each of Probs. 8 to 11 is 100.

8 $1 + 2 + 3 + 4 + 5 + 6 + 7 + (8)(9)$ **9** $1 + 2 + 34 - (5 - 67 + 8 - 9)$

10 $12 + 3 - (4 - 5 - 67 - 8 - 9)$

11 $123 + 4 - (5 - 67 + 89)$

Verify the calculations in Probs. 12 to 16.

12 $12^2 + 5^2 = 11^2 + 7^2 - 1^2$

13 $3^3 + 4^3 + 5^3 = 6^3$

14 $1^3 + 12^3 - (9^3 + 10^3) = 0$

15 $\dfrac{428{,}571}{142{,}857} = \dfrac{857{,}142}{285{,}714} = 3$

16 $(2^2 - 3^2)(4^2 - 1^2) = 5^2 - 10^2$

17   Is $\frac{3}{11}$ rational?

18   Is $\frac{3}{1}$ an integer?

19   Is it true that $.121221222 \cdots + .212112111 \cdots = \frac{2}{6}$?

20   Is it true that $\dfrac{4}{-3} = \dfrac{-4}{3} = -\dfrac{4}{3}$?

21   Show that $-[-(-3)] = -3$.

22   Find $|(1)(-2)(-3)|$.

23   Find $|(1)(-2) + (-3)|$.

Express each rational number in Probs. 24 to 26 as a repeating decimal.

24 $\frac{4}{9}$ **25** $\frac{5}{11}$ **26** $\frac{7}{15}$

Simplify the expressions in Probs. 27 to 34.

27 $(2^3)(2^4)$ **28** $(3^5)(3^3)$ **29** $4^5/4^3$ **30** $(3^2)(2^2)$

31 $6^3/2^3$ **32** $12^4/4^4$ **33** $(3^4)^5$ **34** $(5^4)^2$

35   Show that $[8, 12](8, 12) = 96$.

Perform the indicated operations in Probs. 36 to 42.

36 $\dfrac{4}{5} - \dfrac{2}{5} + \dfrac{3}{5}$ **37** $\dfrac{2}{5} + \dfrac{1}{8}$ **38** $\dfrac{4}{3} - \dfrac{3}{4}$ **39** $\dfrac{2}{3} + \dfrac{4}{5} + \dfrac{2}{15}$

40 $\dfrac{4}{15}\left(\dfrac{-3}{10}\right)$ **41** $\dfrac{6}{7} \times \dfrac{35}{33} \times \dfrac{11}{4}$ **42** $\dfrac{18}{35} \div \dfrac{2}{5}$

43   Show that if $b \neq 0$ then $(b^{-1})^{-1} = b$.

44   The following argument shows why division by zero is not defined; justify each step.

$$\frac{a}{0} = q \qquad \text{means} \qquad a = (0)(q)$$

Thus, if $a = 0$, then every q satisfies $a = (0)(q)$, while if $a \neq 0$, then no q satisfies $a = (0)(q)$.

45   Show that $\dfrac{a}{b} + \dfrac{c}{d} = \dfrac{ad + bc}{bd}$ (even though bd may not be the least common multiple of b and d).

46   Show that the rationals are a field.   Note that some of the field properties (such as associativity and commutativity) are "inherited" from the reals, while others (such as closure and identities) need to be proved for rationals.

47   Is it true that $(a^m)^n = a^{(m^n)}$?   Verify your answer for $a = 4$, $m = 3$, $n = 2$.

48 Show that the field of Example 3 of Sec. 1.4 may be ordered in more than one way by showing that both

$$P_1 = \{a + b\sqrt{2} \,|\, a + b\sqrt{2} > 0\} \qquad \text{and} \qquad P_2 = \{a + b\sqrt{2} \,|\, a - b\sqrt{2} \in P_1\}$$

satisfy the definition of an ordered field.

49 Show that the field of Example 4 of Sec. 1.4 cannot be ordered. *Hint*: Assume there is a set P satisfying the definition and consider the element $(0, 1)$. Either it or its negative is in P. Note that $(0, 1) \cdot (0, 1) = -[(0, 1) \cdot (0, 1) \cdot (0, 1) \cdot (0, 1)]$.

2
polynomials, products, and factoring

In this chapter we discuss the addition, subtraction, multiplication, and division of numbers and variables. Factoring, in which sums are expressed as products, is also extensively treated. Each expression in this chapter represents a real number; thus, the properties of real numbers presented in Chap. 1 may all be used.

2.1 ALGEBRAIC EXPRESSIONS

Algebraic Expression

If one or more numbers or variables (symbols for numbers) are combined by the four fundamental operations, the result is called an *algebraic expression*, or simply an expression.

Example 1 Each of the following is an expression.

$$2xy^2 \tag{1}$$
$$-\tfrac{4}{3}y^5 \tag{2}$$
$$6 \tag{3}$$
$$4x^3y^2 + 3x^2y^3 \tag{4}$$
$$2zy - 6x^2z + xy^4 \tag{5}$$
$$x^4yz^2 - 2xz^3 + x^6 - yz \tag{6}$$
$$8x^4 - 7x^2 + \tfrac{2}{3}x - 5 \tag{7}$$
$$\frac{8x}{y+z} \tag{8}$$

Monomial

An algebraic expression involving only multiplications of real numbers and positive integer powers of variables is called a *monomial*. In Example 1, (1), (2), and (3) are monomials, while neither (8) nor $2x/y$ is one.

Terms The monomials in an expression, together with their respective signs, are called *terms* of the expression. The terms of (5) are $2zy$, $-6x^2z$, and xy^4.

Binomial A sum of two monomials is called a *binomial*, and a sum of
Trinomial three monomials is called a *trinomial*. Thus, (4) is a binomial and (5) is a trinomial.

Polynomial A (finite) sum of monomials is called a *polynomial*. Each of the above except (8) is a polynomial. Expressions (2) and (7) are polynomials in one variable, while (1), (4), (5), and (6) are polynomials in several variables.

Degree One way to classify polynomials is by their degree. The *degree of a monomial in one variable* is the exponent of the variable. Thus, the degree of (2) is 5. The *degree of a polynomial in one variable* is the largest degree of any of its terms. It follows that the degree of (7) is 4. The degree of a constant, like (3), is zero, except that the degree of the constant 0 is not defined.

A polynomial in several variables may also be considered as a polynomial in any combination of its variables. Thus, (6) may be considered as a polynomial in x, y, and z together, or in x and y, or in x and z, or in y and z, or in x, or in y, or in z. Its degree in x is 6, in y is 1, and in z is 3.

The *degree of a monomial in several variables* is the sum of its degrees in each of the variables separately. Thus, the degree of (1) is 3 since $1 + 2 = 3$; the degree of each term in (4) is 5; and the degree of the first term in (6) is 7.

The *degree of a polynomial in several variables* is the largest degree of any of its terms. Hence, the degree of (4) is 5, and the degree of (6) is 7. The degree of (6) is 6 if we consider it as a polynomial in x and y.

The degree of (5) is 5 in x, y, and z; it is 5 in x and y; it is 3 in x and z; and it is 2 in x, 4 in y, and 1 in z.

If a monomial is expressed as the product of two or more sym-
Coefficient bols, each of the symbols is called the *coefficient* of the product of the others.

Example 2 In $3ab$, 3 is the coefficient of ab; a is the coefficient of $3b$; and b is the coefficient of $3a$. In $3ab$, 3 is the numerical coefficient. Usually, when we refer to the coefficient in a monomial, we mean the *numerical coefficient*.

Similar Two monomials or two terms are *similar* if they differ only in
Terms their numerical coefficients.

Example 3 The monomials $3a^2b$ and $-2a^2b$ are similar, and the terms are similar in

$$4\left(\frac{3a}{5b}\right) + 2\left(\frac{3a}{5b}\right)$$

2.2   ADDITION OF MONOMIALS AND POLYNOMIALS

In order to add similar monomials we may use the distributive law

$$a(b + c) = ab + ac$$

By the commutative law, this may be rewritten as

$$(b + c)a = ab + ac$$

This may, of course, be extended to more than two terms, say

$$a(b - c + d) = ab - ac + ad$$

Example 1   Combine terms in $4ab + 6ab - 3ab$ into a single term.

Solution $4ab + 6ab - 3ab = (4 + 6 - 3)ab$ **by distributive axiom**

$$= 7ab \qquad \text{since } 4 + 6 - 3 = 7$$

Example 2   Find the sum of $3a^2b$, $6a^2b$, $-8a^2b$, and $-5a^2b$.

Solution   We express this sum and proceed as follows:

$$3a^2b + 6a^2b + (-8a^2b) + (-5a^2b)$$
$$= [(3 + 6) + (-8 - 5)]a^2b \qquad \text{\textbf{by associative and distributive axioms}}$$
$$= [9 + (-13)]a^2b$$
$$= -4a^2b$$

If a polynomial contains two or more sets of similar terms, we use the commutative and distributive axioms to combine the terms in each set, as illustrated in Example 3.

Example 3 $7y^2 + 3xy + 4x^2 + 5x^3 - 2xy + 2x^2 - 6x^3 - xy - x^2 - 2y^2$

$$= 7y^2 - 2y^2 + 3xy - 2xy - xy + 4x^2 + 2x^2 - x^2 + 5x^3 - 6x^3$$
by commutative axiom
$$= (7 - 2)y^2 + (3 - 2 - 1)xy + (4 + 2 - 1)x^2 + (5 - 6)x^3$$
by distributive axiom and the fact that
$$-xy = -1 \cdot xy \text{ and } -x^2 = -1 \cdot x^2$$
$$= 5y^2 + 0xy + 5x^2 - x^3 \qquad \text{adding each coefficient}$$
$$= 5y^2 + 5x^2 - x^3 \qquad \text{since } 0(xy) = 0$$

The process of adding two or more polynomials makes use of the commutative axiom of addition and the distributive axiom. The commutative axiom enables us to rearrange the terms in the sum so that similar terms are together, and the distributive axiom enables us to combine the similar terms.

Example 4   Find the sum of $3x^2 - 2xy + y^2$, $2xy - 3y^2 - 2x^2$, and $4y^2 - 5x^2 + 4xy$.

Solution　We first write the sum of three trinomials and then rearrange and combine terms as follows:

$$3x^2 - 2xy + y^2 + 2xy - 3y^2 - 2x^2 + 4y^2 - 5x^2 + 4xy$$
$$= 3x^2 - 2x^2 - 5x^2 - 2xy + 2xy + 4xy + y^2 - 3y^2 + 4y^2$$
$$= x^2(3 - 2 - 5) + xy(-2 + 2 + 4) + y^2(1 - 3 + 4)$$
$$= x^2(-4) + xy(4) + y^2(2)$$
$$= -4x^2 + 4xy + 2y^2$$

This process justifies the following procedure, which is often used: Rewrite the expressions so that each one after the first is below the preceding, and at the same time rearrange terms so that those containing the same variables and powers thereof form vertical columns. Finally, draw a horizontal line below the last expression; then combine like terms and write the result below the line. When this is done, we have

$$
\begin{array}{r}
3x^2 - 2xy + y^2 \\
-2x^2 + 2xy - 3y^2 \\
-5x^2 + 4xy + 4y^2 \\
\hline
-4x^2 + 4xy + 2y^2
\end{array}
$$

Example 5　Find the sum of $3a^2 - 2a - 2b^2$, $2ab - 3b - 2a^2$, and $3b^2 - 4a + 4a^2 - 2b$.

Solution　We proceed in this problem as we did in Example 4. However, since the first expression contains neither a b term nor an ab term and the second contains neither an a term nor a b^2 term, when we write the second trinomial below the first, we leave the spaces under $-2a$ and $-2b^2$ blank and write $-3b$ and $2ab$ at the right. Thus, we have

$$
\begin{array}{l}
3a^2 - 2a - 2b^2 \\
-2a^2 \qquad\qquad\quad -3b + 2ab \\
\underline{4a^2 - 4a + 3b^2 - 2b} \\
5a^2 - 6a + \ b^2 - 5b + 2ab
\end{array}
$$

According to the definition of subtraction, $a - b = a + (-b)$. Consequently, we have the following rule for algebraic subtraction:

In order to subtract one number (or one polynomial) from another, we change the sign (or the signs) of the subtrahend and then proceed as in addition.

Example 6　Subtract $8a^2$ from $6a^2$.

Solution   In accordance with the above rule, the solution is

$$6a^2 + (-8a^2) = 6a^2 - 8a^2 = -2a^2$$

Example 7   Subtract $-4xy$ from $-8xy$.

Solution   If we change the sign of the subtrahend and add, we have

$$-8xy - (-4xy) = -8xy + (+4xy) = -8xy + 4xy = -4xy$$

Example 8   Subtract $3a^2 - 2a + 4ab + 3b^2$ from $4a^2 - 2ab - b^2 + 2b$.

Solution   We first write the minuend and then place the subtrahend below it so that like terms in the two expressions are together; thus,

$$4a^2 - 2ab - \ \ b^2 + 2b \qquad \text{minuend}$$
$$\underline{3a^2 + 4ab + 3b^2 \qquad\quad - 2a} \qquad \text{subtrahend}$$
$$\text{difference}$$

Mentally, we now change the sign of each term in the subtrahend, add it to the like term in the minuend, and write the result. The completed problem thus appears as

$$4a^2 - 2ab - \ \ b^2 + 2b$$
$$\underline{3a^2 + 4ab + 3b^2 \qquad\quad - 2a}$$
$$a^2 - 6ab - 4b^2 + 2b + 2a$$

2.3   SYMBOLS OF GROUPING

The symbols of grouping, ( ), [ ], { }, are used to make the meanings of certain expressions clear and to indicate the order in which operations are performed.

Frequently, it is desirable to remove the symbols of grouping from an expression, and we use (1.6) and the distributive axiom for this purpose. We illustrate the procedure with two examples.

Example 1   Remove the parentheses from

$$3x - (2x + 3y) + 2(3x - 4y)$$

and combine similar terms.

Solution $-(2x + 3y) = -2x - 3y$ \qquad **by (1.6)**

and

$2(3x - 4y) = 6x - 8y$ \qquad **by distributive axiom**

Consequently,

$$3x - (2x + 3y) + 2(3x - 4y) = 3x - 2x - 3y + 6x - 8y$$
$$= 3x - 2x + 6x - 3y - 8y$$
by commutative axiom
$$= 7x - 11y$$

This example illustrates the fact that if a pair of grouping symbols preceded by a minus sign is removed, the sign of every term enclosed by the symbols must be changed. However, a pair of grouping symbols preceded by a plus sign can be removed without affecting the signs of the enclosed terms.

Frequently, one or more pairs of grouping symbols are enclosed in another pair. In such cases it is advisable to remove the innermost symbols first.

Example 2   Remove the symbols of grouping from

$$3x^2 - \{3x^2 - xy - [5(x^2 - xy) - 3(x^2 - y^2)] + 4xy\} - 3y^2$$

Solution   We first apply the distributive axiom to the expression in the brackets and get

$$3x^2 - \{3x^2 - xy - [5x^2 - 5xy - (3x^2 - 3y^2)] + 4xy\} - 3y^2$$
$$= 3x^2 - \{3x^2 - xy - [5x^2 - 5xy - 3x^2 + 3y^2] + 4xy\} - 3y^2$$
removing parentheses
$$= 3x^2 - \{3x^2 - xy - [2x^2 - 5xy + 3y^2] + 4xy\} - 3y^2$$
combining terms in brackets
$$= 3x^2 - \{3x^2 - xy - 2x^2 + 5xy - 3y^2 + 4xy\} - 3y^2$$
removing brackets
$$= 3x^2 - \{x^2 + 8xy - 3y^2\} - 3y^2 \quad \text{combining terms in braces}$$
$$= 3x^2 - x^2 - 8xy + 3y^2 - 3y^2 \quad \text{removing braces}$$
$$= 2x^2 - 8xy \quad \text{combining similar terms}$$

It is sometimes beneficial to introduce symbols of grouping instead of removing them.

Example 3   All the following expressions are equal.

$$2x - 4y + 8t + 3p - q$$
$$2x - 4y + 8t - (-3p + q)$$
$$2x - 4y - (-8t - 3p + q)$$
$$2x - (4y - 8t - 3p + q)$$
$$2x - [4y - (8t + 3p - q)]$$

EXERCISE 2.1   Algebraic Expressions

Classify each expression in Probs. 1 to 4 as a monomial, binomial, trinomial, polynomial, or as none of these.

1 $x^2 + 2xy, 2x + 5$

2 $2x + \dfrac{1}{x}, x^4 + x^3 + x^2 + x$

3 $x^2y + xz + 2z, 17y - 2/(x+1)$

4 $z, 8xy^2z$

Classify each polynomial in Probs. 5 to 8 as a monomial, binomial, or trinomial. Give its degree as a polynomial and its degree in y.

5 $63x^2y^3z^7$

6 $x^2 + 2y^2 - yz^2$

7 $7x^4z^4$

8 $xy^2 - yx^3$

Combine similar terms in Progs. 9 to 24.

9 $6a + 4a - 3a$

10 $10y - 5y + 7y$

11 $7xy - 3xy + 2xy + 3xy$

12 $6ac - 5ac - 4ac - 3ac$

13 $7ab - 4ac + 3ab + 6ac$

14 $8x + 2y + 3x - 4x + 3y$

15 $-4xy + 2xz - 3yz + 5xz + 7yz$

16 $-8 + 5ab + 3a - 2 - 5ab$

17 $a + b - c - 3b + 2c - 5a$

18 $ab - 2ac + 3bc + 5bc - 2ac + 3ab$

19 $x + 2xy - 3z + 5x + 8z - 4xy$

20 $2y + 3xz - 4y + 6z - 2xz + 3z$

21 $x + y - 2z + 3w - 2x + y + 2z - 5w$

22 $2a - 3b + 4c + d - a - 2b - 3c + d$

23 $3a - 2b + c - 4d + 2a + 6b - 3d + 4a$

24 $-x + 2y + 3z - x + w - 2y - 2x + 4z$

Remove the symbols of grouping and combine similar terms in Probs. 25 to 36.

25 $x + y + 2(x + 2y) - 3(2x - y)$

26 $2x - y - (x + 3y) + 2(2x + 3y)$

27 $a + 2b + 4(-a - 3b) - 2(2a + 5b)$

28 $a + 2x - 4(x - a) - 3(2a + x)$

29 $x + 3[2y - 3x + 4(-x + 2y)]$

30 $2x - 3y + 6[-x + 3(2x - 5y) + x]$

31 $a + 1 - 2[2a + 4 - 3(-1 + 5a)]$

32 $y - 4 - 3[6 - 2(-1 + 5y) + 3y - 2]$

33 $2x + 2\{y - [4x - (z + 2y)] + z\} - 2y$

34 $3a - \{b - 2[c - 3b + 2(-a + c) + b] + 2a\}$

35 $6d - 4e - \{2f + 2[-d + e - 2(d - f)] + e\} + e$

36 $2g - 3\{h - 4[i + 2(g - h + 2i) - g] + 2h\} + 3i$

Find the indicated sums in Probs. 37 to 48.

37 $(3b + c - d) + (2b + 3c - 4d) + (-4b + 4c - d)$

38 $(3x^2 + x - 1) + (2x^2 + 3x - 4) + (-4x^2 + 4x - 1)$

39 $(2p + q - 2r) + (-3p - 2q - r) + (-2p + 3q - r)$

40 $(2w - 3x + y) + (2w - 3x + y) + (-w + 6x + 3y)$

41
$$6x^3 \qquad\quad + 2x + 1$$
$$\quad - 3x^2 + 4x - 3$$
$$-4x^3 + \ x^2 + 3x$$

42
$$7x^4 \qquad\quad + 2x^2 - x$$
$$\quad - 2x^3 + 3x^2 \qquad - 1$$
$$4x^4 + 2x^3 \qquad\quad - 3x$$

43
$$a - 2b + \ c - \ d$$
$$\quad 3b - 2c + 3d$$
$$2a + \ \ b - 4c$$

44
$$x - \ y + z$$
$$\quad 2y - z + \ w$$
$$3x \qquad + z - 2w$$

45 $2x - 3y$
$\quad\quad 4y + 6z$
$\quad 3x \quad\quad - 4z$

46 $a + c$
$\quad\quad - c - e$
$\quad a \quad\quad - e$

47 $\quad g - \quad e$
$\quad\quad 2e + 3m$
$\quad -2g \quad\quad - 2m$

48 $\quad 2f \quad\quad - i$
$\quad\quad\quad - 2b + i$
$\quad -2f + 3b$

In Probs. 49 to 60, subtract the second expression from the first.

49 $-265, 324$ **50** $28, -35$ **51** $-89, -129$ **52** $-823, -417$

53 $\quad -x + \quad y - 2z$
$\quad\quad 2x + 3y - \quad z$

54 $\quad 2a + \quad b - \quad d$
$\quad -3a + 2b + 2d$

55 $\quad x^2 + 2x - 3$
$\quad -2x^2 - \quad x + 5$

56 $8 + 6a - \quad z$
$\quad 7 - 2a - 4z$

57 $5a^2 + 2b^2 - 3c^2, 2a^2 - 4b^2 - 3c^2$

58 $2x + 3y^2 - 3z^3, y^2 - 2z^3 + 3x$

59 $x^3 + 2x^2y - 3xy^2 - 5y^3, 2x^3 - 5x^2y - 4xy^2 + 2y^3$

60 $a + b - x - y, a - b + x - y$

2.4 PRODUCTS OF MONOMIALS AND POLYNOMIALS

Examples 1 to 4 below illustrate the procedure for obtaining the product of two or more monomials.

Example 1 $(3x^2)(4x^3) = (3)(4)(x^2)(x^3)$ **by associative and commutative axioms**

$\quad\quad\quad\quad\quad\quad = 12x^5$ **since** $x^2 \cdot x^3 = x^5$

Example 2 $(2a)(-3b)(4c) = (2)(-3)(4)a \cdot b \cdot c$

$\quad\quad\quad\quad\quad\quad\quad\quad = (-6)(4)a \cdot b \cdot c$ **by associative axiom and law of signs**

$\quad\quad\quad\quad\quad\quad\quad\quad = -24abc$

Example 3 $(-2a^2)(-8a^3) = (-2)(-8)a^2 \cdot a^3$

$\quad\quad\quad\quad\quad\quad\quad = 16a^5$ **since** $(-2)(-8) = 16$ **and** $a^3 \cdot a^2 = a^5$

Example 4 $(-3x^2y)(-4xy^3)(-5x^4y^2) = (-3)(-4)(-5)(x^2x^1x^4)(y^1y^3y^2)$

$\quad\quad\quad\quad\quad\quad\quad\quad\quad\quad = -60x^{2+1+4} \, y^{1+3+2} = -60x^7y^6$

We use the distributive axiom to obtain the product of a monomial and a polynomial, as illustrated in Examples 5 and 6.

Example 5   Find the product of $2x^3y - 5x^2y^2 + 6xy^3$ and $3x^2y^3$.

Solution   We write the product, use the distributive axiom, and proceed as follows:

$3x^2y^3(2x^3y - 5x^2y^2 + 6xy^3)$

$\quad = 3(2)x^2x^3y^3y^1 + 3(-5)x^2x^2y^3y^2 + 3(6)x^2x^1y^3y^3$

by distributive axiom and commutative axiom

$\quad = 6x^5y^4 - 15x^4y^5 + 18x^3x^6$ **by law of signs and law of exponents**

Example 6   Perform the multiplication in

$$(2a^3b^2 - 4a^2b^3 + 7ab^4)3a^2b^3 - 2ab^4(4a^4b - 6a^3b^2 + a^2b^3) \qquad (1)$$

Solution   By use of the distributive axiom, the associative and commutative axioms, and the law of signs, we have

$$(2a^3b^2 - 4a^2b^3 + 7ab^4)3a^2b^3 = 6a^5b^5 - 12a^4b^6 + 21a^3b^7$$

Similarly,

$$-2ab^4(4a^4b - 6a^3b^2 + a^2b^3) = -8a^5b^5 + 12a^4b^6 - 2a^3b^7$$

Consequently, expression (1) is equal to

$$6a^5b^5 - 12a^4b^6 + 21a^3b^7 - 8a^5b^5 + 12a^4b^6 - 2a^3b^7$$
$$= 6a^5b^5 - 8a^5b^5 - 12a^4b^6 + 12a^4b^6 + 21a^3b^7 - 2a^3b^7$$
by associative and commutative axioms

$$= -2a^5b^5 + 19a^3b^7 \qquad \text{**combining similar terms**}$$

The following example illustrates the use of the distributive axioms for removing symbols of grouping:

Example 7   Remove the symbols of grouping from

$$3x^3 - \{2x^3 - x^2y - 3x[x(x - y) - y(2x - y)] + 4xy^2\} - 3xy^2 \qquad (2)$$

and combine similar terms.

Solution   As stated earlier, it is advisable to remove the innermost symbols first. We therefore proceed as follows:
Expression (2) is equal to

$$3x^3 - \{2x^3 - x^2y - 3x[x^2 - xy - 2xy + y^2] + 4xy^2\} - 3xy^2$$
removing parentheses by distributive axiom

$$= 3x^3 - \{2x^3 - x^2y - 3x[x^2 - 3xy + y^2] + 4xy^2\} - 3xy^2$$
combining terms in brackets

$$= 3x^3 - \{2x^3 - x^2y - 3x^3 + 9x^2y - 3xy^2 + 4xy^2\} - 3xy^2$$
removing brackets by distributive axiom

$$= 3x^3 - \{-x^3 + 8x^2y + xy^2\} - 3xy^2 \qquad \text{**combining terms in braces**}$$
$$= 3x^3 + x^3 - 8x^2y - xy^2 - 3xy^2 \qquad \text{**removing braces by use of (1.6)**}$$
$$= 4x^3 - 8x^2y - 4xy^2 \qquad \text{**combining similar terms**}$$

2.5   THE PRODUCT OF TWO POLYNOMIALS

We obtain the product of two polynomials by repeated applications of the distributive axiom and the commutative axiom. The method is illustrated in the following example.

Example 1   Obtain the product of $3x^3 - 4y^3 - 6x^2y + 2xy^2$ and $2xy - 5x^2 + 3y^2$

Solution $(3x^3 - 4y^3 - 6x^2y + 2xy^2)(2xy - 5x^2 + 3y^2)$

$= (3x^3 - 4y^3 - 6x^2y + 2xy^2)(2xy) + (3x^3 - 4y^3 - 6x^2y + 2xy^2)(-5x^2)$

$+ (3x^3 - 4y^3 - 6x^2y + 2xy^2)(3y^2)$ **by distributive axiom**

$= 6x^4y - 8xy^4 - 12x^3y^2 + 4x^2y^3 - 15x^5$

$+ 20x^2y^3 + 30x^4y - 10x^3y^2 + 9x^3y^2 - 12y^5 - 18x^2y^3 + 6xy^4$

by distributive axiom

$= -15x^5 + 30x^4y + 6x^4y - 10x^3y^2 - 12x^3y^2$

$+ 9x^3y^2 + 20x^2y^3 + 4x^2y^3 - 18x^2y^3 - 8xy^4 + 6xy^4 - 12y^5$

by commutative axiom for addition

$= -15x^5 + 36x^4y - 13x^3y^2 + 6x^2y^3 - 2xy^4 - 12y^5$

combining similar terms

The above process can be abbreviated considerably by use of the vertical method that we shall next explain. We first arrange the terms in each polynomial so that the exponents of one of the variables are in descending numerical order. In this case, we base our arrangement on the exponents of x and proceed as follows:

$3x^3 - 6x^2y + 2xy^2 - 4y^3$

$\underline{\quad\; - 5x^2 \;\; + 2xy + 3y^2\quad}$

$-15x^5 + 30x^4y - 10x^3y^2 + 20x^2y^3$ $-5x^2(3x^3 - 6x^2y + 2xy^2 - 4y^3)$

$+ \;\; 6x^4y - 12x^3y^2 + \;\; 4x^2y^3 - 8xy^4$ $2xy(3x^3 - 6x^2y + 2xy^2 - 4y^3)$

$9x^3y^2 - 18x^2y^3 + 6xy^4 - 12y^5$ $3y^2(3x^3 - 6x^2y + 2xy^2 - 4y^3)$

$\overline{-15x^5 + 36x^4y - 13x^3y^2 + \;\; 6x^2y^3 - 2xy^4 - 12y^5}$

For convenience in adding, the terms in the partial products are arranged so that similar terms form columns.

Example 2   To multiply $2a + 3b$ and $5a - 7b$ vertically, we write

$2a + 3b$

$\underline{5a - 7b}$

$10a^2 + 15ab$

$\underline{\qquad\quad - 14ab - 21b^2}$

$10a^2 + \quad ab - 21b^2$

EXERCISE 2.2   Products of Polynomials

Find the indicated products.

1   $(5x^3y^2)(2x^2y^4)$

2   $(2a^2b)(-3ab^4)$

3   $(6xy^3z^2)(3xy^4z^2)$

4   $(-8ax^4)(-3a^3x^5)$

5   $5x^2(x^3 + 2x)$

6   $-6x^4(2x^5 - x^2)$

7   $8xy^2(2xy + 4xy^2)$

8   $3ab^5(-2a^4b + 3a^2b^2)$

9   $6xy^2z(2x^2yz^2 - 3x^2y^2z)$

10   $-10x^2yz^2(6xyz^2 - 4xy)$

11 $(3x^2yz - 4xy^2z^3)(2xy^2z^4)$.

12 $(4x^2y - 2xy^2z^4)(-3xz^5)$

13 $3x^4y(2x^3y^2z^2 - 5x^4y^4z^3) - 5x^3y^2z(x^4yz - 3x^5y^3z^2)$

14 $3a^3b^4c^2(6a^2b^3 - 5a^4c^5 + 4b^2c^3) - 6a^2b^4c(3a^3b^3c + 2ab^2c^4)$

15 $8p^3q^4r^5(4p^2q^3 - 3q^5r^6) - 4p^3q^7r^4(8p^2r - 3p^3q - 6q^2r^7)$

16 $4a^3b^2c^4(2 - 3a^2b^5) + 3a^5b^6c^4(4b + 3a^3c) - 2a^3bc^2(4bc^2 - 3a^5b^5c^3)$

17 $3 + x[-2 + x(4 + x)]$

18 $1 + x[-5 + x(-3 + 2x)]$

19 $-4 + x\{2 + x[-1 + x(2 + 3x)]\}$

20 $4 + x\{-6 + x[3 + x(-4 - 2x)]\}$

21 $2x(3x + 1) - 3x(2x - 5)$

22 $3x(-2x^2 + 4x - 1) + 4x(x^2 + 6x - 5)$

23 $2x[8x - 3x(x + 3) + 2x^2] - 3x^2[2 - x(2 + 3x) - x^2 + 3x]$

24 $4x[2 + 5x(-x + 1) - 2x] + 3x[-x + 2(2 + 3x) - 2x^2]$

25 $(2x + 5)(x - 3)$

26 $(3b - 1)(1 + 4b)$

27 $(a - 2b)(2a - b)$

28 $(3x + y)(2x - 3y)$

29 $(x - y)(x + y)$

30 $(x + y)(x + y)$

31 $(x - y)(x - y)$

32 $(x - a)(x - b)$

33 $(x + y)^3$

34 $(x - y)^3$

35 $(x + y)(x^2 - xy + y^2)$

36 $(x - y)(x^2 + xy + y^2)$

37 $(x - 2y)(x^2 + xy - y^2)$

38 $(a + 4b)(2a^2 - 3ab - 2b^2)$

39 $(x - 1)(x^2 + 2x - 5)$

40 $(2x + 3)(3x^2 - 3x + 2)$

Multiply the two polynomials in each of Probs. 41 to 52.

41 $2x^2 - 3xy + y^2$
      $3x^2 + 2xy - 2y^2$

42 $3x^2 - xy + 2y^2$
      $2x^2 + xy - 3y^2$

43 $5x^2 + 3xy - y^2$
      $3x^2 - xy + 2y^2$

44 $2x^2 - 3xy + y^2$
      $x^2 + 3xy - y^2$

45 $5x^2 + x - 1$
      $2x^2 + 3x + 2$

46 $4x^2 + x - 2$
      $3x^2 + 2x + 1$

47 $2x^2 - x - 4$
      $-x^2 + 2x + 3$

48 $2x^2 - x + 1$
      $-x^2 - 3x - 3$

49 $2a + 3b - 2c$
      $4a - 2b + c$

50 $a^4 - a^3 + a^2 - a + 1$
      $a + 1$

51 $x^4 - 2x^3 + 4x^2 - 8x + 16$
      $x + 2$

52 $x^3 - x^2y + 2xy^2 + 2y^3$
      $x^2 + xy - y^2$

2.6 PRODUCTS OF SPECIAL BINOMIALS

Section 2.5 showed how to multiply any two polynomials together. The next three sections will show how to do so more quickly in certain special cases. These special cases come up often in mathematics, so it is worthwhile to be able to handle them.

The corresponding terms of the binomials $ax + by$ and $cx + dy$

are similar. We obtain the product of these two binomials by the procedure shown here in which we make use of the distributive axiom.

$$(ax + by)(cx + dy) = ax(cx + dy) + by(cx + dy)$$
$$= acx^2 + adxy + bcxy + bdy^2 \quad \text{by distributive and commutative axioms}$$
$$= acx^2 + (ad + bc)xy + bdy^2 \quad \text{since } adxy + bcxy = (ad + bc)xy$$

Hence, we have

Product of Two Binomials

$$(ax + by)(cx + dy) = acx^2 + (ad + bc)xy + bdy^2 \qquad (2.1)$$

We see, by observing the product on the right, that we obtain the product of two binomials with corresponding terms similar by performing the following steps:

1 Multiply the first terms in the binomials to obtain the first term in the product.

2 Add the products obtained by multiplying the first term in each binomial by the second term in the other. This yields the second term in the product.

3 Multiply the second terms in the binomials to get the third term in the product.

Ordinarily, the computation required by these three steps can be done mentally, and the result can be written with no intermediate steps. This is illustrated by the following example.

Example 1 Obtain the product of $2x - 5y$ and $4x + 3y$.

Solution We write the product as shown here, proceed as directed following the problem, and record the results in the positions indicated by the flow lines.

$$(2x - 5y)(4x + 3y) = 8x^2 - 14xy - 15y^2$$

Get these products mentally:

1 $2x \cdot 4x = $

2 $(2x \cdot 3y) + (-5y \cdot 4x) = 6xy - 20xy = $

3 $-5y \cdot 3y = $

Equation (2.1) is given in two variables, but it also applies to one variable if we take $y = 1$. In fact, it then becomes

$$(ax + b)(cx + d) = acx^2 + (ad + bc)x + bd \qquad (2.2)$$

Example 2 Find $(x + 2)(3x - 5)$.

Solution   By (2.2), we have

$$(x + 2)(3x - 5) = (1)(3)(x^2) + [(1)(-5) + (2)(3)]x + (2)(-5)$$
$$= 3x^2 + x - 10$$

The square of the sum of two numbers x and y is expressed as $(x + y)^2$. Since $(x + y)^2 = (x + y)(x + y)$, we may use Formula (2.1) and get

$$(x + y)^2 = (x + y)(x + y)$$
$$= x^2 + (xy + xy) + y^2 \qquad \text{by (2.1)}$$
$$= x^2 + 2xy + y^2$$

Consequently,

Square of a Sum $$(x + y)^2 = x^2 + 2xy + y^2 \tag{2.3}$$

Similarly,

Square of a Difference $$(x - y)^2 = x^2 - 2xy + y^2 \tag{2.4}$$

Therefore we have the following rule:

The square of the sum or of the difference of two numbers is the square of the first term, plus or minus twice the product of the first term and the second term, plus the square of the second term.

The product of the sum and the difference of the numbers a and b is expressed as $(a + b)(a - b)$. If we apply Formula (2.1) to this product, we get

$$(a + b)(a - b) = a^2 + ab - ab - b^2 = a^2 - b^2$$

Consequently,

Product of the Sum and the Difference of the Same Two Numbers $$(a + b)(a - b) = a^2 - b^2 \tag{2.5}$$

Therefore we have the following rule:

The product of the sum and the difference of the same two numbers is equal to the difference of their squares.

We shall illustrate the application of Formulas (2.3), (2.4), and (2.5) with two examples.

Example 3   By use of Formulas (2.3) and (2.4), obtain the square of $2a + 5b$ and the square of $3x - 4y$.

Solution $(2a + 5b)^2 = (2a)^2 + 2(2a)(5b) + (5b)^2$ **by (2.3)**

$\qquad\qquad = 4a^2 + 20ab + 25b^2$ **by commutative and associative axioms**

$\qquad (3x - 4y)^2 = (3x)^2 - 2(3x)(4y) + (4y)^2$ **by (2.4)**

$\qquad\qquad = 9x^2 - 24xy + 16y^2$

Example 4   By use of Formula (2.5), obtain the product of $3x + 5y$ and $3x - 5y$, and the product of 104 and 96.

Solution $(3x + 5y)(3x - 5y) = (3x)^2 - (5y)^2$ **by (2.5)**

$\qquad\qquad = 9x^2 - 25y^2$

To get the product of 104 and 96, we use the fact that $104 = 100 + 4$ and $96 = 100 - 4$.   Therefore,

$$(104)(96) = (100 + 4)(100 - 4)$$
$$= 100^2 - 4^2 \quad \text{\textbf{by (2.5)}}$$
$$= 10,000 - 16$$
$$= 9984$$

In order to find $(x + y)^3$, we could write it as $(x + y)^2 \cdot (x + y)$ and use (2.3), or we could refer to Prob. 33 in Exercise 2.2.   In any case, we find that

Cube of a Sum $(x + y)^3 = x^3 + 3x^2y + 3xy^2 + y^3$ (2.6)

By writing $x - y = x + (-y)$, we see that

Cube of a Difference $(x - y)^3 = x^3 - 3x^2y + 3xy^2 - y^3$ (2.7)

Two related products, which may be verified by multiplication, are

Sum of Two Cubes $(x + y)(x^2 - xy + y^2) = x^3 + y^3$ (2.8)

Difference of Two Cubes $(x - y)(x^2 + xy + y^2) = x^3 - y^3$ (2.9)

Example 5   (a) $(2a + b)^3 = (2a)^3 + 3(2a)^2(b) + 3(2a)(b^2) + b^3$

$\qquad\qquad = 8a^3 + 12a^2b + 6ab^2 + b^3$

(b) $(5a - 2)^3 = (5a)^3 - 3(5a)^2(2) + 3(5a)(2^2) - 2^3$

$\qquad\qquad = 125a^3 - 150a^2 + 60a - 8$

(c) $(2a + 3c)(4a^2 - 6ac + 9c^2) = 8a^3 + 27c^3$

2.7   PRODUCTS INVOLVING TRINOMIALS

The square of a trinomial may be obtained by suitably grouping the terms and then applying (2.3) or (2.4).   We shall illustrate the procedure with two examples.

Example 1   Obtain the square of $y + z + w$ by use of (2.3).

Solution   We shall treat $z + w$ as a single number, and we indicate our intention by the use of parentheses, writing $y + (z + w)$. Then $[y + (z + w)]^2$ is the square of the sum of two numbers, and we may obtain the square by two applications of (2.3) as is now shown.

$$(y + z + w)^2 = [y + (z + w)]^2$$
$$= y^2 + 2y(z + w) + (z + w)^2 \qquad \text{by (2.3)}$$
$$= y^2 + 2yz + 2yw + z^2 + 2zw + w^2$$

Square of a Trinomial
$$(y + z + w)^2 = y^2 + z^2 + w^2 + 2yz + 2yw + 2zw \qquad (2.10)$$

Example 2   Obtain the square of $2a - 3b - 5c$.

Solution   (a)   In this solution we enclose the first two terms in parentheses and then apply (2.4) to obtain

$$(2a - 3b - 5c)^2 = [(2a - 3b) - 5c]^2$$
$$= (2a - 3b)^2 - 2(2a - 3b)(5c) + (-5c)^2 \qquad \text{by (2.4)}$$
$$= 4a^2 - 12ab + 9b^2 - 20ac + 30bc + 25c^2$$

(b)   In this solution, we use (2.10) with $y = 2a$, $z = -3b$, and $w = -5c$. Then

$$(2a - 3b - 5c)^2 = (2a)^2 + (-3b)^2 + (-5c)^2$$
$$+ 2(2a)(-3b) + 2(2a)(-5c) + 2(-3b)(-5c)$$
$$= 4a^2 + 9b^2 + 25c^2 - 12ab - 20ac + 30bc$$

In some cases it is possible to group the terms in two trinomials so that one of them is the sum of two numbers, and the other is the difference of the same two numbers. Then the product can be obtained by the use of (2.5) and (2.3) or (2.4). The following examples illustrate such situations.

Example 3   Obtain the product of $3x + 2y + 5z$ and $3x + 2y - 5z$.

Solution   If we enclose the first two terms in each trinomial in parentheses, we obtain

$$(3x + 2y) + 5z \qquad \text{and} \qquad (3x + 2y) - 5z$$

and we have the sum and the difference of the same two numbers. Hence, we can obtain the product by first using (2.5) and then complete the problem by using (2.3). Thus, we get

$$[(3x + 2y) + 5z][(3x + 2y) - 5z]$$
$$= (3x + 2y)^2 - (5z)^2 \qquad \text{by (2.5)}$$
$$= 9x^2 + 12xy + 4y^2 - 25z^2$$

Example 4 Obtain the product $(3a + 4b + c)(3a - 4b - c)$.

Solution We first notice that if we group the first two terms in each trinomial together, we do not obtain the product of the sum and difference of the same two numbers. If, however, we group the terms as $[3a + (4b + c)][3a - (4b + c)]$, we see that the expressions in the first and second brackets are, respectively, the sum and the difference of the same two numbers. Note that the parentheses in the second trinomial were inserted after a minus sign, and so the signs of all enclosed terms were changed. We may now complete the solution as

$$(3a + 4b + c)(3a - 4b - c) = [3a + (4b + c)][3a - (4b + c)]$$
$$= (3a)^2 - (4b + c)^2 \qquad \text{by (2.5)}$$
$$= 9a^2 - (16b^2 + 8bc + c^2) \qquad \text{by (2.3)}$$
$$= 9a^2 - 16b^2 - 8bc - c^2 \qquad \substack{\text{removing} \\ \text{parentheses}}$$

2.8 THE SQUARE OF A POLYNOMIAL

We may use (2.3) and (2.10) to obtain the square of a polynomial containing four terms. The method is illustrated in Example 1.

Example 1 Obtain the square of $x + y + z + w$.

Solution
$$(x + y + z + w)^2 = [x + (y + z + w)]^2$$
$$= x^2 + 2x(y + z + w) + (y + z + w)^2 \qquad \text{by (2.3)}$$
$$= x^2 + 2xy + 2xz + 2xw + y^2 + z^2 \qquad \substack{\text{by distributive law} \\ \text{and (2.10)}}$$
$$+ w^2 + 2yz + 2yw + 2zw$$
$$= x^2 + y^2 + z^2 + w^2 + 2xy + 2xz$$
$$+ 2xw + 2yz + 2yw + 2zw$$

The preceding example and (2.10) illustrate the following rule for obtaining the square of a polynomial:

Square of a Polynomial The square of a polynomial is equal to the sum of (1) the squares of the separate terms and (2) twice the product of each term and the sum of all terms that follow it.

At present we are not in a position to prove that this rule is true for polynomials containing more than four terms. The usefulness of the rule, however, justifies its inclusion here. We shall illustrate the application of the rule with the following example.

Example 2 Obtain the square of $2x + 3y - 4z - 2w$.

Solution $(2x + 3y - 4z - 2w)^2 = (2x)^2 + (3y)^2 + (-4z)^2 + (-2w)^2$
$$+ 2(2x)(3y - 4z - 2w)$$
$$+ 2(3y)(-4z - 2w) + 2(-4z)(-2w)$$
$$= 4x^2 + 9y^2 + 16z^2 + 4w^2 + 12xy - 16xz - 8xw$$
$$- 24yz - 12yw + 16zw$$

EXERCISE 2.3   Special Products

Find the products in Probs. 1 to 68.

1 $(x - 1)(x + 3)$ **2** $(x + 2)(x - 4)$ **3** $(x + 3)(x + 2)$

4 $(x - 5)(x - 3)$ **5** $(2x + 3)(3x - 2)$ **6** $(3x - 4)(x + 5)$

7 $(4x - 1)(-2x + 3)$ **8** $(2x - 3)(3x - 4)$ **9** $(2x - y)(3x + y)$

10 $(x + 2y)(x - 4y)$ **11** $(3x + 5y)(2x - 3y)$ **12** $(2x + y)(x - 3y)$

13 $(8k - 3m)(9k - 5m)$ **14** $(6c - 11d)(2c - 5d)$ **15** $(3a - 10b)(4a + 7b)$

16 $(7h + 3k)(4h - 7k)$ **17** $(a + 2b)^2$ **18** $(3h - k)^2$

19 $(2x + 3)^2$ **20** $(3r + t)^2$ **21** $(5m + 2n)^2$

22 $(4x - 5)^2$ **23** $(10x + 5)^2$ **24** $(50 + x)^2$

25 $35^2 = (30 + 5)^2$ **26** $45^2 = (40 + 5)^2$ **27** $75^2 = (70 + 5)^2$

28 $95^2 = (90 + 5)^2$ **29** $48^2 = (50 - 2)^2$ **30** $53^2 = (50 + 3)^2$

31 $41^2 = (50 - 9)^2$ **32** $68^2 = (50 + 18)^2$ **33** $(x - 3)(x + 3)$

34 $(a - 4)(a + 4)$ **35** $(2x + 1)(2x - 1)$ **36** $(b + 2)(b - 2)$

37 $(3x - 2y)(3x + 2y)$ **38** $(5a + 2b)(5a - 2b)$ **39** $(4c - 7d)(4c + 7d)$

40 $(6m + 5p)(6m - 5p)$ **41** $(102)(98) = (100 + 2)(100 - 2)$ **42** $(62)(58)$

43 $(47)(53)$ **44** $(26)(34)$ **45** $(23)(37)$

46 $(35)(45)$ **47** $(84)(76)$ **48** $(93)(67)$

49 $(a + 2b)^3$ **50** $(2a - b)^3$ **51** $(x + 3y)^3$

52 $(5x - y)^3$ **53** $(5x + 2)^3$ **54** $(3a - 4)^3$

55 $(2x - 3)^3$ **56** $(3x + 5)^3$ **57** $(x + y - 1)^2$

58 $(2x + y + 3)^2$ **59** $(a - 2b + 3c)^2$ **60** $(2a - b + 2c)^2$

61 $(2x + y + z)(2x + y - z)$ **62** $(x - 3y + 2z)(x - 3y - 2z)$ **63** $(a - b + c)(a + b - c)$

64 $(a + 2b - c)(a - 2b + c)$ **65** $(a + b - c + d)^2$ **66** $(x - y + z - 1)^2$

67 $(2x + y - a + 3)^2$ **68** $(x - 2a + 4b - 5)^2$ **69** Verify Eq. (2.4).

70 Verify Eq. (2.6). **71** Verify Eq. (2.9).

72 Verify the result of Example 1 of Sec. 2.8 by writing $(x + y + z + w)^2 = [(x + y) + (z + w)]^2$ and expanding.

2.9   COMMON FACTORS

Factors   A number is *factored* if it is expressed as the product of two or more other numbers.   Several such expressions may be possible.

Prime Factors

For example, $6 = 6 \cdot 1 = 3 \cdot 2 = 9 \cdot \frac{2}{3}$. In this section, however, we shall consider only *prime* factors. A *prime number* is an integer greater than 1 that has no factors except itself and 1. Therefore, the only prime factors of 6 are 3 and 2. We shall limit our discussion in this section to polynomials in which the numerical coefficients are integers. Examples of such polynomials are $3x^2 + 2x + 1$; $4x^3 + 2x^2y - xy^2 + 2y^3$; and $4ab - 3bc +$

Prime Polynomial

$3ac + 4cd$. A polynomial with integral coefficients is said to be *prime* or *irreducible* if it has no factors of the same type except itself and 1.

Common Factor

Each term of a polynomial may be divisible by the same monomial. This monomial is called the *common factor*. This polynomial can be factored by expressing it as the product of the common factor and the sum of the quotients obtained by dividing each term of the polynomial by the common factor. This procedure is justified by the distributive axiom. If the factors thus obtained are not prime, we continue factoring. For example,

$$204 = 3 \cdot 68 = 3 \cdot 17 \cdot 4 = 3 \cdot 17 \cdot 2^2$$
$$ax + ay - az = a(x + y - z)$$
$$6a^3b + 3a^2b^2 - 18ab^3 = 3ab(2a^2 + ab - 6b^2)$$
$$= 3ab(2a - 3b)(a + 2b)$$

This method can be extended to include polynomials in which the terms are polynomials that have a common factor that is not a monomial. For example,

$$(a + b)(a - b) + 2(a + b) = (a + b)(a - b + 2)$$
$$(x - 1)(x + 2) - (x - 1)(2x - 3) = (x - 1)[(x + 2) - (2x - 3)]$$
$$= (x - 1)(x + 2 - 2x + 3)$$
$$= (x - 1)(-x + 5)$$

2.10   FACTORING BY GROUPING

Frequently, the terms of a polynomial can be grouped in such a way that each group has a common factor, and then the method of common factors can be applied. We shall illustrate the method by three examples.

Example 1   Factor $ax + bx - ay - by$.

Solution   We notice that the first two terms have the common factor x, and the third and fourth have the common factor y. Hence, we group the terms in this way,

$$(ax + bx) - (ay + by)$$

and then proceed as follows:

$$ax + bx - ay - by = (ax + bx) - (ay + by)$$
$$= x(a + b) - y(a + b)$$
$$= (a + b)(x - y)$$

with x as common factor of first group, y as common factor of second group, and with $a + b$ as common factor of $x(a + b)$ and $y(a + b)$

We could also have written

$$ax + bx - ay - by = ax - ay + bx - by$$
$$= a(x - y) + b(x - y)$$
$$= (x - y)(a + b)$$

commutative law

common factors

distributive law

Example 2   Factor $2x^2 + 10x - xy - 5y$.

Solution $2x^2 + 10x = 2x(x + 5)$ and $-xy - 5y = -y(x + 5)$, so
$$2x^2 + 10x - xy - 5y = 2x(x + 5) - y(x + 5)$$
$$= (x + 5)(2x - y)$$

Example 3   Factor $a^2 + ab - 2b^2 + 2a - 2b$.

Solution   Since
$$a^2 + ab - 2b^2 = (a + 2b)(a - b) \quad \text{and} \quad 2a - 2b = 2(a - b)$$
we proceed as indicated here:
$$a^2 + ab - 2b^2 + 2a - 2b = (a^2 + ab - 2b^2) + (2a - 2b)$$
$$= (a + 2b)(a - b) + 2(a - b)$$
$$= (a - b)(a + 2b + 2) \quad \text{with } a - b \text{ as common factor}$$

Example 4   Factor $4c^2 - a^2 + 2ab - b^2$.

Solution $4c^2 - a^2 + 2ab - b^2 = 4c^2 - (a^2 - 2ab + b^2)$
$$= (2c)^2 - (a - b)^2 \quad \text{by (2.4)}$$
$$= [2c + (a - b)][2c - (a - b)] \quad \text{by (2.5)}$$
$$= (2c + a - b)(2c - a + b)$$

EXERCISE 2.4   Factoring

In Probs. 1 to 4, factor each number into prime factors.

1   18

2   40

3   30

4   84

Factor each expression in Probs. 5 to 44.

5    $2x + 2y$

6    $3x - 3a$

7    $7x + 7t$

8    $-3a + 3b$

9 $4x + 8y - 16t$	**10** $6x^2 + 3x + 15$
11 $6xy - 2xz + 8yz$	**12** $5x^2 + 5xy - 15y^2$
13 $x^2 + xy$	**14** $a^3 - 2a$
15 $x^2y - xy^2$	**16** $2b^2 - 5b$
17 $x^{2n} - x^n$	**18** $a^{n+2} - a^{n-1}$
19 $3a^2b - 12ab^2 + 9ab$	**20** $x^2yz - 2xy^2z - 3xz$
21 $3(a - b) - x(a - b)$	**22** $4a(x + 2y) - b(x + 2y)$
23 $(a + 3b)(2x) + (a + 3b)(3y)$	**24** $(a + b)(2x + y) - (a + b)(x - 3y)$
25 $ax + bx + ay + by$	**26** $2ax - 2ay - bx + by$
27 $2ax + ay - 6bx - 3by$	**28** $6ax - 12ay + 4bx - 8by$
29 $ax + ay - az - x - y + z$	**30** $3ax + 3ay - 3a - 6bx - 6by + 6b$
31 $2ax - 4ay + x - 2y - 3z - 6az$	**32** $3ax - 2a - 3bx + 2b - 8 + 12x$
33 $2x^2 + 5x - 2xy - 5y$	**34** $3x^2 - 12x + xy - 4y$
35 $2a + 2b - ab - a^2$	**36** $2ax + 4a^2 + 2a + x$
37 $x^2 - x + 3x - 3$	**38** $2x^2 - x + 4x - 2$
39 $2 - 2x + 3x - 3x^2$	**40** $4 + 3x - 6x^2 - 8x$
41 $x^3 + x^2 + x + 1$	**42** $x^3 - 2x^2 - 3x + 6$
43 $x^4 - x^3 + 2x^2 - 2x$	**44** $-2x^4 + 6x^3 + 10x^2 - 30x$

2.11 FACTORS OF A QUADRATIC TRINOMIAL

A trinomial of the type $px^2 + qxy + ry^2$, where p, q, and r are integers, is a *quadratic trinomial with integral coefficients*. In this section, we shall discuss the method for finding the two binomial factors of such a trinomial if the factors exist. Sine we shall use (2.1) for this purpose, we rewrite it here with the members interchanged.

$$acx^2 + (ad + bc)xy + bdy^2 = (ax + by)(cx + dy) \qquad (2.1)$$

Example 1 To use (2.1) in factoring $3x^2 - 10xy - 8y^2$, we must find four numbers a, b, c, and d such that $ac = 3$, $bd = -8$, and $ad + bc = -10$. The only possibilities for a and c are ± 3† and ± 1, and these numbers must have the same sign since $ac > 0$. The possibilities for b and d are ± 4 and ∓ 2 or ± 8 and ∓ 1, where the double sign indicates that if one of the two numbers is positive, the other must be negative. If we let $a = 3$ and $c = 1$, then $ad + bc = 3d + b = -10$, and this is true only when $d = -4$ and $b = 2$. Therefore,

$$3x^2 - 10xy - 8y^2 = (3x + 2y)(x - 4y)$$

†This symbol means $+3$ or -3.

We call attention to the fact that $-3x - 2y$ and $-x + 4y$ are also factors of $3x^2 - 10xy - 8y^2$, as the reader may verify. By the distributive and commutative axioms, however,

$$
\begin{aligned}
(-3x - 2y)(-x + 4y) &= -1(3x + 2y)(-1)(x - 4y) \\
&= (-1)(-1)(3x + 2y)(x - 4y) \\
&= (3x + 2y)(x - 4y)
\end{aligned}
$$

Hence, if the first term of the trinomial is positive, we may choose positive values for a and c. Furthermore, if we had let $a = 1$ and $c = 3$, then $ad + bc = -10$ would have been $d + 3b = -10$, which is satisfied by $b = -4$ and $d = 2$; so the factors would be $(x - 4y)(3x + 2y)$, which are the same as obtained above except in reverse order.

Usually, if a trinomial is factorable, the factors may be found after relatively few trials. If the first and last terms of the trinomial can be factored in more than one way, several combinations may be tried before the correct one is found. If the correct factors are not readily seen, it is advisable to list the possible corresponding values of a and c and of b and d, and then systematically try each possibility until the correct combination is found.

Quadratic trinomials of one variable, $px^2 + qx + r$, may be treated in a similar manner.

Example 2   In order to write $10x^2 - 11x - 6$ as $(ax + b)(cx + d)$, we need $ac = 10$, $ad + bc = -11$, and $bd = -6$. If we let $a = 5$ and $c = 2$, then we need $ad + bc = 5d + 2b = -11$, and $bd = -6$. These hold if $b = 2$ and $d = -3$, and so $10x^2 - 11x - 6 = (5x + 2)(2x - 3)$.

In Example 2, if we had chosen $a = 2$ and $c = 5$, we would have gotten $b = -3$ and $d = 2$, giving the factors $2x - 3$ and $5x + 2$, which are the same as above except for order. We can cut down the number of actual possibilities we need to consider: $ac = 10$ would originally give $a = 10, 5, 2, 1, -1, -2, -5, -10$. However, taking a and c positive leaves only $10, 5, 2, 1$ for a, and then using only 10 or 5 gives all possibilities except for order.

Even after cutting down the number of possibilities, there may be considerable trial and error involved. Some polynomials, such as $x^2 + 3x + 1$, do not have factors $ax + b$ and $cx + d$ with a, b, c, and d all being integers. If we know this beforehand, we can avoid the futile search for factors. By methods in Chap. 6, it can be shown that a quadratic trinomial $px^2 + qx + r$, where p, q, and r are integers, may be factored into terms with integer coefficients if and only if

$q^2 - 4pr$ is a perfect square   (1)

In Example 2, $q^2 - 4pr = (-11)^2 - 4(10)(-6) = 121 + 240 = 361 = 19^2$, so factors with integer coefficients do exist (and we found them). For $x^2 + 3x + 1$, we have $q^2 - 4pr = 3^2 - 4(1)(1) = 9 - 4 = 5$, which is not a perfect square. Thus, no factors with integer coefficients exist, and we need not waste time looking for any. In Example 1, $q^2 - 4pr = (-10y)^2 - 4(3)(8y)^2 = 100y^2 + 96y^2 = 196y^2 = (14y)^2$, so factors exist. We may also consider the trinomial in Example 1 as a polynomial in y, namely, $-8y^2 - 10xy + 3x^2$. In this case, $q^2 - 4pr = (-10x)^2 - 4(-8)(3x^2) = 100x^2 + 96x^2 = 196x^2 = (14x)^2$; it makes no difference whether we consider x or y as the variable, since (1) will hold both ways or neither way.

Example 3 Factor $6x^2 + 47xy + 15y^2$.

Solution We refer to (2.1) and see that a, c, b, and d must have values that satisfy $ac = 6$, $bd = 15$, and $ad + bc = 47$. The corresponding values of a and c are

a	6	3
c	1	2

and those of b and d are

b	±3	±5	±15	±1
d	±5	±3	±1	±15

where the corresponding values must have the same sign. Since, however, $ad + bc = 47$ is positive, and a and c are also positive, we may rule out the negative signs for b and d. Now, using $a = 6$ and $c = 1$, we have $ad + bc = 6d + b = 47$. We readily verify that no one of the above pairs of corresponding values of b and d satisfies $6d + b = 47$. Hence, we use $a = 3$ and $b = 2$ and have $3d + 2c = 47$, and we see at once that $d = 15$ and $c = 1$ satisfy this equation. Consequently, we have

$$6x^2 + 47xy + 15y^2 = (3x + y)(2x + 15y)$$

Example 4 Factor $12x^2 + 71xy - 60y^2$.

Solution Referring to (2.1), we see that a, c, b, and d must be chosen so that

$$ac = 12 \qquad bd = -60 \qquad ad + bc = 71$$

Considering the possibilities $(12x \quad)(x \quad)$, $(6x \quad)(2x \quad)$, and $(4x \quad)(3x \quad)$ leads to

$$12x^2 + 71xy - 60y^2 = (4x - 3y)(3x + 20y)$$

2.12   TRINOMIALS THAT ARE PERFECT SQUARES

If a trinomial is the square of a binomial, we know by Formulas (2.3) and (2.4) that two of its terms are perfect squares and hence are positive, and that the other term is twice the product of the square roots of these two. Furthermore, such a trinomial is the square of a binomial composed of the square roots of the two perfect-square terms of the trinomial connected by the sign of the other term.

The trinomial $px^2 + qx + r$ is the square of $ax + b$, where a, b, p, q, and r are all integers, if and only if

$$q^2 - 4pr = 0 \tag{1}$$

This corresponds to Eq. (1) in Sec. 2.11. Thus, $px^2 + qx + r$ is factorable if $q^2 - 4pr$ is a perfect square, and $px^2 + qx + r$ is a perfect square if $q^2 - 4pr = 0$.

Example   Factor each of the following:
(a) $4x^2 - 12xy + 9y^2$
(b) $9a^2 + 24ab + 16b^2$
(c) $(2a - 3b)^2 - 8(2a - 3b) + 16$

Solution   Since in (a), $q^2 - 4pr = (-12y)^2 - (4)(4)(9y^2) = 144y^2 - 144y^2 = 0$, we see that (a) is a perfect square. Also

$$4x^2 = (2x)^2 \qquad 9y^2 = (3y)^2 \qquad 12xy = 2(2x)(3y)$$

so we have

$$4x^2 - 12xy + 9y^2 = (2x)^2 - 2(2x)(3y) + (3y)^2 = (2x - 3y)(2x - 3y)$$
$$= (2x - 3y)^2 \tag{1}$$

For (b),

$$9a^2 + 24ab + 16b^2 = (3a)^2 + 2(3a)(4b) + (4b)^2 = (3a + 4b)^2 \tag{2}$$

For (c), we may consider it as a polynomial in $2a - 3b$. Then $q^2 - 4pr = (-8)^2 - 4(1)(16) = 64 - 64 = 0$. Thus it is a perfect square, and

$$(2a - 3b)^2 - 8(2a - 3b) + 16 = (2a - 3b)^2 - 2(2a - 3b)(4) + 4^2$$
$$= [(2a - 3b) - 4]^2$$
$$= (2a - 3b - 4)^2 \tag{3}$$

Note   A trinomial that is a perfect square can be factored by the method given in Sec. 2.11. However, if a trinomial is recognized as a perfect square, it can be factored more quickly by this method.

EXERCISE 2.5   Factoring Trinomials

In Probs. 1 to 16, state whether or not the trinomial is factorable into factors with integer coefficients.   If it is also a perfect square, write PS.   Do not factor.

1 $x^2 + 2x + 1$

2 $x^2 + 2x - 1$

3 $x^2 - x - 12$

4 $x^2 + x - 12$

5 $2x^2 - 5x - 3$

6 $3x^2 - 7x + 5$

7 $4x^2 - 6x + 3$

8 $5x^2 + 9x - 3$

9 $x^2 + 8xy + 14y^2$

10 $x^2 - 18xy + 81y^2$

11 $x^2 - 7xy + 12y^2$

12 $x^2 - 2xy - 42y^2$

13 $9a^2 + 24ab + 16b^2$

14 $12p^2 + 7pq - 12q^2$

15 $12b^2 - 8bc - 15c^2$

16 $18m^2 + 19mt - 12t^2$

Factor each of the following expressions.

17 $x^2 + 2x - 15$

18 $x^2 + 8x + 12$

19 $x^2 - 12x + 32$

20 $x^2 + x - 12$

21 $2y^2 + y - 3$

22 $3y^2 - y - 4$

23 $2y^2 + 9y + 10$

24 $4y^2 - 23y + 15$

25 $8a^2 - 2a - 3$

26 $4a^2 + 28a + 49$

27 $10a^2 - 29a + 10$

28 $28a^2 + 57a + 14$

29 $6b^2 + 25b + 21$

30 $64p^2 - 16p + 1$

31 $54r^2 - 147r + 65$

32 $40y^2 + 22y - 35$

33 $x^2 - 4xy + 4y^2$

34 $2x^2 - 5xy + 2y^2$

35 $4a^2 - 20ab + 25b^2$

36 $4m^2 + 11mt - 3t^2$

37 $16y^2 - 32yz + 15z^2$

38 $30x^2 + 7xy - 15y^2$

39 $30a^2 - 31ab - 12b^2$

40 $16r^2 + 46rs - 35s^2$

41 $36x^2 - 60xy + 25y^2$

42 $96a^2 - 28ab - 55b^2$

43 $81c^2 + 144cd + 64d^2$

44 $72x^2 - 37xy - 24y^2$

45 $x^2 + y^2 + 1 + 2xy + 2x + 2y =$
$(x + y)^2 + 2(x + y) + 1$

46 $a^2 - 2ab + b^2 + 2a - 2b + 1$

47 $4x^2 + 4xy + y^2 - 8x - 4y + 4$

48 $x^2 - 4xy + 4y^2 + 8x - 16y + 16$

49 $(y - 4)^2 - 5(y - 4) + 6$

50 $(b + 1)^2 - 3(b + 1) - 10$

51 $(a + b)^2 + (a + b) - 2$

52 $(x + y)^2 + 3z(x + y) - 10z^2$

53 $2x^2 - xy - y^2 - x + y$

54 $2x^2 + 3xy + y^2 + 6x + 3y$

55 $a^2 - 6ab + 9b^2 + ac - 3bc$

56 $2a^2 + ab - 10b^2 + 6ac + 15bc$

57 $12a^2 + 24ab + 9b^2 - 6a - 9b$

58 $4x^2 - 8xy + 4y^2 + 4xz - 4yz$

59 $x^3 + 4x^2y + 4xy^2 - 2x^2 - 4xy$

60 $3b^3 - 3b + 3ab^2 + 3ab$

2.13 FACTORS OF A BINOMIAL

If we interchange the members of formula (2.5), we obtain

The Difference of Two Squares

$$a^2 - b^2 = (a + b)(a - b) \tag{2.5}$$

Consequently we have the following rule for factoring the difference of the squares of two numbers:

The difference of the squares of two numbers is equal to the product of the sum and the difference of the two numbers.

We illustrate the application of this rule with the following example.

Example 1 Factor

(a) $49a^2 - 16b^2$
(b) $(a + 3b)^2 - 4$
(c) $x^2 - (y + z)^2$

Solution

(a) $49a^2 - 16b^2 = (7a)^2 - (4b)^2$
$$= (7a + 4b)(7a - 4b)$$

(b) $(a + 3b)^2 - 4 = (a + 3b)^2 - 2^2$
$$= (a + 3b + 2)(a + 3b - 2)$$

(c) $x^2 - (y + z)^2 = [x + (y + z)][x - (y + z)]$
$$= (x + y + z)(x - y - z)$$

The expression $a^2 + b^2$ cannot be factored using real numbers. The sum and difference of two cubes can be expressed as $x^3 + y^3$ and $x^3 - y^3$, respectively. Rewriting Eq. (2.8) gives

Sum of Two Cubes

$$x^3 + y^3 = (x + y)(x^2 - xy + y^2) \tag{2.8}$$

Similarly,

Difference of Two Cubes

$$x^3 - y^3 = (x - y)(x^2 + xy + y^2) \tag{2.9}$$

Hence, we have the following two rules:

If a binomial is expressed as the sum of the cubes of two numbers, one factor is the sum of the two numbers. The other factor is the square of the first number minus the product of the two numbers plus the square of the second number.

If a binomial is expressed as the difference of the cubes of two numbers, one factor is the difference of the two numbers. The other factor is the square of the first number plus the product of the two numbers plus the square of the second number.

Example 2 Factor

(a) $8x^3 + 27y^3$

(b) $27a^3 - 64b^6$

Solution (a) $8x^3 + 27y^3 = (2x)^3 + (3y)^3$

$$= (2x + 3y)[(2x)^2 - (2x)(3y) + (3y)^2]$$
$$= (2x + 3y)(4x^2 - 6xy + 9y^2)$$

(b) $27a^3 - 64b^6 = (3a)^3 - (4b^2)^3$

$$= (3a - 4b^2)[(3a)^2 + (3a)(4b^2) + (4b^2)^2]$$
$$= (3a - 4b^2)(9a^2 + 12ab^2 + 16b^4)$$

Note 1 In the case of the sum of two cubes, the sign between the two terms of the first factor is plus, and the sign of the middle term of the second factor is minus.

Note 2 In the case of the difference of two cubes, the sign between the two terms of the first factor is minus, and the sign of the middle term of the second factor is plus.

Note 3 In each case, the middle term of the second factor is the product of the two terms of the first factor (not twice the product).

Frequently, the factors obtained by use of Formulas (2.5), (2.8), and (2.9) can be further factored by a repeated application of one or more of these formulas.

Example 3 Factor

(a) $x^6 - y^6$

(b) $a^8 - y^8$

Solution (a) $x^6 - y^6 = (x^3)^2 - (y^3)^2$

$$= (x^3 - y^3)(x^3 + y^3) \qquad \text{by (2.5)}$$
$$= (x - y)(x^2 + xy + y^2)(x + y)(x^2 - xy + y^2) \qquad \text{by (2.9) and (2.8)}$$

(b) $x^8 - y^8 = (x^4)^2 - (y^4)^2$

$$= (x^4 - y^4)(x^4 + y^4) \qquad \text{by (2.5)}$$
$$= (x^2 - y^2)(x^2 + y^2)(x^4 + y^4) \qquad \text{by (2.5)}$$
$$= (x - y)(x + y)(x^2 + y^2)(x^4 + y^4) \qquad \text{by (2.5)}$$

2.14   TRINOMIALS REDUCIBLE TO THE DIFFERENCE OF TWO SQUARES

Frequently, it is possible to convert a trinomial into the difference of two squares by adding and subtracting a monomial that is a perfect square. For example, if we add $4a^2b^2$ to $a^4 + 2a^2b^2 + 9b^4$ and then subtract $4a^2b^2$, we have

$$a^4 + 2a^2b^2 + 9b^4 = a^4 + 2a^2b^2 + 9b^4 + 4a^2b^2 - 4a^2b^2$$
$$= a^4 + 6a^2b^2 + 9b^4 - 4a^2b^2$$
$$= (a^2 + 3b^2)^2 - (2ab)^2$$

This process is possible only when the trinomial becomes a perfect square when a perfect-square monomial is *added* to it.

After a trinomial has been converted into the difference of two squares, it can be factored by use of (2.5).

Example 1   Factor $4x^4 - 21x^2y^2 + 9y^4$.   (1)

Solution   In this trinomial, $4x^4$ and $9y^4$ are perfect squares, and twice the product of their square roots is

$$2(2x^2)(3y^2) = 12x^2y^2$$

Since $-12x^2y^2 = -21x^2y^2 + 9x^2y^2$, we get a perfect square if we add $9x^2y^2$ to the trinomial (1). Then we must also subtract $9x^2y^2$, and get

$$4x^2 - 21x^2y^2 + 9y^4 = 4x^4 - 21x^2y^2 + 9y^4 + 9x^2y^2 - 9x^2y^2$$
$$= 4x^4 - 12x^2y^2 + 9y^4 - 9x^2y^2$$
$$= (2x^2 - 3y^2)^2 - (3xy)^2$$
$$= [(2x^2 - 3y^2) + 3xy][(2x^2 - 3y^2) - 3xy]$$

   by (2.5)

$$= (2x^2 + 3xy - 3y^2)(2x^2 - 3xy - 3y^2)$$

Example 2   In Example 3a in the previous section, we factored $x^6 - y^6$ as a difference of two squares. It may also be treated as a difference of cubes by writing

$$x^6 - y^6 = (x^2)^3 - (y^2)^3$$
$$= (x^2 - y^2)(x^4 + x^2y^2 + y^4) \qquad \text{\textbf{by (2.9)}}$$
$$= (x - y)(x + y)[(x^4 + 2x^2y^2 + y^4) - x^2y^2] \qquad \text{\textbf{by (2.5), and adding and subtracting } } x^2y^2$$
$$= (x - y)(x + y)[(x^2 + y^2)^2 - (xy)^2] \qquad \text{\textbf{the square of a sum}}$$
$$= (x - y)(x + y)(x^2 + y^2 - xy)(x^2 + y^2 + xy) \qquad \text{\textbf{by (2.5)}}$$

EXERCISE 2.6   Factoring Special Binomials

Verify the equations in Probs. 1 to 8.

1 $(3^2 - 2^2)(4^2 - 1^2) = 14^2 - 11^2$

2 $(4^2 - 2^2)(5^2 - 2^2) = 24^2 - 18^2$

3 $(7^2 - 4^2)(3^2 - 1^2) = 25^2 - 19^2$

4 $(8^2 - 7^2)(6^2 - 5^2) = 83^2 - 82^2$

5 $(3^2 + 2^2)(4^2 + 1^2) = 10^2 + 11^2$

6 $(7^2 + 4^2)(3^2 + 1^2) = 17^2 + 19^2$

7 $(6^2 + 5^2)(4^2 + 3^2) = 9^2 + 38^2$

8 $(2^2 + 5^2)(2^2 + 3^2) = 11^2 + 16^2$

Factor the following expressions.

9 $a^2 - 4$

10 $b^2 - 36$

11 $x^2 - 4y^2$

12 $y^2 - 64z^2$

13 $25x^2 - 36y^2$

14 $121a^2 - 289b^2$

15 $9x^4 - 64y^2$

16 $81c^6 - 49d^4$

17 $121h^{12} - 4t^6$

18 $100x^{100} - 64y^{64}$

19 $64x^8 - 25z^{10}$

20 $256a^{20} - 36b^4$

21 $a^3 - 8$

22 $b^3 + 27$

23 $8a^3 - 1$

24 $b^3 + 8$

25 $8x^3 - 27y^3$

26 $64a^3 + 27b^3$

27 $125y^3 + 8x^3$

28 $216a^3 - 125c^3$

29 $x^6 + 1$

30 $y^9 + 8$

31 $b^{15} - 8c^9$

32 $27m^{12} + t^{21}$

33 $27a^{27} + 216b^{216}$

34 $343x^9 + 27y^{12}$

35 $512r^{24} - 27s^6$

36 $216x^{27} - 343y^{18}$

37 $x^4 - y^4$

38 $81x^8 - 1$

39 $a^9 + b^9$

40 $a^{12} - b^6$

41 $x^2 + 2xy + y^2 - 9$

42 $a^2 + 4ab + 4b^2 - 16$

43 $x^2 - 6x + 9 - y^2$

44 $4a^2 + 12ab + 9b^2 - x^2$

45 $x^3 + 3x^2y + 3xy^2 + y^3 + z^3$

46 $x^3 + 3x^2y + 3xy^2 + y^3 - 8a^3$

47 $x^2 + 2xy + y^2 - a^2 - 2ab - b^2$

48 $x^2 + 4xy + 4y^2 - a^2 + 6a - 9$

49 $2x^4 - 8$

50 $a^3 - 4ab^2$

51 $12c^4 - 3c^2d^2$

52 $x^4y - x^2y^3$

53 $2a^3 - 16b^3$

54 $81 + 3a^6$

55 $a^4 - ab^3$

56 $x^4y + 8xy^4$

57 $a^2 - b^2 + a - b$

58 $x^2 - 4y^2 - x - 2y$

59 $4x^2 - y^2 + 4x - 2y$

60 $a^3 + b^3 + a + b$

61 $a^4 - 3a^2 + 1$

62 $x^4 - 5x^2 + 4$

63 $y^8 + 5y^4 + 9$

64 $x^4 + 4x^2 + 16$

65 $a^4 + a^2b^2 + b^4$

66 $x^4 - 8x^2y^2 + 4y^4$

67 $9m^4 - 7m^2n^2 + n^4$

68 $a^4 - 12a^2b^2 + 16b^4$

2.15   A POLYNOMIAL DIVIDED BY A MONOMIAL

We obtain the quotient of two monomials that have powers of the same variable by first expressing it as a product and then using the laws of exponents.

Example 1 $\dfrac{48a^4b^2}{12a^2b^2} = \dfrac{48}{12}\cdot\dfrac{a^4}{a^2}\cdot\dfrac{b^2}{b^2}$ **written as a product**

$\qquad\qquad = 4a^{4-2}b^{2-2}$ **by law of exponents**

$\qquad\qquad = 4a^2b^0$

$\qquad\qquad = 4a^2$ **since $b^0=1$**

If the dividend is a polynomial and the divisor is a monomial, we use the distributive law in obtaining the quotient:

Division of a Polynomial by a Monomial $\dfrac{a+b-c}{d} = \dfrac{a}{d}+\dfrac{b}{d}-\dfrac{c}{d}$

Therefore, *if the dividend is a polynomial and the divisor is a monomial, the quotient is the algebraic sum of the quotients obtained by dividing each term in the dividend by the divisor.*

Example 2 Divide $6x^4 + 4x^3y^3 - 3x^2y^2 - 2x^2$ by $3x^2$.

Solution $\dfrac{6x^4 + 4x^3y^3 - 3x^2y^2 - 2x^2}{3x^2} = \dfrac{6x^4}{3x^2}+\dfrac{4x^3y^3}{3x^2}-\dfrac{3x^2y^2}{3x^2}-\dfrac{2x^2}{3x^2}$

$\qquad\qquad = 2x^2 + \tfrac{4}{3}xy^3 - x^0y^2 - \tfrac{2}{3}x^0$

$\qquad\qquad = 2x^2 + \tfrac{4}{3}xy^3 - y^2 - \tfrac{2}{3}$ **since $x^0=1$**

2.16 QUOTIENT OF TWO POLYNOMIALS

In this section, we shall discuss the procedure for obtaining the quotient of two polynomials. We shall discuss only polynomials in which each term is the product of an integer and one or more integral powers of a variable. The *degree of a polynomial* in any variable is the number that is equal to the greatest exponent of that variable in the polynomial. For example, $3x^4 + 2x^3y + 5x^2y^2 + xy^3$ is a polynomial of degree 4 in x and of degree 3 in y.

Degree of a Polynomial

Before discussing the procedure for dividing one polynomial by another, we shall consider the quotient of 231 and 5. By methods of arithmetic, we have

$\begin{array}{r} 46 \\ 5)\overline{231} \\ 20 \\ \hline 31 \\ 30 \\ \hline 1 \end{array}$ or $231 = 5(46) + 1$

or $\tfrac{231}{5} = 46 + \tfrac{1}{5}$

Hence, if we divide 231 by 5, we obtain the integer 46 and the remainder 1. Likewise,

$\dfrac{6x^2+4}{3x} = \dfrac{6x^2}{3x}+\dfrac{4}{3x} = 2x + \dfrac{4}{3x}$ or $6x^2+4 = (3x)(2x)+4$

and we have the quotient $2x$ and the remainder 4.

We may readily verify that in each of the above examples, the following relation is satisfied:

Dividend = (divisor)(quotient) + remainder (2.11)

since

$$231 = 5(46) + 1 \quad \text{and} \quad 6x^2 + 4 = (3x)(2x) + 4$$

In order to divide one polynomial by another, we first arrange the terms in each polynomial so that they are in the order of the descending powers of some variable that appears in each. Then we seek the quotient that is a polynomial and that satisfies the relation (2.11), where the degree of the remainder in the variable chosen as the basis of the arrangement of terms is less than the degree of the divisor in that variable.

The formal steps in the process of dividing one polynomial by another are the following:

1 Arrange the terms in both the dividend and divisor in the order of the descending powers of a variable that appears in each.
2 Divide the first term in the dividend by the first term in the divisor to get the first term in the quotient.
3 Multiply the divisor by the first term in the quotient, and subtract the product from the dividend.
4 Treat the remainder obtained in step 3 as a new dividend, and repeat steps 2 and 3.
5 Continue this process until a remainder is obtained that is of lower degree than the divisor in the variable that is chosen as a basis for the arrangement in step 1.

The quotient can be checked by the use of the relation

Dividend = (divisor)(quotient) + remainder

We shall illustrate the process by the following examples.

Example 1 Find the quotient obtained by dividing $6x^2 + 5x - 1$ by $2x - 1$.

Solution Here the dividend is $6x^2 + 5x - 1$, the divisor is $2x - 1$, and we seek the quotient that satisfies the relation

$$6x^2 + 5x - 1 = (2x - 1)(\text{quotient}) + \text{remainder} \quad (1)$$

Since the degree of the dividend is 2, the degree of the divisor is 1, and the degree of the remainder is less than 1, it follows that the degree of the quotient is 1. Hence we write the quotient in the form $ax + b$, substitute this expression in Eq. (1) and get

$$6x^2 + 5x - 1 = (2x - 1)(ax + b) + \text{remainder} \tag{2}$$

We shall now divide $6x^2 + 5x - 1$ by $2x - 1$ by use of the usual long division process. If this process is examined closely, it may be seen that it is a condensation of the procedure outlined above in steps 1 to 5.

$$
\begin{array}{r}
3x + 4 \\
2x - 1 \overline{)6x^2 + 5x - 1} \\
\underline{6x^2 - 3x} \\
8x - 1 \\
\underline{8x - 4} \\
3
\end{array}
$$

divisor — (above $2x - 1$)
quotient — $3x + 4$
dividend — $6x^2 + 5x - 1$
$(2x - 1)(3x)$ — $6x^2 - 3x$
subtracting — $8x - 1$
$(2x - 1)4$ — $8x - 4$
remainder — 3

Example 2 Divide $x^3 - 2x^2 - x + 2$ by $x - 1$.

Solution We follow the general procedure by writing

$$
\begin{array}{r}
x^2 - x - 2 \\
x - 1 \overline{)x^3 - 2x^2 - x + 2} \\
\underline{x^3 - x^2} \\
-x^2 - x \\
\underline{-x^2 + x} \\
-2x + 2 \\
\underline{-2x + 2} \\
0
\end{array}
$$

$(x - 1)(x^2)$

subtracting

$(x - 1)(-x)$

subtracting

$(x - 1)(-2)$

remainder is 0

Notice that since the remainder is 0, (2.11) says that

$$x^3 - 2x^2 - x + 2 = (x - 1)(x^2 - x - 2)$$

and thus $x - 1$ is a factor of $x^3 - 2x^2 - x + 2$.

Example 3 Divide $2x^2 + 5xy - 4y^2$ by $2x - y$.

Solution

$$
\begin{array}{r}
x + 3y \\
2x - y \overline{)2x^2 + 5xy - 4y^2} \\
\underline{2x^2 - xy} \\
6xy - 4y^2 \\
\underline{6xy - 3y^2} \\
-y^2
\end{array}
$$

$(2x - y)(x)$

subtracting

$(2x - y)(3y)$

remainder

Equation (2.11) says $2x^2 + 5xy - 4y^2 = (2x - y)(x + 3y) - y^2$.

Example 4 Divide $6x^4 - 6x^2y^2 - 3y^4 + 5xy^3 - x^3y$ by $-2y^2 + 2x^2 + xy$.

Solution   We shall arrange the terms in the dividend and divisor in the order of the descending powers of x and proceed as follows:

$$
\begin{array}{r}
3x^2 - 2xy + y^2 \\
\hline
2x^2 + xy - 2y^2\,)\,6x^4 - x^3y - 6x^2y^2 + 5xy^3 - 3y^4 \\
6x^4 + 3x^3y - 6x^2y^2 \\
\hline
-4x^3y + 5xy^3 - 3y^4 \\
-4x^3y - 2x^2y^2 + 4xy^3 \\
\hline
2x^2y^2 + xy^3 - 3y^4 \\
2x^2y^2 + xy^3 - 2y^4 \\
\hline
-y^4
\end{array}
$$

divisor / quotient / dividend

$(2x^2 + xy - 2y^2)3x^2$

subtracting

$(2x^2 + xy - 2y^2)(-2xy)$

subtracting

$(2x^2 + xy - 2y^2)y^2$

remainder

EXERCISE 2.7   Division of Polynomials

Find the quotients in Probs. 1 to 20.

1 $x^6 \div x^2$

2 $a^8 \div a^3$

3 $-z^7 \div z^4$

4 $b^6 \div -b^2$

5 $8a^4 \div 2a^3$

6 $27x^5y^2 \div 9x^3y^2$

7 $-32a^2b^3c^4 \div 4ab^3c^2$

8 $-16x^3y^5z^3 \div -4x^2y^2z^2$

9 $45c^4d^5e^3 \div -9c^4d^2e^3$

10 $-36r^7s^5t^3 \div 9r^5s^2t$

11 $-42a^{10}b^8c^5 \div -7a^3b^5c^2$

12 $56x^{12}y^{10}z^8 \div -8x^8y^6z^2$

13 $(6a^6 + 8a^4) \div 2a^2$

14 $(9b^7 - 6b^5) \div 3b^3$

15 $(12x^8 - 9x^4) \div -3x^4$

16 $(-24a^6 - 16a^4) \div -4a^3$

17 $(6x^4y^3 - 4x^3y^4 + 2xy^5) \div 2xy^3$

18 $(12a^{12}b^6 - 8a^8b^4 - 4a^4b^2) \div 4a^4b^2$

19 $(20r^7s^5t^3 - 25r^4s^4t^4 - 35rs^2t^5) \div -5rs^2t^3$

20 $(-12x^8y^2z^5 - 18x^6y^3z^4 + 24x^2y^4z^5) \div -6x^2y^2z^4$

In Probs. 21 to 36, show that the second expression is a factor of the first by dividing the first by the second and making sure the remainder is zero.

21 161; 7

22 962; 13

23 19,175; 59

24 16,928; 46

25 $2x^2 - 5x + 3;\ x - 1$

26 $6a^2 - 5a - 4;\ 2a + 1$

27 $12b^2 + 25bc + 12c^2;\ 4b + 3c$

28 $3y^2 - 10xy + 3x^2;\ 3y - x$

29 $6x^3 - 13x^2 + 8x - 3;\ 2x - 3$

30 $6a^3 - 2a^2 + 4a - 16;\ 3a - 4$

31 $6x^3 - 13x^2y + 8xy^2 - 3y^3;\ 2x - 3y$

32 $5w^3 + 23w^2z + 14wz^2 + 8z^3;\ w + 4z$

33 $2a^3 + 3a^2 - 5a - 6;\ 2a^2 - a - 3$

34 $4x^4 + x^3 - 4x^2 + 6x - 3;\ x^2 + x - 1$

35 $6y^3 - 11y^2d + 7yd^2 - 6d^3;\ 3y^2 - yd + 2d^2$

36 $2a^4 + a^3b - 4a^2b^2 + 6ab^3 - 3b^4;\ 2a^2 + 3ab - 3b^2$

In Probs. 37 to 52, find the quotient and remainder when the first expression is divided by the second.

37 1488; 53

38 670; 8

39 17,618; 41

40 62,905; 72

41 $3a^2 - a - 2; a + 1$ **42** $2x^2 - x - 7; x - 3$

43 $6b^2 + 17bd + 19d^2; 2b + 3d$ **44** $16x^2 - 34xy - y^2; 8x + 3y$

45 $2x^3 - 3x^2 - 5x + 11; x - 2$ **46** $14x^3 + 15x^2 + 10x - 2; 2x + 1$

47 $10x^3 + 24x^2y - 27xy^2 - 14y^3; 5x + 2y$ **48** $12a^3 - 28a^2b + 13ab^2 - b^3; 6a + b$

49 $6x^3 + x^2 - 19x + 6; 2x^2 + x - 6$

50 $15t^4 - 7t^3 + 9t^2 - 7t + 19; 5t^2 - 4t + 1$

51 $24y^3 - 23y^2z - 52yz^2 - 9z^3 + 2z^2; 3y^2 - 4yz - 5z^2$

52 $27x^4 - 24x^3y - 14x^2y^2 + 31xy^3 - 10y^4 + 7; 9x^2 + 7xy - 2y^2$

In Probs. 53 to 56, (a) consider both expressions as being in the variable x, (b) consider both expressions as being in the variable y. For both (a) and (b), find the quotient and remainder when the first expression is divided by the second.

53 $6x^2 + 11xy + 11y^2; 2x + y$

54 $8x^2 - 2xy - 9y^2; 2x - 3y$

55 $8x^4 - 2x^3y + x^2y^2 + 2xy^3 + y^4; 2x^2 - xy + y^2$

56 $6x^4 + x^3y + 15x^2y^2 + 2xy^3 + 12y^4; 3x^2 + 2xy + 4y^2$

2.17 SUMMARY

In Chap. 1 we worked with real numbers mainly as constants. In Chap. 2 we consider real numbers in expressions involving both constants and variables. In either form, the field axioms and other properties of real numbers are true. Addition, subtraction, multiplication, and division of monomials and polynomials in general are treated in detail, along with removing and inserting symbols of grouping. The distributive law is especially useful since it involves both addition and multiplication. Multiplication of general polynomials is followed by products of special types of binomials and trinomials. Many of the multiplication formulas are used as formulas for factoring when read "backward." Common factors are included, as well as rules telling when trinomials can be factored—how to factor them is still a matter of trial and error, although certain rules decrease the number of possibilities which need to be considered. Division of two polynomials is discussed, showing how to get the quotient and remainder. If the remainder is zero, the divisor is a factor.

When working with real numbers as expressions, in this chapter and throughout the book, the student should try to see the parallel situation of real numbers as constants.

EXERCISE 2.8 Review

1 What is the degree of $2x^2yz - 5xz^6 + 8x^4y^4$? What is its degree in x? In y? In z?

Combine similar terms in Probs. 2 to 4.

2 $6x + 2y - 3z + 8y - 2x + 3z - 4y$

3 $3x - 2y + 4[-x + 3(y - 2x) - 3y] - 2x$

4 $3a + 2\{b - 4[a + 2(2a - b) + b] - a\} + 2b$

Perform the indicated operations in Probs. 5 to 18.

5 $(6x^3 + x^2 - 1) + (2x^2 + 3x - 4) - (4x^3 - x - 7)$

6 $(a + 3b - 2ab) - 4(2a - b - ab)$

7 $(9x^2yz - 4x^2y^4z^3)(-3xy^4)$

8 $3a^2b(4ab^4 - 2a^3b^3) - 2ab^2(5a^4b^2 - 3a^2b^3)$

9 $3x + 2 - x\{4x + 3[2 - 5x(x + 1) - x] + 3 - 2x\}$

10 $(2x - 5y)(7x^2 + 4xy - y^2)$ **11** $(2a + 5b + 8c)(8a - 7b - 6c)$

12 $(2x + 7y)(8x - 5y)$ **13** $(3x - 2y)^2$

14 $(3x - 2y)^3$ **15** $(3x - 2y)^4$

16 $62^2 = (50 + 12)^2$ **17** $(62)(78) = (70 - 8)(70 + 8)$

18 $(x^2 - 3x + 1)^2$

Factor the expressions in Probs. 19 to 37.

19 $18x^2y - 6xyz^2 + 9xz^2$ **20** $4x^2 + 8xy + x + 2y$

21 $6x(a - 2b) + 5(2b - a)$ **22** $2ax - 3a - 28by + 7ay - 8bx + 12b$

23 $(a - b)(3x + 2y) + (a - b)(5x - 4y)$ **24** $3x^3 + 3x^2 - 6x$

25 $2x^3 - 10x^2 + 3x - 15$ **26** $24x^2 - 10x - 25$

27 $24x^2 - 2xy - 15y^2$ **28** $9x^2 - 24xy + 16y^2$

29 $9x^2 + 6xy + y^2 - 6x - 2y + 1$ **30** $x^3 + 2x^2y + xy^2 - 2x^2 - 2xy$

31 $32x^3 - 98xy^2$ **32** $64a^8 - b^{12}$

33 $64a^9 - b^{12}$ **34** $x^9 - 8$

35 $x^8 - 9$ **36** $x^2 + 6x + 9 - y^2$

37 $x^4 + 3x^2y^2 + 4y^4$

In Probs. 38 to 41, find the quotient and remainder when the first expression is divided by the second.

38 $2x^2 + 7x + 2, 2x + 3$ **39** $10x^2 + 31xy + 12y^2, 2x + 5y$

40 $x^3 - 2x^2y - xy^2 + 14y^3, x^2 - 4xy + 7y^2$

41 $x^4 - 3x^3 + 5x^2 - 12x + 1, x^2 - 3x + 1$

42 Suppose p, q, and r are integers. Show that $px^2 + qx + r$ is factorable (with integer coefficients) if and only if $px^2 - qx + r$ is factorable.

3
rational expressions

Fractions such as $\frac{1}{2}$, $\frac{3}{4}$, $\frac{2}{3}$, and $\frac{5}{7}$ commonly occur in arithmetic and are used constantly in everyday living. Algebraic fractions are equally important in mathematics and in all fields in which algebra is applied. Skill in the operations that involve fractions is essential for progress in any of these fields. In this chapter we consider the basic operations dealing with fractions.

3.1 DEFINITIONS

In Chapter 1 we dealt with real numbers and their properties. We shall be doing the same thing in this chapter, except that the real numbers will appear in more varied forms than they did earlier.

Rational Number   A *rational number* was defined to be m/n, where m and n are integers and $n \neq 0$. We define a *rational expression* to be a

Rational Expression   quotient of polynomials P/Q, where Q is not 0. Examples are

$$\frac{2}{x} \qquad \frac{4x^2 - 3}{2 - 5x} \qquad \frac{1 - ab}{a^3 + b^3} \qquad \frac{3}{5} \qquad \frac{4x^4 - x^3y + 2y^2}{3}$$

Numerator   In a rational expression P/Q, P is called the *numerator* and Q the
Denominator   *denominator*. Since Q is never allowed to be 0, then in the first example above, $2/x$, we allow x to be any real number except 0. In $(4x^2 - 3)/(2 - 5x)$, we allow x to be any real number except $2/5$, and in the third expression we require that $b \neq -a$.

Fraction   A *fraction* is another name for either a rational number or a rational expression.

Signs of a Fraction   There are three *signs in a fraction*—the sign of the numerator, the sign of the denominator, and the sign of the fraction itself. Each sign may be $+$ or $-$, giving rise to eight possibilities. The fractions

$$+\frac{+P}{+Q} \qquad +\frac{-P}{-Q} \qquad -\frac{+P}{-Q} \qquad -\frac{-P}{+Q} \tag{3.1}$$

are all equal to each other. For example, the second and third fractions are equal since

$$+\frac{-P}{-Q} = (+1)\frac{(-1)(P)}{(-1)(Q)} = (-1)(-1)\frac{(-1)(P)}{(-1)(Q)} = (-1)\left(\frac{-1}{1}\right)\frac{(-1)P}{(-1)Q}$$

$$= -\frac{(-1)(-1)P}{(1)(-1)Q} = -\frac{(1)P}{(-1)Q} = -\frac{+P}{-Q}$$

The other four,

$$-\frac{-P}{-Q} \qquad -\frac{+P}{+Q} \qquad +\frac{-P}{+Q} \qquad +\frac{+P}{-Q} \tag{3.2}$$

are also equal to each other and are the negatives of the fractions in (3.1).

Example 1 $\quad \dfrac{2}{-3} = -\dfrac{2}{3} = \dfrac{-2}{3} = -\dfrac{-2}{-3}$

Example 2 $\quad \dfrac{x-1}{2x^2+5} = -\dfrac{-(x-1)}{2x^2+5} = -\dfrac{1-x}{2x^2+5}$

Example 3 $\quad \dfrac{-ab+b^2}{-1-b^3} = \dfrac{ab-b^2}{1+b^3}$

3.2 THE FUNDAMENTAL PRINCIPLE OF FRACTIONS

Fundamental Principle of Fractions

Equation (1.26) is called the *fundamental principle of fractions*. It states that, for $b \neq 0$ and $d \neq 0$,

$$\frac{a}{b} = \frac{c}{d} \qquad \text{if and only if} \qquad ad = bc \tag{3.3}$$

Here a/b and c/d are fractions, and a, b, c, and d are polynomials.

The main consequence of the fundamental principle of fractions is Eq. (1.27), which states that

$$\frac{a}{b} = \frac{ae}{be} = \frac{a/f}{b/f} \tag{3.4}$$

If this is written as

$$\frac{a \cdot e}{b \cdot e} = \frac{a}{b}$$

Law of Cancellation

we have the *law of cancellation*, which is valid for $b \neq 0$ and $e \neq 0$.

To reduce a fraction to lowest terms, we divide the numerator and denominator by every factor that is common to both. If the members of the fractions are polynomials, it is advisable to factor each as a first step in the reduction.

Example 1 $\dfrac{8}{12} = \dfrac{2 \cdot 4}{3 \cdot 4} = \dfrac{2}{3}$

Example 2 $\dfrac{x - y}{x^2 - y^2} = \dfrac{(x - y)(1)}{(x - y)(x + y)} = \dfrac{1}{x + y}$

Example 3 $\dfrac{a^2 b^5}{3abc} = \dfrac{ab^4}{3c}$

Example 4 $\dfrac{(a^3 + b^3)(2a^2 + 5ab - 3b^2)}{(2a^2 + ab - b^2)(a + 3b)}$

$$= \dfrac{(a + b)(a^2 - ab + b^2)(2a - b)(a + 3b)}{(a + b)(2a - b)(a + 3b)}$$

$$= a^2 - ab + b^2$$

Notice that to reduce a fraction by cancellation, the factor common to the numerator and denominator must be a factor of the whole numerator and the whole denominator, not just a part of either. Thus

$$\dfrac{2x + 2y + z}{3x + 3y - w} = \dfrac{2(x + y) + z}{3(x + y) - w} \neq \dfrac{2 + z}{3 - w}$$

since $x + y$ is a factor of only part of the numerator and denominator. Also,

$$\dfrac{4x^2 + x - 3}{4x^2 - 2x + 1} \neq \dfrac{x - 3}{-2x + 1}$$

3.3   MULTIPLICATION AND DIVISION OF FRACTIONS

To multiply two fractions, or rational expressions, we use (1.30). Thus,

$$\dfrac{a}{b} \cdot \dfrac{c}{d} = \dfrac{ac}{bd} \tag{3.5}$$

This rule may be extended to include the product of more than two fractions:

The product of two or more fractions is a fraction whose numerator is the product of the numerators and whose denominator is the product of the denominators.

The application of this rule is illustrated by the following examples.

Example 1   Obtain the product of a/b, c/d, and $(x - y)/(x + y)$.

Solution $$\frac{a}{b} \cdot \frac{c}{d} \cdot \frac{x - y}{x + y} = \frac{ac(x - y)}{bd(x + y)}$$

Frequently, the numerator and denominator of a product have a common factor. In such cases the fraction should be reduced to lowest terms by dividing both members by the common factor. It is advisable to factor the numerators and denominators of the fractions, if possible, before the final product is written; then the factors that are common to the numerator and denominator of the product can be more easily detected.

Example 2   Obtain the product of $\frac{2}{3}$, $\frac{6}{7}$, and $\frac{21}{8}$.

Solution $$\frac{2}{3} \cdot \frac{6}{7} \cdot \frac{21}{8} = \frac{2}{3} \cdot \frac{3 \cdot 2}{7} \cdot \frac{3 \cdot 7}{2 \cdot 2 \cdot 2} \qquad \text{factoring 6, 21, and 8}$$

$$= \frac{2 \cdot 2 \cdot 3 \cdot 3 \cdot 7}{2 \cdot 2 \cdot 2 \cdot 3 \cdot 7} \qquad \text{by (3.5)}$$

$$= \frac{3}{2} \qquad \text{dividing numerator and denominator by } 2 \cdot 2 \cdot 3 \cdot 7$$

Example 3   Obtain the product of

$$\frac{a^2 - 4b^2}{2a^2 - 7ab + 3b^2} \qquad \frac{6a - 3b}{2a + 4b} \qquad \text{and} \qquad \frac{a^2 - 4ab + 3b^2}{a^2 - ab - 2b^2}$$

Solution $$\frac{a^2 - 4b^2}{2a^2 - 7ab + 3b^2} \cdot \frac{6a - 3b}{2a + 4b} \cdot \frac{a^2 - 4ab + 3b^2}{a^2 - ab - 2b^2}$$

$$= \frac{(a - 2b)(a + 2b)}{(2a - b)(a - 3b)} \cdot \frac{3(2a - b)}{2(a + 2b)} \cdot \frac{(a - b)(a - 3b)}{(a + b)(a - 2b)}$$

factoring

$$= \frac{3(a - 2b)(a + 2b)(2a - b)(a - 3b)(a - b)}{2(a - 2b)(a + 2b)(2a - b)(a - 3b)(a + b)}$$

by (3.5)

$$= \frac{3(a - b)}{2(a + b)}$$

dividing numerator and denominator by
(a − 2b)(a + 2b)(2a − b)(a − 3b)

Division of two fractions is carried out as a multiplication according to (1.31). It is

$$\frac{a/b}{c/d} = \frac{a}{b} \cdot \frac{d}{c} = \frac{ad}{bc} \qquad\qquad (3.6)$$

Therefore, *to obtain the quotient of two fractions, multiply the dividend by the reciprocal of the divisor.*

Example 4   Divide $3x^2/4a$ by $6x^3/5a^2$

Solution $\dfrac{3x^2}{4a} \div \dfrac{6x^3}{5a^2} = \dfrac{3x^2}{4a} \cdot \dfrac{5a^2}{6x^3} = \dfrac{15a^2x^2}{24ax^3} = \dfrac{5a}{8x}$

Example 5   Divide

$$\dfrac{x^2 - y^2}{x + 3y} \qquad \text{by} \qquad \dfrac{x - y}{x^2 + 3xy}$$

Solution $\dfrac{x^2 - y^2}{x + 3y} \div \dfrac{x - y}{x^2 + 3xy} = \dfrac{x^2 - y^2}{x + 3y} \cdot \dfrac{x^2 + 3xy}{x - y}$ inverting divisor and multiplying

$$= \dfrac{(x^2 - y^2)(x^2 + 3xy)}{(x + 3y)(x - y)}$$ by (3.5)

$$= \dfrac{(x - y)(x + y)(x)(x + 3y)}{(x + 3y)(x - y)}$$ factoring

$$= (x + y)x$$ dividing numerator and denominator

$$= x^2 + xy$$ by $(x + 3y)(x - y)$

EXERCISE 3.1   Reducing and Multiplying Fractions

Reduce the fractions to lowest terms in Probs. 1 to 20.

1   $\dfrac{15}{21}$ 2 $\dfrac{34}{510}$ 3 $\dfrac{48}{78}$ 4 $\dfrac{57}{266}$

5   $\dfrac{x^2y}{xy^3}$ 6 $\dfrac{a^2bc^2}{ab^2c}$ 7 $\dfrac{xy^4}{x^2y^2z^2}$ 8 $\dfrac{c^2d^5}{bcd}$

9   $\dfrac{4x - 4y}{x^2 - y^2}$ 10 $\dfrac{x^3 - x^2}{x - 1}$ 11 $\dfrac{5xy - 10xy^2}{1 - 4y^2}$ 12 $\dfrac{1 + 4b + 4b^2}{3a + 6ab}$

13   $\dfrac{x^2 - x - 2}{2x^2 - 3x - 2}$ 14 $\dfrac{2x^2 - xy - 6y^2}{4x^2 + 8xy + 3y^2}$

15   $\dfrac{(a - 2b)(a^2 + 4ab + 3b^2)}{(a + 3b)(2a^2 + ab - b^2)}$ 16 $\dfrac{(c + 3d)(9c^2 - 9cd + 2d^2)}{(2c + d)(3c^2 + 7cd - 6d^2)}$

17   $\dfrac{2ax - 2ay + bx - by}{3ax - bx - 3ay + by}$ 18 $\dfrac{2ax + 4ay + 3bx + 6by}{3ax + 2bx + 4by + 6ay}$

19   $\dfrac{a - 1}{(2a - 1)a - 1}$ 20 $\dfrac{x + 2}{(x + 3)x + 2}$

In Probs. 21 to 28, fill in the blank space correctly.

21   $\dfrac{x - 2y}{2x + y} = \dfrac{3x^2 - 6xy}{}$ 22 $\dfrac{2a + 5b}{a - 4b} = \dfrac{}{2a^2b - 8ab^2}$ 23 $\dfrac{a + b}{b^2 - a^2} = \dfrac{}{a^2 - b^2}$

24   $\dfrac{2x - 5y}{x - 3y} = \dfrac{5xy - 2x^2}{}$ 25 $\dfrac{15xy^2}{18x^3y^4} = \dfrac{}{6x^2y^2}$ 26 $\dfrac{32x^2y^3z^4}{48x^5y^4z^5} = \dfrac{}{3x^3yz}$

27   $\dfrac{x^2 - 4}{x^2 - x - 2} = \dfrac{}{x + 1}$ 28 $\dfrac{a^2 - 4b^2}{a^2 - ab - 2b^2} = \dfrac{}{a + b}$

Label each statement in Probs. 29 to 36 as true or false.

29 $\dfrac{2x + y}{x - 3y} = -\dfrac{-2x + y}{x + 3y}$

30 $\dfrac{a^2 - ab^2}{-a - 3x} = \dfrac{ab^2 - a^2}{a + 3x}$

31 $\dfrac{ax - bx + by}{ay + bx - cy} = -\dfrac{bx - ax - by}{bx + ay - cy}$

32 $-\dfrac{2a^2 + b^3}{-a + b} = \dfrac{2a^2 + b^3}{b - a}$

33 $\dfrac{ax + b}{ax + d} = \dfrac{b}{d}$

34 $\dfrac{b(c + d) - 4}{b(c - d) + 4} = \dfrac{c + d - 4}{c - d + 4}$

35 $\dfrac{3x^2 + 5x + 6}{3x^2 + 5x + 3} = 2$

36 $\dfrac{26}{65} = \dfrac{2}{5}$

Perform the indicated operations in the following problems.

37 $\dfrac{4}{5} \cdot \dfrac{3}{8} \cdot \dfrac{10}{9}$

38 $\dfrac{8}{7} \cdot \dfrac{45}{36} \cdot \dfrac{21}{40}$

39 $\dfrac{18}{35} \cdot \dfrac{7}{108} \div \dfrac{5}{72}$

40 $\dfrac{78}{98} \div \left(\dfrac{18}{70} \cdot \dfrac{52}{21}\right)$

41 $\dfrac{2a^3b}{3a^2b^5} \cdot \dfrac{5a^4b^2}{4a^5b^3}$

42 $\dfrac{8x^5y^7}{12x^7y^5} \cdot \dfrac{9x^4y^9}{14x^8y^5}$

43 $\dfrac{x^7y^8}{8xy^5} \div \dfrac{9x^3y^3}{12x^4y^{10}}$

44 $\dfrac{5b^5c^5}{8b^8c^8} \div \dfrac{4b^4c^4}{16b^3c^7}$

45 $\dfrac{x^5y}{xy^2} \cdot \dfrac{x^2y^4}{x^5y^2} \cdot \dfrac{x^8y^3}{x^3y^4}$

46 $\dfrac{xy^2z^5}{x^2y^5z^4} \cdot \dfrac{x^4z^4}{x^2y^2z^6}$

47 $\dfrac{a^4b^5}{a^4b^6} \cdot \dfrac{a^5b^2}{a^3b^8} \div \dfrac{a^2b^2}{a^4b^7}$

48 $\dfrac{a^2b^4c^2}{a^4b^3c^4} \div \dfrac{a^2b^5c^3}{a^5b^4}$

49 $\dfrac{2x - 4y}{6x + 3y} \cdot \dfrac{2x^2 + xy}{4x + 8y}$

50 $\dfrac{ab + ac}{ab + bc} \cdot \dfrac{ab - bc}{bc + c^2}$

51 $\dfrac{3x - 9y}{8x + 4y} \div \dfrac{12x + 12y}{2x^2 + xy}$

52 $\dfrac{2x^2 + 3xy}{xy - 3y^2} \div \dfrac{3x^2 + 6xy}{2x - 6y}$

53 $\dfrac{a^2}{a - b} \cdot \dfrac{b^3}{a + b} \cdot \dfrac{a^2 - b^2}{(ab)^2}$

54 $\dfrac{x^3 - y^3}{x} \cdot \dfrac{y - x}{x^3 + x^2y + xy^2} \cdot \dfrac{xy}{(x - y)^2}$

55 $\dfrac{xy}{x^2 - y^2} \cdot \dfrac{x - y}{x^3 - y^3} \div \dfrac{x^2 - xy + y^2}{x^2 + xy + y^2}$

56 $\dfrac{2a - b}{a - b} \cdot \dfrac{(a - b)^3}{2ab} \div \dfrac{4a^2 - b^2}{4a + 2b}$

57 $\dfrac{x^2 - y^2}{2x^2 - 3xy + y^2} \cdot \dfrac{2x^2 + 5xy - 3y^2}{x^2 + 4xy + 3y^2} \cdot \dfrac{x^2 - 2xy - 3y^2}{x^2 - 4xy + 3y^2}$

58 $\dfrac{4x^2 - y^2}{2x^2 + 5xy + 2y^2} \cdot \dfrac{x^2 - 4y^2}{6x^2 - 5xy + y^2} \cdot \dfrac{9x^2 - y^2}{3x^2 - 5xy - 2y^2}$

59 $\dfrac{2x^2 - 5xy - 3y^2}{x^2 - 9y^2} \cdot \dfrac{2x^2 + 5xy - 3y^2}{4x^2 - y^2} \cdot \dfrac{x^3 - y^3}{x^2 - xy}$

60 $\dfrac{a^2 - 16b^2}{a^2 + 7ab + 12b^2} \cdot \dfrac{a^2 - 9b^2}{a^2 - 2ab - 8b^2} \cdot \dfrac{a^2 - 4b^2}{a^2 - 6ab + 9b^2}$

61 $\dfrac{3a^2 + 10ab + 3b^2}{3a^2 + ab} \cdot \dfrac{2a^2 + 5ab + 2b^2}{2ab + b^2} \div \dfrac{2a^2 + 3ab - 2b^2}{2a^2b - ab^2}$

62 $\dfrac{x^2 + 2xy - xz - 2yz}{x^2 + 5xy + 6y^2} \cdot \dfrac{x^2 + 2xy - 3y^2}{xy - xz - yz + z^2} \div \dfrac{y^2 - xy + yz - xz}{y^2 - z^2}$

63 $\dfrac{(x + 2)x - 3}{(x - 3)x + 2} \cdot \dfrac{(x - 1)x - 2}{(x + 4)x + 3} \div \dfrac{(x + 4)x + 4}{(x - 1)x - 6}$

64 $\dfrac{(x - 3)x - 4}{(x - 2)x - 3} \cdot \dfrac{(x - 2)x + x - 2}{(x - 4)x + 2(x - 4)} \div \dfrac{(x + 1)x - 6}{(x + 1)x - 2}$

3.4 THE LEAST COMMON MULTIPLE

As we saw in Chap. 1, we can find the sum of two or more fractions only when the denominators of the fractions are equal. Hence, if the denominators of the fractions to be added are different, we must change each fraction to an equal fraction with the denominators of the resulting fractions all the same. The denominator should be the least common multiple (lcm) of the denominators; it is called the least common denominator (lcd) of the fractions.

A common multiple of a set of integers is an integer that is divisible by each integer in the set. For example, 48 is a common multiple of 2, 3, and 8. Also, 24 is a common multiple of 2, 3, and 8, and there is no integer less than 24 that is divisible by each of these three numbers. Hence, 24 is the least common multiple of 2, 3, and 8. In arithmetic the lcm of a set of integers is defined to be the least positive integer that is divisible by each integer in the set. In this chapter, however, we shall be dealing with polynomials, and we cannot define the lcm of a set of polynomials in this way, since the adjective "least" has no meaning when applied to polynomials. Therefore, we shall define the lcm of a set of polynomials in this way:

The least common multiple (lcm) of a set of polynomials is the polynomial P such that P is divisible by each polynomial in the set, and furthermore, every polynomial that is divisible by each member of the set is also divisible by P.

The adjective "least" refers to the degree of the lcm.

In order to obtain the lcm of a set of polynomials, we first express each polynomial of the set as the product of powers of its prime factors. Then, by definition, the following statement is true.

The lcm of a set of polynomials must have as factors the highest power of each prime factor that appears in any polynomial of the set and must have no other factors.

Example 1 Find the lcm of $(a - 1)^2$, $(a + 1)^2$, and $(a - 2)(a + 1)$.

Solution The different prime factors are $a - 1$, $a + 1$, and $a - 2$. Multiplying the highest power of each that occurs gives us the lcm:

$$(a - 1)^2(a + 1)^2(a - 2)$$

Example 2 Find the lcm of the five polynomials $x^2 - 2xy + y^2$, $x^2 + 2xy + y^2$, $x^2 - y^2$, $x^2 - 3xy + 2y^2$, and $2x^2 + 3xy + y^2$.

Solution We first write each of these polynomials in the factored form shown here:

$$x^2 - 2xy + y^2 = (x - y)^2$$
$$x^2 + 2xy + y^2 = (x + y)^2$$
$$x^2 - y^2 = (x - y)(x + y)$$
$$x^2 - 3xy + 2y^2 = (x - 2y)(x - y)$$
$$2x^2 + 3xy + y^2 = (2x + y)(x + y)$$

The prime factors which appear are $(x - y)$, $(x + y)$, $(x - 2y)$, and $(2x + y)$. However, $(x - y)$ and $(x + y)$ have exponents 2 in the first and the second polynomials, respectively. Hence, the lcm is $(x - y)^2(x + y)^2(x - 2y)(2x + y)$.

Example 3 Find the lcd of

$$\frac{x + 1}{2x + 1} \qquad \frac{x - 1}{(3x - 1)^2} \qquad \frac{2x + 1}{6x^2 + x - 1} \qquad \text{and} \qquad \frac{x^2 + 1}{(x + 1)^2(2x + 1)^3}$$

Solution The denominators shown in factored form are $2x + 1$, $(3x - 1)^2$, $(2x + 1)(3x - 1)$, and $(x + 1)^2(2x + 1)^3$. Thus, the lcd of the fractions is

$$(2x + 1)^3(3x - 1)^2(x + 1)^2$$

3.5 ADDITION AND SUBTRACTION OF FRACTIONS

We saw in Chap. 1 that if fractions have the same denominator, we may use the distributive law to add and subtract them. For example,

$$\frac{a}{D} + \frac{b}{D} - \frac{c}{D} = \frac{a + b - c}{D}$$

Sum of Fractions Hence, *the sum of two or more fractions with identical denominators is the fraction that has the sum of the given numerators as the numerator, and the common denominator as the denominator.*

Example 1 $\dfrac{3a}{2xy} + \dfrac{5a}{2xy} - \dfrac{c}{2xy} = \dfrac{3a + 5a - c}{2xy} = \dfrac{8a - c}{2xy}$

Example 2 $\dfrac{x + y}{x + 3y} + \dfrac{x - y}{x + 3y} - \dfrac{2x + y}{x + 3y} = \dfrac{(x + y) + (x - y) - (2x + y)}{x + 3y}$

adding numerators

$$= \frac{x + y + x - y - 2x - y}{x + 3y}$$

removing parentheses

$$= \frac{-y}{x + 3y}$$

combining like terms

If the denominators of the fractions to be added are different, we convert each fraction to an equal fraction with the lcd as the new denominator and proceed as in the above examples.

Example 3   Express

$$\frac{1}{6x} + \frac{1}{3y} - \frac{3x + 2y}{12xy}$$

as a single fraction.

Solution   The lcd of the given fractions is $12xy$. To convert the given fractions to equal fractions with $12xy$ as a denominator, we use (3.4) and multiply each member of the first fraction by $2y$, and each member of the second by $4x$. We thereby obtain

$$\frac{1}{6x} + \frac{1}{3y} - \frac{3x + 2y}{12xy} = \frac{2y}{12xy} + \frac{4x}{12xy} - \frac{3x + 2y}{12xy} \qquad \text{by (3.4)}$$

$$= \frac{2y + 4x - (3x + 2y)}{12xy} \qquad \text{adding numerators}$$

$$= \frac{2y + 4x - 3x - 2y}{12xy} \qquad \text{removing parentheses}$$

$$= \frac{x}{12xy} \qquad \text{combining similar terms}$$

$$= \frac{1}{12y} \qquad \text{dividing numerator and denominator by x}$$

Example 4   Combine

$$\frac{3x + y}{x^2 - y^2} - \frac{2y}{x(x - y)} - \frac{1}{x + y}$$

into a single fraction.

Solution   The denominators are $x^2 - y^2 = (x + y)(x - y)$; $x(x - y)$; and $x + y$. Therefore the lcd is $x(x + y)(x - y)$. Consequently, we multiply the numerator and denominator of the first, the second, and the third fraction by x, $x + y$, and $x(x - y)$, respectively, and complete the computation as:

$$\frac{3x + y}{(x + y)(x - y)} - \frac{2y}{x(x - y)} - \frac{1}{x + y}$$

$$= \frac{x(3x + y)}{x(x + y)(x - y)} - \frac{2y(x + y)}{x(x + y)(x - y)} - \frac{x(x - y)}{x(x + y)(x - y)} \qquad \text{by (3.4)}$$

$$= \frac{x(3x + y) - 2y(x + y) - x(x - y)}{x(x + y)(x - y)} \qquad \text{adding numerators}$$

$$= \frac{3x^2 + xy - 2xy - 2y^2 - x^2 + xy}{x(x+y)(x-y)} \qquad \text{removing parentheses}$$

$$= \frac{2x^2 - 2y^2}{x(x+y)(x-y)} \qquad \text{collecting similar terms}$$

$$= \frac{2(x+y)(x-y)}{x(x+y)(x-y)} \qquad \text{factoring numerator}$$

$$= \frac{2}{x} \qquad \text{canceling}$$

In adding fractions, any common denominator may be used, but the lcd minimizes the process of reducing to lowest terms.

EXERCISE 3.2 Adding Fractions

Find the lcd of the fractions in each of Probs. 1 to 12. Do not add.

1 $\dfrac{2}{xy} - \dfrac{3x+1}{x^2z} + \dfrac{x+y}{2yz^4}$

2 $\dfrac{3a}{bc} + \dfrac{3a+1}{abc} - \dfrac{3a+2}{ac}$

3 $\dfrac{5x}{8y} + \dfrac{8y}{9xz} + \dfrac{3x+1}{18x^2y}$

4 $\dfrac{2}{ab^2c} - \dfrac{2a+5}{4ab^2c} - \dfrac{a-3b}{6a^2b}$

5 $\dfrac{8}{x^2-y^2} + \dfrac{12}{x-y} + \dfrac{13x}{(x-y)^2}$

6 $\dfrac{8x+3y}{x^2+y^2} - \dfrac{3x-5xy}{(x-y)(x+y)^2} + \dfrac{2}{(x+y)(x-y)^2}$

7 $\dfrac{c+3}{c+d} - \dfrac{c+2d}{(c-d)(c^2-d^2)} - \dfrac{5d}{(c-3d)(c+d)^3}$

8 $\dfrac{-4}{a+2b} + \dfrac{5}{a^2-4b^2} + \dfrac{6}{(a-2b)^2}$

9 $\dfrac{-3}{(x^2-y^2)(x+2y)} - \dfrac{17}{(x^2+3xy+2y^2)(x+2y)} + \dfrac{2x}{x^3+3x^2y+3xy^2+y^3}$

10 $\dfrac{4a}{(a-2b)(a^2+4ab+4b^2)} + \dfrac{2b}{(a-2b)(a^2-3ab+2b^2)} + \dfrac{5a}{(a-b)(a^2-4ab+4b^2)}$

11 $\dfrac{a^2+4ab}{(a^2+ab-2b^2)(a+b)} + \dfrac{a-2b^2}{(a^2-3ab+2b^2)(a+2b)} + \dfrac{-1+2a}{(a^2-b^2)(a-2b)}$

12 $\dfrac{4}{(a^2-2ab+b^2)(a+b)} - \dfrac{-2+a}{(a^2+4ab+4b^2)(a-b)} + \dfrac{ab}{(a^2+2ab+b^2)(a+2b)}$

Add the following fractions.

13 $\frac{1}{12} + \frac{13}{12} - \frac{7}{12} + \frac{17}{12}$

14 $\frac{3}{7} - \frac{10}{21} + \frac{3}{14} - \frac{2}{3}$

15 $\frac{2}{5} - \frac{3}{10} + \frac{7}{20} + \frac{1}{8}$

16 $\frac{1}{6} + \frac{1}{10} - \frac{2}{15} - \frac{7}{30}$

17 $\dfrac{a+4}{18} - \dfrac{2a-3}{18} + \dfrac{3a+17}{18}$

18 $\dfrac{b-7}{5} + \dfrac{4b-3}{7} - \dfrac{3b+1}{25}$

19 $\dfrac{x-y}{4} + \dfrac{2x+y}{9} - \dfrac{-x+2y}{12} - \dfrac{x}{3}$

20 $\dfrac{x+2y}{4} + \dfrac{x-y}{5} - \dfrac{2x-y}{6} - \dfrac{x-3y}{8}$

21 $\dfrac{x}{6yz} + \dfrac{3y}{10xz} - \dfrac{2z}{15xy}$

22 $\dfrac{3c}{7de} + \dfrac{5d}{14ec} - \dfrac{e}{4cd}$

23 $\dfrac{2p-r}{p^2r} + \dfrac{3p-4r}{2pr^2} - \dfrac{p^2-3r^2}{3p^2r^2}$

24 $\dfrac{x-4y}{2y^3} + \dfrac{6x-y}{3xy^2} - \dfrac{y-2x}{6x^2y}$

25 $\dfrac{4x^2 - 6xy + 3y^2}{2x(x + y)} - \dfrac{2x - 3y}{x + y}$

26 $\dfrac{a^2 - 2ab - b^2}{b(a - b)} + \dfrac{a + b}{a - b}$

27 $\dfrac{6z^2 + 3zw - w^2}{3z(z - w)} - \dfrac{2z + w}{z - w}$

28 $\dfrac{2r^2 - 7rs - 12s^2}{2r(3r - 4s)} + \dfrac{2r + 4s}{3r - 4s}$

29 $\dfrac{2a^2 - 6ab}{a^2 - 2ab - 3b^2} + \dfrac{a + 2b}{a + b}$

30 $\dfrac{b^2}{a^2 - 3ab + 2b^2} - \dfrac{a - b}{a - 2b}$

31 $\dfrac{x - y}{x + y} - \dfrac{2y^2 - 4xy}{x^2 - xy - 2y^2}$

32 $\dfrac{3z^2 + 6zw}{z^2 + zw - 6w^2} - \dfrac{z + 2w}{z - 2w}$

33 $\dfrac{2}{b} - \dfrac{1}{a + b} + \dfrac{1}{a - b}$

34 $\dfrac{3}{3c - 2d} + \dfrac{7}{c + 2d} - \dfrac{2}{2c + d}$

35 $\dfrac{1}{r - 2t} - \dfrac{4}{r + t} + \dfrac{3}{r + 2t}$

36 $\dfrac{3}{x - y} - \dfrac{1}{x + y} - \dfrac{4}{2x - y}$

37 $\dfrac{1}{p + r} - \dfrac{4}{2p - r} + \dfrac{1}{p - 2r}$

38 $\dfrac{3}{h + 1} - \dfrac{8}{2h + 1} + \dfrac{1}{h - 1}$

39 $\dfrac{1}{a - b} - \dfrac{2}{2a + b} + \dfrac{3}{a + b}$

40 $\dfrac{9}{x + 3y} - \dfrac{5}{x - y} + \dfrac{16}{x - 2y}$

41 $\dfrac{4c + d}{c^2 - cd - 2d^2} - \dfrac{1}{c + d} - \dfrac{2}{c - 2d}$

42 $\dfrac{3}{x + 2y} + \dfrac{x + 6y}{x^2 - 4y^2} - \dfrac{2}{x - 2y}$

43 $\dfrac{a - b}{a + b} - \dfrac{6b^2}{a^2 - ab - 2b^2} + \dfrac{a}{a - 2b}$

44 $\dfrac{14pr + 10r^2}{p^2 + 7pr + 10r^2} + \dfrac{4p}{p + 5r} - \dfrac{3p}{p + 2r}$

45 $\dfrac{r - 2t}{r + 2t} - \dfrac{r - 2t}{r - t} + \dfrac{2r^2 - 5t^2}{r^2 + rt - 2t^2}$

46 $\dfrac{2z^2 - 3wz}{2w^2 - 5wz - 3z^2} - \dfrac{3w + z}{2w + z} + \dfrac{w - 2z}{w - 3z}$

47 $\dfrac{5ab - 2a^2}{a^2 + ab - 2b^2} + \dfrac{a - 2b}{a - b} - \dfrac{a + 5b}{a + 2b}$

48 $\dfrac{z + w}{2z - w} - \dfrac{7zw - w^2}{2z^2 + 3zw - 2w^2} - \dfrac{z - w}{z + 2w}$

49 $\dfrac{2x}{x^2 - y^2} - \dfrac{1}{x + y} - \dfrac{y}{x^2 - xy}$

50 $\dfrac{c}{2cd + 3d^2} - \dfrac{9d}{4c^2 - 9d^2} + \dfrac{3}{2c - 3d}$

51 $\dfrac{6}{b^2 - 1} - \dfrac{3}{b^2 - b} + \dfrac{3}{b + 1}$

52 $\dfrac{3a}{a^2 + ab - 2b^2} + \dfrac{2b}{a^2 + 3ab + 2b^2} - \dfrac{2}{a + b}$

53 $\dfrac{x - y}{2x^2 + 3xy + y^2} - \dfrac{x + y}{2x^2 - xy - y^2} + \dfrac{8}{3(2x + y)}$

54 $\dfrac{3a + b}{4(2a + b)(a + b)} - \dfrac{5(a + b)}{8(3a - b)(2a + b)} + \dfrac{1}{8(2a + b)}$

55 $\dfrac{1}{3c - 2d} + \dfrac{3c + 2d}{12(3c - 2d)(c - d)} - \dfrac{3c - 2d}{12(3c + 2d)(c - d)}$

56 $\dfrac{f}{(f + g)(f + 3g)} - \dfrac{1}{f - g} + \dfrac{f}{(f + g)(f - g)}$

57 $\dfrac{3w^2}{(2w + z)(w - z)} + \dfrac{w^2 + wz - z^2}{(2w - z)(w - z)} - \dfrac{4w^2}{(2w + z)(2w - z)}$

58 $\dfrac{11h + 2k}{2(3h + k)(2h - k)} - \dfrac{7h}{2(3h + k)(2h + 3k)} - \dfrac{6k}{(2h + 3k)(2h - k)}$

59 $\dfrac{32r - 2s}{6r^2 - 19rs + 10s^2} + \dfrac{r - 3s}{6r^2 - rs - 2s^2} - \dfrac{20r + 4s}{4r^2 - 8rs - 5s^2}$

60 $\dfrac{a^2 + b^2}{2a^2 + 5ab + 3b^2} - \dfrac{a^2 - 11b^2}{2a^2 - ab - 3b^2} + \dfrac{10ab - 24b^2}{4a^2 - 9b^2}$

3.6 COMPLEX FRACTIONS

A complex fraction is a fraction in which the numerator, the denominator, or both numerator and denominator contain fractions. For example,

$$\frac{1 + \frac{1}{3}}{1 - \frac{1}{4}} \qquad \frac{\dfrac{2x - y}{y}}{x + \dfrac{2}{y}} \qquad \frac{\dfrac{a + b}{a - b} - \dfrac{a}{2b}}{2 - \dfrac{1}{a^2 b^2}}$$

are complex fractions.

We simplify a complex fraction by converting it to an equal fraction that has no fractions in either the numerator or the denominator. A complex fraction can be simplified by first simplifying the numerator and the denominator and then finding their quotient. Usually, however, the most efficient method consists of the following steps:

1 Find the lcm of the denominators of the fractions that appear in the complex fraction.
2 Multiply the members of the complex fraction by the lcm found in step 1.
3 Simplify the result obtained in step 2.

Example 1 Simplify

$$\frac{4 - \dfrac{1}{x}}{16 - \dfrac{1}{x^2}}$$

Solution 1 Since the lcm of the denominators is x^2, we will proceed as follows:

$$\frac{4 - \dfrac{1}{x}}{16 - \dfrac{1}{x^2}} = \frac{x^2\left(4 - \dfrac{1}{x}\right)}{x^2\left(16 - \dfrac{1}{x^2}\right)} \qquad \text{multiplying numerator and denominator by } x^2$$

$$= \frac{4x^2 - x}{16x^2 - 1} \qquad \text{performing indicated operations}$$

$$= \frac{x(4x - 1)}{(4x + 1)(4x - 1)} \qquad \text{factoring}$$

$$= \frac{x}{4x + 1} \qquad \text{dividing numerator and denominator by } 4x - 1$$

Solution 2 If we first simplify the numerator and denominator separately, we get

$$4 - \frac{1}{x} = \frac{4x-1}{x} \qquad \text{and} \qquad 16 - \frac{1}{x^2} = \frac{16x^2-1}{x^2}$$

thus

$$\frac{4 - \dfrac{1}{x}}{16 - \dfrac{1}{x^2}} = \frac{4x-1}{x} \div \frac{16x^2-1}{x^2} = \frac{4x-1}{x} \cdot \frac{x^2}{16x^2-1}$$

$$= \frac{(4x-1) \cdot x^2}{x(4x-1)(4x+1)} = \frac{x}{4x+1}$$

Example 2   Simplify

$$\frac{2 - \dfrac{3}{a+2}}{\dfrac{1}{a-1} + \dfrac{1}{a+2}}$$

Solution   The lcm of the denominators is $(a-1)(a+2)$. Therefore, we proceed as follows:

$$\frac{2 - \dfrac{3}{a+2}}{\dfrac{1}{a-1} + \dfrac{1}{a+2}} = \frac{(a-1)(a+2)\left(2 - \dfrac{3}{a+2}\right)}{(a-1)(a+2)\left(\dfrac{1}{a-1} + \dfrac{1}{a+2}\right)}$$ multiplying each member by $(a-1)(a+2)$

$$= \frac{2(a-1)(a+2) - 3(a-1)}{a+2+a-1}$$ performing indicated operations

$$= \frac{2(a^2 + a - 2) - 3(a-1)}{2a+1}$$

$$= \frac{2a^2 + 2a - 4 - 3a + 3}{2a+1}$$ removing parentheses

$$= \frac{2a^2 - a - 1}{2a+1}$$ combining terms

$$= \frac{(2a+1)(a-1)}{2a+1}$$ factoring

$$= a - 1$$ dividing numerator and denominator by $2a+1$

Example 3   In the complex fraction

$$\frac{1 - \dfrac{1}{1 - \dfrac{1}{a}}}{\dfrac{1}{1 - \dfrac{a}{3}} - 1}$$

the numerator and denominator are themselves complex fractions. In such cases the member or members that contain complex fractions should be simplified first. If the fraction thus obtained is complex, the process is continued until a fraction is obtained in which no fractions appear in either the numerator or the denominator. We shall simplify the above complex fraction.

$$\frac{1 - \dfrac{a}{a-1}}{\dfrac{3}{3-a} - 1}$$ multiplying members of complex fraction in numerator by a and members of complex fraction in denominator by 3

$$= \frac{\dfrac{a-1-a}{a-1}}{\dfrac{3-3+a}{3-a}}$$ adding fractions in numerator and denominator

$$= \frac{\dfrac{-1}{a-1}}{\dfrac{a}{3-a}}$$ combining terms

$$= \frac{a-3}{a(a-1)}$$ by (3.6)

EXERCISE 3.3   Complex Fractions

Simplify the fractions in Probs. 1 to 40.

1. $\dfrac{3}{2 + \dfrac{1}{5}}$

2. $\dfrac{4}{1 - \dfrac{3}{7}}$

3. $\dfrac{6 + \dfrac{4}{3}}{3}$

4. $\dfrac{1 - \dfrac{5}{6}}{2}$

5. $\dfrac{\dfrac{3}{4} - \dfrac{1}{3}}{\dfrac{5}{12} - \dfrac{3}{4}}$

6. $\dfrac{\dfrac{5}{6} + \dfrac{4}{9}}{\dfrac{4}{3} + \dfrac{3}{2}}$

7. $\dfrac{\dfrac{8}{5} - \dfrac{7}{4}}{\dfrac{3}{10} + \dfrac{2}{5}}$

8. $\dfrac{\dfrac{1}{2} + \dfrac{2}{5}}{\dfrac{2}{3} + \dfrac{7}{10}}$

9. $\dfrac{2 + \dfrac{1}{x}}{4 - \dfrac{1}{x^2}}$

10. $\dfrac{a + \dfrac{1}{a^2}}{a - 1 + \dfrac{1}{a}}$

11. $\dfrac{2 - \dfrac{1}{x}}{\dfrac{1}{x}}$

12. $\dfrac{1 - \dfrac{1}{x^3}}{1 - \dfrac{1}{x}}$

13. $\dfrac{x + \dfrac{1}{3x}}{\dfrac{1}{x} - \dfrac{2}{3x}}$

14. $\dfrac{\dfrac{1}{2} - \dfrac{4}{d}}{\dfrac{1}{d} + \dfrac{3}{2d}}$

15. $\dfrac{\dfrac{a}{3b} - \dfrac{1}{3}}{\dfrac{a}{2b} + \dfrac{b}{3}}$

16. $\dfrac{\dfrac{1}{4} - \dfrac{1}{2x}}{\dfrac{1}{12x} - \dfrac{1}{6x^2}}$

17. $\dfrac{2h + 4}{1 + \dfrac{1}{2h+3}}$

18. $\dfrac{a + 2b}{2 - \dfrac{2b}{a+3b}}$

19. $\dfrac{3 - \dfrac{x+3y}{x+y}}{1 + \dfrac{y}{x-y}}$

20. $\dfrac{1 - \dfrac{2b}{a-b}}{2 - \dfrac{a+b}{a-b}}$

21 $\dfrac{a - \dfrac{2}{a - 1}}{\dfrac{1}{a - 1} + \dfrac{2}{(a - 1)^2}}$

22 $\dfrac{1 + \dfrac{4b}{a - b}}{3 + \dfrac{8ab}{a^2 - b^2}}$

23 $\dfrac{x - \dfrac{xy}{x - y}}{x - \dfrac{2y^2}{x - y}}$

24 $\dfrac{h + \dfrac{2hk - k^2}{h - 2k}}{h - \dfrac{hk - 2k^2}{h - 2k}}$

25 $\dfrac{x + 2 - \dfrac{2}{x - 1}}{x - 1 - \dfrac{2}{x + 2}}$

26 $\dfrac{\dfrac{x}{2} - \dfrac{1}{x + 1}}{\dfrac{x}{2} + \dfrac{2x + 3}{x + 1}}$

27 $\dfrac{x - \dfrac{4x - 3}{2x - 1}}{2x - \dfrac{x - 3}{x - 2}}$

28 $\dfrac{3x + \dfrac{x - 5}{x - 1}}{x - \dfrac{5}{3x - 2}}$

29 $\dfrac{1 + \dfrac{xy + 2y^2}{x^2 - y^2}}{\dfrac{1}{x - y}}$

30 $\dfrac{1 - \dfrac{6d}{2c + 3d}}{2 + \dfrac{d}{c - 2d}}$

31 $\dfrac{1 - \dfrac{6b}{a + 2b}}{\dfrac{2b}{a - 3b} - 2}$

32 $\dfrac{\dfrac{6x}{x + 2y} - 2}{\dfrac{3x}{x - 2y} + 3}$

33 $\dfrac{\dfrac{w}{w + 1} - \dfrac{w^2}{w^2 - 1}}{\dfrac{1}{w - 1} + 1}$

34 $\dfrac{\dfrac{2a + 3}{a + 2} - \dfrac{2a}{a + 1}}{\dfrac{a}{a + 2} - 1}$

35 $\dfrac{\dfrac{3}{x} - \dfrac{4}{x + y}}{\dfrac{-3}{2x} + \dfrac{1}{x - y}}$

36 $\dfrac{\dfrac{4}{a + 2b} - \dfrac{1}{a}}{\dfrac{2}{a - 2b} + \dfrac{1}{a}}$

37 $\dfrac{1}{1 - \dfrac{1}{1 - \dfrac{1}{a}}}$

38 $\dfrac{a}{\dfrac{1}{1 - \dfrac{a}{3}} - 1}$

39 $\dfrac{2 - \dfrac{1}{\dfrac{x}{y} - 1}}{\dfrac{2}{1 + \dfrac{y}{x}} - \dfrac{3}{\dfrac{x}{y} + 1}}$

40 $\dfrac{a - \dfrac{b}{1 - \dfrac{a}{a + b}}}{1 + \dfrac{b}{a - b}}$

In each of Probs. 41 to 44, show that the first number is larger than the second.

41 $2 + \dfrac{1}{2 + \dfrac{1}{2 + \dfrac{1}{2 + \dfrac{1}{2}}}}; \quad 2 + \dfrac{1}{2 + \dfrac{1}{2}}$

42 $2 + \dfrac{1}{2}; \quad 2 + \dfrac{1}{2 + \dfrac{1}{2 + \dfrac{1}{2}}}$

43 $4 + \dfrac{1}{3 + \dfrac{1}{4 + \dfrac{1}{3 + \dfrac{1}{4}}}}; \quad 4 + \dfrac{1}{3 + \dfrac{1}{4}}$

44 $4 + \dfrac{1}{3}; \quad 4 + \dfrac{1}{3 + \dfrac{1}{4 + \dfrac{1}{3}}}$

3.7 SUMMARY

We treat quotients of polynomials (rational expressions or fractions) just as we treated quotients of integers (rational numbers) in Chap. 1. The fundamental principle of fractions gives us the conditions under which two fractions are equal. This leads to the cancellation law. In the remainder of the chapter we multiply, divide, add, and subtract fractions and simplify the results. More complicated fractions, called complex fractions, are also discussed.

EXERCISE 3.4   Review

Reduce the fractions in Probs. 1 to 3.

1 $\dfrac{x^2 y^2 z^4}{x y^3 z^6}$

2 $\dfrac{6x^2 - xy - y^2}{3x^2 - 11xy - 4y^2}$

3 $\dfrac{ax - 2ay + 4by - 2bx}{ax - 2ay + bx - 2by}$

State whether the equation in each of Probs. 4 to 6 is true or false.

4 $\dfrac{x - 1}{2 - x} = \dfrac{1 - x}{x - 2}$

5 $\dfrac{5x + y - 2}{-3x + y + 1} = \dfrac{5x - 2}{-3x + 1}$

6 $\dfrac{19}{95} = \dfrac{1}{5}$

Perform the indicated operations in Probs. 7 to 10.

7 $\dfrac{7a^2 b^5}{12abc^4} \cdot \dfrac{8a^4 b^4 c}{21a^4 b^6 c^2}$

8 $\dfrac{4a + 6b}{a^2 - ab} \div \dfrac{2a^2 + 5ab + 3b^2}{a^2 - b^2}$

9 $\dfrac{x^2 + xy - 2y^2}{4x^2 + 4xy - 3y^2} \cdot \dfrac{6x^2 - 5xy + y^2}{3x^2 - 4xy + y^2} \cdot \dfrac{6x^2 + 5xy - 6y^2}{3x^2 + 5xy - 2y^2}$

10 $\dfrac{x^2 - xy - 12y^2}{3x^2 + 4xy} \div \left(\dfrac{xy - 4y^2}{2x^2 - 5xy} \cdot \dfrac{2x^3 + x^2 y - 15xy^2}{3xy + 4y^2} \right)$

Find the lcd of the fractions in each of Probs. 11 to 13.

11 $\dfrac{3x}{4} + \dfrac{5x - 1}{2xy} + \dfrac{6x^2 + 5}{3x^2}$

12 $\dfrac{2x + 1}{(x - 1)^2 (x + 2)} + \dfrac{x - 1}{(2x + 1)^3 (x + 2)^4} + \dfrac{x + 2}{(x - 1)^3 (2x + 1)}$

13 $\dfrac{4}{(x - 1)(x^2 + 4x + 4)} + \dfrac{5}{(x^2 + x - 2)(x + 3)} - \dfrac{6}{x^3 + 5x^2 + 6x}$

Add the fractions in Probs. 14 to 19.

14 $\dfrac{x^2 + 3}{15} - \dfrac{2x - 3}{21} + \dfrac{2x^2 - 5x}{35}$

15 $\dfrac{8}{ab} - \dfrac{5}{ac} + \dfrac{3}{bc}$

16 $\dfrac{5a^2 + 2ab + 6b^2}{4a(a - b)} - \dfrac{a + 4b}{a - b}$

17 $\dfrac{4}{x - y} + \dfrac{12}{2x - y} - \dfrac{5}{3x - y}$

18 $\dfrac{a + 3}{6a^2 + 13a - 5} - \dfrac{3a + 1}{2a^2 - a - 15} + \dfrac{4}{3a - 1}$

19 $\dfrac{2a - b}{(4a + b)(a + 3b)} + \dfrac{4a - b}{(a + 3b)(2a + b)} - \dfrac{20a}{(2a + b)(4a + b)}$

Simplify the fractions in Probs. 20 to 25.

20 $\dfrac{\dfrac{5}{6} + \dfrac{6}{11}}{\dfrac{3}{22} - \dfrac{5}{3}}$

21 $\dfrac{x - \dfrac{2}{x}}{x + \dfrac{2}{x}}$

22 $\dfrac{\dfrac{4a}{b} - \dfrac{1}{2}}{\dfrac{3a}{2b} + \dfrac{b}{5}}$

23 $\dfrac{3 + \dfrac{a}{a - 5b}}{4 - \dfrac{3}{a - 5b}}$

24 $\dfrac{2 + \dfrac{3x}{x - 2y}}{5 + \dfrac{xy}{x^2 - 4y^2}}$

25 $\dfrac{\dfrac{x - y}{2x + 5y} - \dfrac{x}{x - 3y}}{\dfrac{x + 2y}{2x + 5y} - 3}$

26 Show that $3 + \dfrac{1}{3 + \dfrac{1}{3}} < 3 + \dfrac{1}{3 + \dfrac{1}{3 + \dfrac{1}{3}}}$.

27 Show that $3 + \dfrac{1}{3 + \dfrac{1}{3 + \dfrac{1}{3 + \dfrac{1}{3}}}} < 3 + \dfrac{1}{3 + \dfrac{1}{3 + \dfrac{1}{3}}}$.

4

linear and fractional equations

Heretofore, we have been concerned with formal operations that followed prescribed rules of procedure. In this chapter we shall investigate conditions under which two algebraic expressions are equal. For example, we shall explain methods for finding the replacement for x so that a statement such as $\frac{1}{2}(x + 3) = \frac{1}{4}(x - 2)$ is true. A statement of this type is called an *equation*.

The equation is a powerful tool in mathematics and is essential in the development and understanding of problems in the physical sciences, engineering, life sciences, and social sciences.

4.1 OPEN SENTENCES

In this section we shall consider statements of these types:

$$5 + x = 9 \tag{1}$$

x is an integer between 2 and 6 (2)

x is a color in the United States flag (3)

No one of these statements is true as it stands. However, (1) is true if x is replaced by 4; (2) is true if x is replaced by an element of $\{3, 4, 5\}$; and (3) is true if x is replaced by a color of the set $\{\text{red, white, blue}\}$. Furthermore, each of the statements (1), (2), and (3) is false if x is replaced by any number or word other than the replacements specified above.

In statements (1) and (2), x stands for a number; and in (3), x stands for a word. As defined in Chap. 1, the letter x is called *Variable* a *variable*, and it may be any element of a specified set. The *Replacement* specified set is called the *replacement set*. *Set*

The assertions (1), (2), and (3) are called *open sentences* and illustrate the following definition:

Open Sentence   An *open sentence* is an assertion, containing a variable, that is neither true nor false but that becomes a true or false statement if the variable is replaced by an element chosen from the replacement set.

Truth Set   If an open sentence becomes a true statement when the variable is replaced by each element of a set T and no other elements, then T is called the *truth set* for the open sentence.

According to this definition, the truth sets for the open sentences (1), (2), and (3) are $\{4\}$, $\{3, 4, 5\}$, and $\{$red, white, blue$\}$, respectively.

Other examples of open sentences and their truth sets follow:

Open Sentence	Truth Set
x is a state of the United States larger than California	$\{$Alaska, Texas$\}$
x is a color in a rainbow	$\{$violet, indigo, blue, green, yellow, orange, red$\}$
x is a woman who has been president of the United States	$\varnothing$
x is a positive odd integer less than 10.	$\{1, 3, 5, 7, 9\}$
x is an element of $\{a, b, c, d\} \cap \{a, c, d, e, f\}$	$\{a, c, d\}$
$x - 8 = 2$	$\{10\}$

4.2   EQUATIONS

Open sentences of the types

$$3x - 5 = 4 + 2x \tag{1}$$

$$\frac{2x - 3}{x + 2} = \frac{x - 1}{4} \tag{2}$$

$$x + 1 = x + 3 \tag{3}$$

$$\frac{x - 1}{3} + \frac{x + 1}{2} = \frac{5x + 1}{6} \tag{4}$$

are called equations and illustrate the following definition.

Equation   An *equation* is an open sentence which states that two expressions are equal. Each of the two expressions is called a *member* of the equation.

Root
Solution

If an equation is a true statement after the variable is replaced by a specific number, then that number is called a *root* or a *solution* of the equation and is said to *satisfy* it.

For example, 9 is a root of Eq. (1) since $3(9) - 5 = 4 + 2(9)$. Furthermore, 2 and 5 are roots of Eq. (2), since each member of (2) is equal to $\frac{1}{4}$ if x is replaced by 2, and each member is equal to 1 if x is replaced by 5.

Solution Set   The set of all roots of an equation is called the *solution set* of the equation.

Solving the
Equation

A procedure for finding the solution set of an equation is called *solving the equation*.

As stated above, the solution sets of Eqs. (1) and (2) are {9} and {2, 5}, respectively. Equation (3), however, is a false statement for every replacement for x, since the sum of a number and 1 is not equal to the sum of the same number and 3. Therefore, the solution set of Eq. (3) is the empty set $\varnothing$.

Equation (4), however, is a true statement for every replacement for x, since if we add the fractions at the left of the equality sign, we get the fraction on the right. Therefore, the solution set of Eq. (4) is the set of all numbers. Equations (1), (2), and (3) are called *conditional equations*, and Eq. (4) is an *identity*. These illustrate the following definitions:

Conditional
Equation

A *conditional equation* is an equation whose solution set is a proper subset of the replacement set.

Identity

An *identity* is an equation whose solution set is the replacement set. $(x + 3)^2 = x^2 + 6x + 9$ is an identity, and so is $6x/3 = 2x$. Furthermore,

$$\frac{x - 2}{x - 1} + x = \frac{x^2 - 2}{x - 1}$$

is an identity, but 1 is not a part of the replacement set since division by 0 is not a permissible operation. Thus the replacement set is $\{x \mid x \text{ is real and not } 1\}$.

We use set notation to show that {9} is the solution set of $3x - 5 = 4 + 2x$ by writing $\{x \mid 3x - 5 = 4 + 2x\} = \{9\}$ but could just say that 9 is the solution of the equation. Similarly, we can write

$$\left\{ x \ \middle| \ \frac{2x-3}{x+2} = \frac{x-1}{4} \right\} = \{2, 5\}$$

or simply say that the solutions are 2 and 5.

We now introduce the notation $f(x)$, read "f of x," which is very important in mathematics. In this chapter we shall use $f(x)$ to stand for an algebraic expression in x. This notation will be discussed more fully in a later chapter where it will be given a broader interpretation.

In this book we shall usually deal with conditional equations and the methods for solving them. The variable in an equation *Unknown*   is often called the *unknown*, and we shall frequently refer to it in this way.

4.3   EQUIVALENT EQUATIONS

The objective in solving an equation is to find a replacement for the variable that satisfies the equation. The simpler the equation is in form, the easier it is to solve. For example, consider the equations

$$7x - 45 = 5x - 43 \tag{1}$$

$$2x = 2 \tag{2}$$

At this stage, the only way that we can find a root of Eq. (1) is to guess at a number, substitute it for x, and see if it satisfies the equation. In Eq. (2), however, it is obvious that the root is 1. Now, if we substitute 1 for x in Eq. (1) and combine terms, we get $-38 = -38$. Therefore 1 is also a root of Eq. (1). Even though Eqs. (1) and (2) are different statements, each statement is true if x is replaced by 1. Two equations of this type are *equivalent* and illustrate the following definition:

Equivalent   Two equations are *equivalent* if their solution sets are equal.

The procedure for solving a given equation is to obtain a succession of equivalent equations, each simpler in form than the preceding, until we ultimately obtain one whose solution set can be easily found. The following theorems are essential for this purpose.

Equivalent   **If $f(x)$, $g(x)$, and $h(x)$ are expressions, then the two equations**
Equations   **$f(x) = g(x)$ and $f(x) + h(x) = g(x) + h(x)$ are equivalent.   (4.1)**

If k is a nonzero constant, then the equations $f(x) = g(x)$ and $k \cdot f(x) = k \cdot g(x)$ are equivalent.   (4.2)

We use this theorem when one or more of the coefficients or constant terms in $f(x) = g(x)$ are fractions. If k is the lcm of the denominators of the fractions in the equation, then $k \cdot f(x) = k \cdot g(x)$ will contain no fractions.

We shall use a modification of (4.2) to solve fractional equations later in this chapter, multiplying by nonconstant expressions which are never zero [see (4.3)].

The proofs of (4.1) and (4.2) depend upon the following facts, which are in reality axioms for the real numbers:

If equals are added to equals, the results are equal.
If equals are subtracted from equals, the results are equal.
If equals are multiplied by equals, the results are equal.
If equals are divided by nonzero equals, the results are equal.

For example if $x = y$, then $x + 2 = y + 2$, and if $x = y$, then $x - w = y - w$. Also, if $x = y$, then $5x = 5y$ [and $(0)(x) = (0)(y)$]. Finally, if $x = y$, then $x/8 = y/8$ (but we may not write $x/z = y/z$ if $z = 0$).

Example 1 If we begin with the equation

$$5x + 3 = 4x - 9 \tag{3}$$

we may add -3 to each member (or subtract 3). This yields the equivalent equations

$$5x + 3 + (-3) = 4x - 9 + (-3)$$
$$5x = 4x - 12$$

Similarly we now add $-4x$ to each member (or subtract $4x$ from each member) and get

$$5x - 4x = 4x - 12 - 4x$$
$$x = -12 \tag{4}$$

Now Eq. (3) is equivalent to Eq. (4), and the solution to (4) is obvious, namely -12. We may check that -12 satisfies (3) by substituting -12 for x in both members of (3) and getting

$$5(-12) + 3 = -60 + 3 = -57$$
and $\quad 4(-12) - 9 = -48 - 9 = -57$

The first step in solving Eq. (3) was to add -3 to each member. The effect was to convert $+3$ on one side of the equation to -3 on the other side. With a little practice we may write directly

$$5x + 3 = 4x - 9$$
$$5x - 4x = -9 - 3$$

Transposing This process of writing an expression on the other side of the equation after changing its sign is called *transposing*. It is nothing more than (4.1).

Example 2 In the equation

$$5x - 7 = 13x + 1$$

we may transpose both $5x$ and 1. This leads to the equivalent equations

$$-7 - 1 = 13x - 5x$$

$$-8 = 8x \qquad \text{combining like terms}$$

$$-1 = x \qquad \text{dividing by 8 (using 4.2)}$$

Example 3 Solve the equation $\frac{1}{2}x + \frac{2}{3} = \frac{1}{4}x - \frac{1}{6}$.

Solution The lcm of the denominators in the equation is 12. Consequently, we proceed as follows:

$$\frac{1}{2}x + \frac{2}{3} = \frac{1}{4}x - \frac{1}{6} \qquad \text{given equation}$$

$$12(\tfrac{1}{2}x + \tfrac{2}{3}) = 12(\tfrac{1}{4}x - \tfrac{1}{6}) \qquad \text{by (4.2) with } k = 12$$

$$6x + 8 = 3x - 2 \qquad \text{multiplying}$$

$$6x - 3x = -2 - 8 \qquad \text{transposing}$$

$$3x = -10 \qquad \text{combining similar terms}$$

$$x = -\tfrac{10}{3} \qquad \text{by (4.2) with } k = \tfrac{1}{3}$$

Hence, the solution of the given equation is $-\frac{10}{3}$. We verify this fact by replacing x in the given equation with $-\frac{10}{3}$ and finding that each member is equal to -1.

EXERCISE 4.1   Equations

Find the truth set for each of the open sentences in Probs. 1 to 12.

1 x is an even, positive integer less than 10.

2 x is a positive integer whose English spelling has exactly three letters.

3 x is an American state admitted to the union since 1920.

4 x is the southernmost American state.

5 x was a United States president some time between Jan. 1, 1960, and Jan. 21, 1977.

6 x was the first person to walk on the moon.

7 x is the name of a bird and a major league baseball team.

8 A swimming x and a x table are both fun.

9 $x \in \{1, 2, 5, 8\} \cap \{2, 3, 4, 5\}$

10 $x = \{1, 2, 5, 8\} \cap \{2, 3, 4, 5\}$

11 $x = 285$

12 $2x = 46$

Is each of Probs. 13 to 20 a conditional equation or an identity?

13 $\dfrac{x}{2} + \dfrac{x}{3} = \dfrac{x}{5}$

14 $(x-2)(x+3) = x^2 - 6$

15 $\dfrac{x^2+1}{x+1} - x = \dfrac{1-x}{1+x}$

16 $(2x+3)^2 = 4x^2 + 9$

17 $(3x-2)^2 = 9x^2 - 6x - 4$

18 $(x+2)^3 = x^3 + 6x^2 + 12x + 8$

19 $\dfrac{x^2+3x+4}{x^2+3x-2} = -2$

20 $x - 5(x+2) = x^2 - 3x - 10$

In Probs. 21 to 28, show that the given number or numbers are roots of the equation that follows.

21 $4; x + 1 = 2x - 3$

22 $5; 2x - 1 = 4x - 11$

23 $6; \dfrac{x}{3} + \dfrac{x}{6} = 3$

24 $2; \dfrac{2x-1}{3} + \dfrac{x+1}{4} = \dfrac{7}{4}$

25 $3, -1; x^2 = 2x + 3$

26 $\dfrac{1}{2}, -4; 2x^2 + 7x = 4$

27 $\dfrac{2}{3}, \dfrac{-3}{2}; 3x + 1 = \dfrac{7}{2x+1}$

28 $-\dfrac{5}{2}, 3; 2x + 3 = \dfrac{9}{x-2}$

For each pair of equations in Probs. 29 to 40, state whether or not they are equivalent.

29 $5x - 1 = 2x + 4$
$3x = 5$

30 $2x + 5 = 7x - 10$
$15 = 9x$

31 $\dfrac{3}{2}x - \dfrac{1}{3} = \dfrac{1}{4}x + \dfrac{2}{5}$
$\dfrac{5x}{4} = \dfrac{11}{15}$

32 $\dfrac{2x}{7} - 1 = \dfrac{x}{3}$
$\dfrac{x}{21} = -1$

33 $6x + 1 = x - 3$
$12x + 2 = 2x - 6$

34 $4x - 3 = 3x + 2$
$12x - 9 = 3x + 2$

35 $\dfrac{x}{2} - \dfrac{1}{3} = \dfrac{3}{4} - \dfrac{2x}{3}$
$6x - 4 = 9 - 8x$

36 $2x - \dfrac{1}{5} = \dfrac{x}{6} + \dfrac{1}{3}$
$30x - 6 = 5x + 10$

37 $\dfrac{x}{4} - \dfrac{x}{5} = 6$
$5x - 4x = 6$

38 $\dfrac{x}{3} - \dfrac{x-2}{5} = \dfrac{1}{15}$
$5x - 3x - 6 = 1$

39 $4x = 9$
$x = 9 - 4$

40 $\dfrac{x}{x-2} = \dfrac{2}{x-2}$
$x = 2$

4.4 LINEAR EQUATIONS

Linear A *linear equation* in one variable is an equation
Equation

$$ax + b = cx + d \qquad\qquad (1)$$

where a, b, c, and d are real numbers with either a or c (or both) $\neq 0$. Examples are

$$3x - 4 = 7x + 1 \qquad 4x - 5 = 7 \qquad 2x = 11$$

We shall first show how to solve a special case of (1), and then after Example 2, explain how to reduce the general case to this special case.

For the equation

$$ax + b = 0 \qquad a \neq 0 \tag{2}$$

we use (4.1) and (4.2) to get the equivalent equations

$$ax = -b \qquad \text{by (4.1)}$$

$$x = -\frac{b}{a} \qquad \text{by (4.2)}$$

So far we have shown that if (2) has a solution, then it must be $-b/a$. We verify that it is a solution by substituting $-b/a$ for x in (2) and getting

$$a\left(-\frac{b}{a}\right) + b = -b + b = 0$$

Thus the one and only solution of (2) is $-b/a$.

Example 1 Solve $5x + 7 = 0$.

Solution Here $a = 5$ and $b = 7$, and the solution is $x = -\frac{7}{5}$.

Example 2 Solve $2x/3 - \frac{1}{2} = 0$.

Solution We write the equivalent equation $2x/3 = \frac{1}{2}$ by adding $\frac{1}{2}$ to each member. Then multiplying by $\frac{3}{2}$ gives

$$x = \frac{3}{2} \cdot \frac{1}{2} = \frac{3}{4}$$

We shall now return to Eq. (1), write it as

$$ax - cx + b - d = 0 \qquad \text{adding } -d - cx \text{ to each member}$$
$$(a - c)x + (b - d) = 0$$

and proceed as in (2) with a replaced by $a - c$, and b by $b - d$.

Example 3 Find the solution set of

$$3x - 5 = 4 - 2x$$

Solution We transpose terms, and thereby get

$$3x + 2x = 4 + 5 \qquad \text{transposing 5 and 2x}$$
$$5x = 9 \qquad \text{combining similar terms}$$
$$x = \tfrac{9}{5} \qquad \text{multiplying by } \tfrac{1}{5}$$

Thus the solution set is $\{\tfrac{9}{5}\}$.

Example 4   Solve $2x/3 - \frac{3}{4} = \frac{5}{6} - x/8$.

Solution 1   Transposing terms gives

$$\frac{2x}{3} + \frac{x}{8} = \frac{5}{6} + \frac{3}{4}$$

$$\frac{16x + 3x}{24} = \frac{10 + 9}{12} \qquad \text{adding fractions}$$

$$\frac{19x}{24} = \frac{19}{12} \qquad \text{collecting terms}$$

$$x = \frac{19}{12} \cdot \frac{24}{19} = 2 \qquad \text{multiplying by } \frac{24}{19}$$

Solution 2   Multiplying by the lcd, which is 24, gives

$$16x - 18 = 20 - 3x$$
$$16x + 3x = 20 + 18 \qquad \text{transposing}$$
$$19x = 38 \qquad \text{collecting terms}$$
$$x = 2 \qquad \text{dividing by 19}$$

Example 5   Solve $|x - 2| = 5$.

Solution   Since $|x - 2|$ is equal to $x - 2$ if $x - 2 > 0$, and $-(x - 2)$ if $x - 2 < 0$, the given equation with absolute values is equivalent to the two equations

$$\begin{array}{ll} x - 2 = 5 & -(x - 2) = 5 \\ x = 5 + 2 & x - 2 = -5 \\ x = 7 & x = -5 + 2 = -3 \end{array}$$

Thus the solutions are 7 and -3, or we may say the solution set is $\{7, -3\}$.

EXERCISE 4.2   Linear Equations

Solve each of the following equations.

1   $2x + 5 = 0$

2   $3x - 2 = 0$

3   $x - 7 = 0$

4   $5x - 15 = 0$

5   $4x - 1 = 11$

6   $-2x + 5 = 7$

7   $3x - 4 = 4x$

8   $-x + 9 = 2x$

9   $4x + 1 = 6x - 7$

10   $3x + 8 = -2x - 17$

11   $7x + 3 = 2x - 4$

12   $6x - 13 = 2x + 5$

13   $2(x - 5) = 3(2x + 1)$

14   $8(2x + 3) = -5(-3x + 2)$

15   $6(3x - 1) = 5(4x + 3)$

16   $4(-3x + 5) = 7(2x + 3)$

17   $\frac{1}{3}(5x - 2) = x + 2$

18   $\frac{2}{5}(7x - 1) = 3x - 1$

19 $\frac{3}{4}(3x - 2) = 3x$

20 $\frac{3}{7}(4x - 7) = 2x - 5$

21 $\frac{6x + 1}{3} = 3x + 2$

22 $\frac{12x + 2}{5} = 3x$

23 $\frac{8x - 3}{9} = -2x + 4$

24 $\frac{15x - 1}{8} = 5x - 2$

25 $\frac{6x + 2}{5} + \frac{4x - 9}{3} = 5$

26 $\frac{4x + 13}{7} - \frac{3x - 5}{10} = 1$

27 $\frac{6x - 5}{11} - \frac{-4x + 5}{3} = -4$

28 $\frac{7x - 10}{4} + \frac{-3x - 4}{5} = -1$

29 $\frac{3}{4}x + 5 = -\frac{x}{2} + 10$

30 $\frac{5x}{3} - \frac{1}{2} = \frac{3x}{2} + \frac{1}{2}$

31 $\frac{4x}{3} + \frac{1}{6} = \frac{10x}{5} - \frac{1}{6}$

32 $\frac{6x}{5} + \frac{3}{10} = \frac{9x}{2} - \frac{19}{10}$

33 $\frac{2x}{3} + \frac{3}{4} - \frac{x}{4} = \frac{x}{2} - \frac{1}{4}$

34 $\frac{5x}{6} + \frac{5}{9} - \frac{2x}{9} = \frac{2x}{3} - \frac{4}{9}$

35 $\frac{5x}{6} + \frac{7}{9} - \frac{3x}{4} = \frac{x}{9} - \frac{2}{9}$

36 $\frac{3x}{8} - \frac{5}{12} + \frac{x}{6} = \frac{x}{2} + \frac{7}{12}$

37 $\frac{x}{a} - b = \frac{x - b^2}{a + b}$

38 $\frac{x}{2b} + b^2 = a + \frac{bx}{2a}$

39 $ax + \frac{x}{b^2} = \frac{1}{a} + b^2$

40 $\frac{x}{b} + ab = \frac{ax}{2b} + 2b$

41 $|x - 3| = 5$

42 $|x - 5| = 3$

43 $|x + 2| = 6$

44 $|x + 3| = 3$

45 $|2x - 1| = 5$

46 $|4x + 3| = 11$

47 $|3x - 2| = 4$

48 $|6x + 1| = 3$

4.5   FRACTIONAL EQUATIONS

Fractional Equation   If at least one fraction with the variable in the denominator appears in an equation, then the equation is a *fractional equation*. For example,

$$\frac{x}{x + 1} + \frac{5}{8} = \frac{5}{2(x + 1)} + \frac{3}{4}$$

is a fractional equation, but

$$\frac{x + 1}{2} - 5x = 7$$

is not. We use a variation of (4.2) in solving a fractional equation:

Equivalent Equations **If $k(x)$ is a nonzero expression, then the equations $f(x) = g(x)$ and $k(x) \cdot f(x) = k(x) \cdot g(x)$ are equivalent.** **(4.3)**

If $f(x) = g(x)$ is a fractional equation, and $k(x)$ is the lcm of the denominators, then $k(x) \cdot f(x) = k(x) \cdot g(x)$ will contain no fractions. If the latter equation is linear, we can solve it by the methods of Sec. 4.4.

Example 1 Solve the equation

$$\frac{x}{x+1} + \frac{5}{8} = \frac{5}{2(x+1)} + \frac{3}{4} \tag{1}$$

Solution We use (4.3) as a first step in the solution. Since the lcm of the denominators is $8(x+1)$, we let $k(x) = 8(x+1)$, assume $x + 1 \neq 0$, and multiply each member of Eq. (1) by $8(x+1)$. We then proceed as follows:

$$8(x+1)\left(\frac{x}{x+1} + \frac{5}{8}\right) = 8(x+1)\left[\frac{5}{2(x+1)} + \frac{3}{4}\right] \tag{2}$$

$$8x + 5(x+1) = 4(5) + 6(x+1) \qquad \text{performing indicated multiplication}$$

$$8x + 5x + 5 = 20 + 6x + 6 \qquad \text{by distributive axiom}$$

$$8x + 5x - 6x = 20 + 6 - 5 \qquad \text{transposing}$$

$$7x = 21 \qquad \text{combining terms}$$

$$x = 3 \qquad \text{multiplying by } \tfrac{1}{7}$$

In (1), we cannot divide by 0, so we assumed $x + 1 \neq 0$, or $x \neq -1$. Thus by (4.3), since $k(x) = 8(x+1)$ is not zero, Eq. (1) and $x = 3$ are equivalent equations. To check that 3 is a root of (1), we replace x by 3 in Eq. (1). If this is done, we find that each member is equal to $\frac{11}{8}$. Hence, the root of Eq. (1) is 3.

Example 2 Find

$$\left\{ x \ \middle| \ \frac{2}{x+1} - 3 = \frac{4x+6}{x+1} \right\}$$

Solution The required set is the solution set of

$$\frac{2}{x+1} - 3 = \frac{4x+6}{x+1} \tag{3}$$

We first use (4.3), assume $x + 1 \neq 0$, multiply each member by $x + 1$, and get

$$2 - 3x - 3 = 4x + 6$$

Then we proceed as follows:

$$-3x - 1 = 4x + 6 \qquad \text{combining terms}$$
$$-3x - 1 + 1 - 4x = 4x + 6 + 1 - 4x \qquad \text{adding } 1 - 4x \text{ to each member}$$
$$-7x = 7 \qquad \text{combining terms}$$
$$x = -1$$

Now $x = -1$ is the only possible root. However, it is excluded since $x + 1 \neq 0$. Hence, we conclude that Eq. (3) has no roots, and that

$$\left\{ x \ \Big| \ \frac{2}{x + 1} - 3 = \frac{4x + 6}{x + 1} \right\} = \varnothing$$

where $\varnothing$ is the empty set.

EXERCISE 4.3 Fractional Equations

Solve the equations in Probs. 1 to 32.

1 $\dfrac{1}{x} = \dfrac{1}{4}$ **2** $\dfrac{1}{x - 1} = \dfrac{2}{3}$ **3** $\dfrac{2}{x + 1} = \dfrac{-1}{5}$

4 $\dfrac{2}{2x + 1} = \dfrac{-3}{7}$ **5** $\dfrac{2}{3x + 1} = \dfrac{1}{x}$ **6** $\dfrac{2}{3x + 1} = \dfrac{5}{8x + 1}$

7 $\dfrac{11}{6x + 1} = \dfrac{2}{x + 1}$ **8** $\dfrac{8}{5x - 4} = \dfrac{5}{3x - 1}$ **9** $\dfrac{x + 1}{x - 2} = \dfrac{x - 1}{x - 3}$

10 $\dfrac{x + 1}{x - 5} = \dfrac{x + 4}{x - 4}$ **11** $\dfrac{x + 5}{x - 1} = \dfrac{x + 2}{x - 2}$ **12** $\dfrac{x + 4}{x - 3} = \dfrac{x + 2}{x - 4}$

13 $\dfrac{2x + 5}{4x + 1} = \dfrac{3x + 5}{6x - 1}$ **14** $\dfrac{4x - 3}{2x - 3} = \dfrac{8x + 5}{4x + 1}$ **15** $\dfrac{2x - 5}{4x - 1} = \dfrac{3x - 4}{6x + 9}$

16 $\dfrac{6x - 8}{9x + 8} = \dfrac{2x - 3}{3x + 2}$ **17** $\dfrac{4}{x - 2} - \dfrac{3}{x + 1} = \dfrac{8}{(x - 2)(x + 1)}$

18 $\dfrac{1}{x + 5} + \dfrac{1}{2x + 9} = \dfrac{2}{(x + 5)(2x + 9)}$ **19** $\dfrac{1}{2x + 3} - \dfrac{3}{x - 3} = \dfrac{3}{(2x + 3)(x - 3)}$

20 $\dfrac{2}{x + 2} + \dfrac{1}{2x - 1} = \dfrac{5}{(x + 2)(2x - 1)}$ **21** $\dfrac{2}{x + 1} + \dfrac{3}{2x - 3} = \dfrac{6x + 1}{2x^2 - x - 3}$

22 $\dfrac{5}{3x - 1} - \dfrac{1}{5x - 7} = \dfrac{11x - 1}{15x^2 - 26x + 7}$ **23** $\dfrac{5}{2x + 1} + \dfrac{4}{x - 1} = \dfrac{12x + 6}{2x^2 - x - 1}$

24 $\dfrac{9}{2x + 3} - \dfrac{2}{x - 1} = \dfrac{x + 9}{2x^2 + x - 3}$ **25** $\dfrac{4}{2x - 3} + \dfrac{5}{5x - 4} = \dfrac{3}{x + 2}$

26 $\dfrac{4}{3x - 2} - \dfrac{1}{2x - 3} = \dfrac{5}{6x + 3}$ **27** $\dfrac{4}{3x - 1} - \dfrac{3}{2x + 3} = \dfrac{-1}{6x - 24}$

28 $\dfrac{3}{x + 3} - \dfrac{2}{2x - 5} = \dfrac{6}{3x - 13}$ **29** $\dfrac{x + 7}{(2x - 3)(x + 1)} = \dfrac{8x - 9}{2x - 3} - \dfrac{4x + 9}{x + 2}$

30 $\dfrac{3x + 8}{(x + 1)(x + 3)} = \dfrac{x + 3}{x + 1} - \dfrac{2x + 3}{2x + 5}$ **31** $\dfrac{5x + 20}{(3x - 5)(x + 1)} = \dfrac{2x - 7}{x - 5} - \dfrac{6x - 6}{3x - 5}$

32 $\dfrac{15x - 79}{(3x + 1)(x - 1)} = \dfrac{x - 5}{x - 1} - \dfrac{x - 6}{x + 3}$

In Probs. 33 to 36, find the value of b for which x = 5 is a solution of the equation.

33 $\dfrac{x-3}{b+2} + \dfrac{x-2}{b+6} = \dfrac{x}{b+4}$

34 $\dfrac{x-3}{b+6} + \dfrac{2x-8}{2b-9} = \dfrac{x-2}{b-1}$

35 $\dfrac{x-3}{b+5} + \dfrac{x+1}{3b+5} = \dfrac{2x-6}{b+3}$

36 $\dfrac{x-2}{3b+7} - \dfrac{x-4}{2b+18} = \dfrac{1}{2b-2}$

Show that the equation in each of Probs. 37 to 44 has no solution.

37 $\dfrac{4x+1}{3x+1} = \dfrac{4}{3}$

38 $\dfrac{6x+1}{5-2x} = -3$

39 $\dfrac{2}{x-1} + \dfrac{3}{2x+1} = \dfrac{7x-2}{2x^2-x-1}$

40 $\dfrac{6}{x+3} - \dfrac{5}{3x-1} = \dfrac{13x-20}{3x^2+8x-3}$

41 $\dfrac{3x-7}{x-4} + 4 = \dfrac{-5}{4-x}$

42 $\dfrac{4x-7}{x-2} = 3 - \dfrac{1}{2-x}$

43 $\dfrac{2}{(x-1)(x+2)} = \dfrac{1}{x-1} - \dfrac{1}{x+2}$

44 $\dfrac{3}{x+3} - \dfrac{3}{x-1} = \dfrac{4}{x^2+2x-3}$

Solve the equation in each of Probs. 45 to 48.

45 $\dfrac{2x+1}{2x+3} = \dfrac{1}{3}$

46 $\dfrac{4x+3}{4x-2} = -\dfrac{3}{2}$

47 $\dfrac{3(x-1)-4}{3(x-1)-2} = 2$

48 $\dfrac{6-5(2x+3)}{-2-5(2x+3)} = -3$

Solve the equation in each of Probs. 49 to 52 by adding the expressions in each member of the equation before multiplying by the lcd.

49 $\dfrac{1}{x+3} - \dfrac{1}{x+1} = \dfrac{1}{x+4} - \dfrac{1}{x+2}$

50 $\dfrac{1}{x+5} - \dfrac{1}{x+8} = \dfrac{1}{x+3} - \dfrac{1}{x+6}$

51 $\dfrac{x-3}{x-1} - \dfrac{x-5}{x-3} = \left(-\dfrac{4}{3}\right)\left(\dfrac{x-1}{x-2} - \dfrac{x-4}{x-5}\right)$

52 $\dfrac{x+3}{x-2} - \dfrac{x-1}{x-6} = \left(-\dfrac{5}{3}\right)\left(\dfrac{x-2}{x+4} - \dfrac{x-4}{x+2}\right)$

4.6 SOLVING STATED PROBLEMS

A stated problem is a word description of a situation that involves both known and unknown quantities and certain relations between these quantities. If the problem is solvable by means of one equation, it must be possible to find two combinations of the quantities in the problem that are equal so that an equation can be written. Furthermore, at least one of the combinations must involve the unknown.

The procedure for solving a stated problem by means of an equation is not always simple, and considerable practice is necessary before one becomes adept at it. The following approach is suggested:

1 Read the problem carefully, and study it until the situation is thoroughly understood.

2 Identify both the known and the unknown quantities that are involved in the problem.

3 Select one of the unknowns, and represent it by a symbol, usually *x*; then express the other unknowns in terms of this symbol.

4 Search the problem for information that tells which quantities or combinations are equal.

5 When the desired combinations are found, set them equal to each other and thus obtain an equation.

6 Solve the equation thus obtained, and check the solution set in the original problem.

It is usually helpful to tabulate the data given in the problem, as is done in the following illustrative examples, which show the methods for solving several types of stated problems.

Problems In-
volving Motion
at a Uniform
Velocity

Problems that involve motion usually state a relation between the velocities (or speeds), between the distances traveled, or between the periods of time involved. The fundamental formulas used in such problems are

$$d = vt \qquad v = \frac{d}{t} \qquad t = \frac{d}{v}$$

where *d* represents distance, *v* represents velocity (or speed), and *t* represents a period of time. When one or more of these formulas is used in the same problem, *d* and *v* must be expressed in terms of the same linear unit, and *v* and *t* must be expressed in the same unit of time.

Example 1 A party of hunters made a trip of 310 miles to a hunting lodge in 7 hours. They traveled 4 hours on a paved highway, and the remainder of the time on a pasture road. If the average speed through the pasture was 25 miles per hour less than that on the highway, find the average speed and the distance traveled on each part of the trip.

Solution The unknown quantities here are the two speeds and the distances traveled on each part of the trip. We shall let

x = speed on highway in miles per hour

Then

x − 25 = speed through pasture

We now tabulate the data of the problem, with the unknown quantities expressed in terms of *x*. Note that the formula $d = vt$ is used to obtain the unknown distances.

	Time t, hours	Velocity v, miles per hour	Distance $d = vt$, miles
On paved highway	4	x	$4x$
On pasture road	$7 - 4 = 3$	$x - 25$	$3(x - 25)$
Total	7		310

The last column gives us two quantities that are equal:

Distance on highway + distance through pasture = total distance

or

$$4x + 3(x - 25) = 310$$

Hence, the desired equation is

$$4x + 3(x - 25) = 310$$

and we solve it as follows:

$4x + 3x - 75 = 310$	distributive axiom
$4x + 3x = 310 + 75$	adding 75 to each member
$7x = 385$	combining terms
$x = 55$	multiplying each member by $\frac{1}{7}$

Thus 55 miles per hour is the velocity on the highway.

$$55 - 25 = 30$$

so that 30 miles per hour is the velocity through the pasture;

$4(55) = 220$ miles traveled on highway

$3(30) = 90$ miles traveled through the pasture

Check $220 + 90 = 310$

Example 2   Airports A, B, and C are located on a north-south line. B is 645 miles north of A, and C is 540 miles north of B.   A pilot flew from A to B, delayed 2 hours, and continued to C.   The wind blew from the south at 15 miles per hour during the first part of the trip, but during the delay it changed to the north, with a velocity of 20 miles per hour.   If each flight required the same period of time, find the airspeed of the plane, that is, the speed delivered by the propeller.

Solution   If we let

$x =$ airspeed of plane

then

$x + 15 =$ groundspeed of plane while wind was from the south

and

$x - 20 =$ groundspeed of plane while wind was from the north

We now use the formula $t = d/v$ to obtain the time required for each flight and tabulate the data:

	Distance d, miles	Airspeed of plane, miles per hour	Speed of wind, miles per hour	Groundspeed v of plane, miles per hour	Period of time $t = d/v$, hours
From A to B	645	x	15, from south	$x + 15$	$\dfrac{645}{x + 15}$
From B to C	540	x	20, from north	$x - 20$	$\dfrac{540}{x - 20}$

According to the statement of the problem, the following quantities are equal:

Time in hours from A to B = time in hours from B to C

$$\frac{645}{x + 15} = \frac{540}{x - 20}$$

Hence, the required equation is

$$\frac{645}{x + 15} = \frac{540}{x - 20}$$

and we solve it as follows:

$$(x - 20)(x + 15)\frac{645}{x + 15} = (x - 20)(x + 15)\frac{540}{x - 20}$$

multiplying each member by lcm of denominators

$$(x - 20)645 = (x + 15)540$$

performing indicated multiplication

$$645x - 12{,}900 = 540x + 8100$$

distributive axiom

$$105x = 21{,}000$$

adding $-540x + 12{,}900$ to each member and combining terms

$$x = 200$$

multiplying each member by $\frac{1}{105}$

Check   The airspeed is 200 miles per hour, the groundspeed from A to B is $200 + 15 = 215$ miles per hour, the groundspeed from B to C is $200 - 20 = 180$ miles per hour, and $\frac{645}{215} = \frac{540}{180} = 3$.

Work Problems   Problems that involve the rate of performance can often be solved by first finding the fractional part of the task done by each individual or agent in one unit of time and then finding a relation between several fractional parts. In this method the unit 1 represents the entire job.

Example 3   A farmer can plow a field in 4 days using a tractor. His hired hand can plow the same field in 6 days using a smaller tractor. How many days will be required for the plowing if they work together?

Solution   We let

x = number of days required to plow the field if they work together.

Now we tabulate the data and complete the solution:

	Farmer	Helper	Together
Days required to plow field	4	6	x
Part plowed in 1 day	$\dfrac{1}{4}$	$\dfrac{1}{6}$	$\dfrac{1}{x}$

Quantities that are equal:

Part done in 1 day by farmer + part done by helper = part done by both

$$\frac{1}{4} + \frac{1}{6} = \frac{1}{x} \qquad \text{desired equation}$$

$$3x + 2x = 12 \qquad \text{multiplying each member by 12x}$$

$$5x = 12$$

$$x = 2\tfrac{2}{5} \text{ days}$$

Check   In $2\tfrac{2}{5}$ days the farmer plows $\tfrac{12}{5}\left(\tfrac{1}{4}\right) = \tfrac{3}{5}$ of the field. The helper plows $\tfrac{12}{5}\left(\tfrac{1}{6}\right) = \tfrac{2}{5}$ of the field, and $\tfrac{3}{5} + \tfrac{2}{5} = 1$, or the entire field.

Example 4   If, in Example 3, the hired hand worked 1 day with the smaller machine and then was joined by the employer, how many days were required to finish the plowing?

Solution   Since the hired hand plowed $\tfrac{1}{6}$ of the field in 1 day, $\tfrac{5}{6}$ of it remained unplowed. We let

x = number of days required to finish the plowing

Then $x/4$ is the part plowed by the farmer, and $x/6$ is the part plowed by the helper. Hence,

$$\frac{x}{4} + \frac{x}{6} = \frac{5}{6}$$

$$3x + 2x = 10 \qquad \text{multiplying each member by 12}$$

$$5x = 10$$

$$x = 2$$

<table>
<tr><td>*Mixture Problems*</td><td>Many problems involve the combination of certain substances of known strengths, usually expressed in percentages, into a mixture of required strength in one of the substances. Others involve the mixing of certain commodities of specified prices. In such problems, it should be remembered that the total amount of any given element in a mixture is equal to the sum of the amounts of that element in the substances combined, and that the monetary value of any mixture is the sum of the values of the substances that are put together.</td></tr>
</table>

Example 5 How many gallons of a liquid that is 74 percent alcohol must be combined with 5 gallons of one that is 90 percent alcohol in order to obtain a mixture that is 84 percent alcohol?

Solution If we let x represent the number of gallons needed of the first liquid, and remember that 74 percent of x is $0.74x$, then the following table showing the data in the problem is self-explanatory.

	Number of gallons	Percentage of alcohol	Number of gallons of alcohol
First liquid	x	74	$0.74x$
Second liquid	5	90	$0.90(5) = 4.5$
Mixture	$x + 5$	84	$0.84(x + 5)$

Quantities that are equal:

Number of gallons of alcohol in first liquid
+ number of gallons of alcohol in second liquid
= number of gallons of alcohol in mixture

Hence, the equation is

$0.74x + 4.5 = 0.84(x + 5)$

and the solution, obtained by the usual method, is $x = 3$.

Other Word Problems There is a wide variety of problems that can be solved by means of equations. The fundamental approach to all of them is the same and entails finding two quantities, one or both of which include the unknown, that are equal.

Example 6 Find three consecutive odd integers whose sum is 69.

Solution An odd integer may be written $2x + 1$, where x is an integer. Then the three consecutive odd integers are $2x + 1$, $(2x + 1) + 2 = 2x + 3$, and $(2x + 1) + 4 = 2x + 5$. Since their sum is 69, we have

$$(2x + 1) + (2x + 3) + (2x + 5) = 69$$
$$6x + 9 = 69$$
$$6x = 60$$
$$x = 10$$

Thus the numbers are $2x + 1 = 2(10) + 1 = 21, 21 + 2 = 23$, and $23 + 2 = 25$.

Example 7   Mrs. Russell has some money invested at 5 percent, and $4000 more than this amount at 7 percent. If her total annual return from these two investments is $1120, how much is invested at each rate?

Solution   If she has x dollars invested at 5 percent, her annual return from this money is $\frac{5}{100}x$. The amount invested at 7 percent is $x + 4000$, and its return is $(\frac{7}{100})(x + 4000)$. The total return is $1120, so

$$\frac{5x}{100} + \frac{7(x + 4000)}{100} = 1120$$
$$5x + 7(x + 4000) = 112,000$$
$$12x = 112,000 - 28,000$$
$$12x = 84,000$$
$$x = 7000$$

She thus has $7000 at 5 percent and $11,000 at 7 percent.

EXERCISE 4.4   Stated Problems

1 Find three consecutive integers whose sum is 72.

2 A saleswoman called on 20 customers in three days. If the second day she called on one more customer than the first day, and the third day she called on three more than the second day, how many calls did she make each day?

3 Two rare coins are worth $90. If the value of one is $1\frac{1}{2}$ times that of the other, what is the value of each?

4 A farmer used 1960 ft of fencing to enclose a rectangular plot of ground. If the width of the plot was three-fourths of the length, what were its dimensions?

5 Fred and Mary have a total of $559 in their bank accounts. If Fred has $213 more in his account than Mary has in hers, how much is in each account?

6 David found that when he completed the 14 hours he was taking that semester, he would have half the hours credit he needed for his degree. If at the beginning of the semester he had had 0.4 of the hours he needed, how many hours were required for graduation?

7 Dean and Lucy caught 21 pounds of fish. If Dean's catch weighed 3.4 pounds more than Lucy's, how many pounds of fish did each one catch?

8 A company safety officer found that of the 62 accidents with company vehicles the previous year, 8 more occurred while the vehicle was backing than all other types of accidents combined. How many backing accidents were there?

9 Sue, Joe, and Jack worked a total of 21 hours overtime. If Sue and Joe together worked 15 hours overtime and Jack worked 2 hours more than Sue, how much overtime did each one work?

10 Mr. Duff got a ticket for speeding. He was driving 17 miles per hour faster than the speed limit allowed. If he had been driving 18 miles per hour faster, he would have been driving twice as fast as the law allowed. What was the speed limit?

11 Three of four poker players lost money. If Mr. Green lost one-fourth of the money, Mr. Brown lost $10 more than Mr. Green, and Mr. White lost $6 more than Mr. Brown, how much did the fourth player, Mr. Black, win?

12 An amusement park charged $1.80 per person for a ticket, but had discount tickets available for $1.55. If $635.90 was collected from the sale of 363 tickets, how many of the tickets were purchased at the discount price?

13 Mrs. Cox counted a total of 27 honor points in her bridge hand from aces, kings, and queens by using the system that allows 4 points per ace, 3 points per king, and 2 points per queen. If she had one more ace than kings and 5 of her 13 cards were not honor cards, how many aces did she have?

14 A man bought a suit at a sale at a 10 percent reduction, and his wife bought a dress at a 15 percent reduction. The original price total for both garments was $160.00, and they paid $141.00. What was the original price of the suit?

15 One month Madge and Anne earned $30.80 by each working 30 hours as a baby sitter. Except for one job on which Anne earned 60 cents an hour, each was paid 50 cents an hour. How many hours did Anne work at the higher rate?

16 A bookstore received $628.75 from the sale of 470 copies of a particular book. If the book was available in both a hardback edition costing $4.00 and a paperback edition costing 75 cents, how many copies of each type of edition were sold?

17 In the annual candy sale, a campfire girl had twice as many boxes of mint sticks to sell as she had boxes of assorted chocolates. After she sold two boxes of each kind of candy, she had three times as many boxes of mint sticks as assorted chocolates. How many boxes of candy remained to be sold?

18 During the first 6 weeks of the summer session of a state college, there were three times as many out-of-state students as there were in-state students enrolled. Of this number, 6000 out-of-state and 1000 in-state students did not enroll for the second summer session, but 9000 new in-state students enrolled. If there were then twice as many in-state students as out-of-state students, what was the enrollment for the first session?

19 In a certain community, there are 400 more registered Republicans than Democrats. If the voting age were changed from 18 to 21, both the Republicans and Democrats would lose 100 eligible voters and there would then be 7/3 as many Republicans as Democrats. How many registered voters over 21 are there in the community?

20 An overweight man weighed twice as much as his wife. After he lost 60 pounds, he weighed $1\frac{1}{2}$ times his wife's weight. How much did his wife weigh?

21 The second number of the combination to unlock a locker is twice the first, and the third number is one-fourth the second. If the sum of the numbers in the combination is 42, what is the combination?

22 The snack bar at a Little League ball game took in $43.20 during one game. If there were 20 one-dollar bills, 20 pennies, 15 more dimes than quarters, and 15 more nickels than dimes, how many nickels, dimes, and quarters were there?

23 If 20 gallons of water at a temperature of 65°F is mixed with 20 gallons of water at a temperature of 81°F, what is the temperature of the mixture?

24 A 5-liter flask designated for 15 percent hydrochloric acid solution was found to be filled with 25 percent hydrochloric acid by mistake. How much should be drained off and replaced with distilled water to provide the desired 15 percent solution?

25 A tour group rode a bus at an average rate of 50 miles per hour to visit a city where they took a walking tour. If they walked at a rate of $\frac{3}{4}$ mile per hour, their day's trip took $7\frac{1}{2}$ hours, and they traveled 178 miles, how long was the walking tour?

26 The tour group of Prob. 25 spent their next day taking a scenic drive along a river to a spot 100 miles from the city. They returned to the city by boat. If the average speed of the boat was five-eighths that of the bus and the boat trip took $1\frac{1}{2}$ hours longer than the bus trip, how long did the group travel that day?

27 A delivery-truck driver drove 30 miles on a freeway from a downtown warehouse to the suburbs where she delivered parcels along a 24-mile route. She spent $2\frac{2}{3}$ as much time delivering parcels as she did driving to the suburbs. Find her average speed on each part of the trip if she drove 35 miles per hour faster on the freeway than in the suburbs.

28 Airport B is 90 miles due north of airport A. One morning a pilot in a small plane flew from A to B, had lunch, and started back to A, but when he had flown the same length of time as he had in the morning, he was only 60 miles of the way back. If there was a wind blowing from the south at a uniform rate of 20 miles per hour all day, find the airspeed of the plane.

29 Two neighbors whose lots are equal in width spent 12 and 16 hours, respectively, building fences across the backs of their yards. How long will it take them to build a fence of equal length between their yards if they work together?

30 A volunteer worker required 2 hours to address a group of envelopes for a fund drive, while a second worker required 3 hours to address a similar group. How long would it take the two volunteers working together to address a third similar group of envelopes?

31 A maintenance man needed 8 hours to wash the windows in a certain building. The next month his assistant took 10 hours to wash the windows. If the two worked together, how much time would be needed to wash the windows?.

32 Joe could read the proof for the school newspaper in 1 hour, while Lois needed $1\frac{1}{4}$ hours to read the same amount of proof. One day they worked together for $\frac{1}{3}$ hours, and then Joe had to leave and Lois finished the proofreading. How long did it take her?

33 Matt and Mary worked together on a jigsaw puzzle for 3 hours. Then Matt left, and

Mary finished the puzzle in 30 minutes. Matt had worked similar puzzles alone in 6 hours. How long would Mary have taken to work the puzzle alone?

34 A political worker could fold enough campaign literature for a mailing list in 10 hours. His wife could do the job in 12 hours, and his daughter in 15 hours. If they all worked together, how long would the job require?

35 Todd, Mark, and Brad had 900 handbills to distribute. Each boy could distribute an average of 120 handbills per hour. Todd started at 8:00 A.M., Mark at 8:30, and Brad at 9:00. What time did they finish?

36 A farmer has a storage tank for irrigation that can be filled by the intake pipe in 10 hours and drained by the outlet in 8 hours. If, at the start of an irrigation job, the tank is full and both pipes are open, how long will it take to drain the tank?

37 Three people could clean the rooms in a certain motel in 8 hours, but two people would need 12 hours to clean all the rooms. One day two of the people started work on time, but the third was 3 hours late. How many hours did it take the three people to finish cleaning the motel?

38 A father and his two sons decided to install a sprinkler system in their backyard. The father could have done the work in 8 hours, the older boy in 12 hours, and the younger in 16 hours. They started work together, but after 2 hours the younger boy left to play baseball, and an hour later the older boy went to band practice. How much longer did the father work alone to finish the sprinkler system?

39 New members of a labor crew were paid $22.50 a day, and experienced members of the crew were paid $25.80 a day. If the average daily wage paid to members of a certain crew was $24 and the total daily wages paid to experienced members of the crew was $129.00, how many new members were on the crew?

40 How many gallons of a 25 percent salt solution must be mixed with 10 gallons of a 15 percent solution to produce a 20 percent salt solution?

41 A 1 percent solution of insecticide was needed to spray a tree, but because of a mixing error, the 8-gallon sprayer was filled with a solution that was 0.6 percent insecticide. How much of the solution must be drawn off and replaced with 8.6 percent solution to provide the proper concentration?

42 A chemist mixed 35 milliliters of 4 percent nitric acid solution with 65 milliliters of 12 percent nitric acid solution. She used a portion of the mixture, and replaced it with distilled water. The new solution tested 6.9 percent nitric acid. How much of the original solution was used?

43 On its last trip of the day, a dump truck carrying a load of dirt traveled 8 miles from an excavation to a construction site, dumped its dirt, and then drove 33 miles to a storage yard. Its loaded speed was two-thirds its empty speed, and the entire traveling time was 1 hour. Find the average speeds of the truck when it was loaded and when it was empty.

44 A commuter drove his car at a rate of 30 miles per hour from his home to a train station where he waited 10 minutes and then boarded a train that averaged 45 miles per hour to the city. If his entire trip was 35 miles and took 1 hour, how far did he live from the train station?

45 Airfield A is 325 miles due north of airfield B, and airfield C is 416 miles due east of airfield B. The pilot of a small plane flew from field A to field B on one day, and on to field C on the next day. The wind was blowing from the south at a rate of 12 miles per hour the first day, and from the west at a rate of 16 miles per hour the second day. Find the airspeed of the plane if both flights took the same length of time.

46 A truck carrying a group of construction workers left the construction camp for the work site. Six minutes later the resident engineer left the camp in his jeep and passed the truck 5 miles from the camp. If traveling time from the camp to the site was 24 minutes for the truck and 12 minutes for the jeep, find the average speed of the truck.

47 Mr. Peterson borrowed $1000 from a loan company and agreed to repay it in equal monthly installments plus interest of 1 percent per month on the outstanding principal. At the end of 5 months, he had paid a total of $45 in interest. Find the amount of his monthly installment.

48 A woman invested $1440 in the common stock of one company and $2160 in the stock of another. The price per share of the second stock was three-fourths that of the first. The next day the price of the more expensive stock declined by $0.75 a share, while the price of the other advanced $1.50 a share; as a result, the value of her investment increased $45. Find the price per share of the more expensive stock.

4.7 SUMMARY

Two equations are equivalent if they have the same solution set. The conditions given in (4.1) and (4.2) are the basic ones in determining whether two equations are equivalent—they state that we may add the same expression to both members of an equation and that both members may be multiplied by the same nonzero expression. Conditional equations and identities are discussed (we normally deal with conditional equations). We solve both linear equations and fractional equations which may be changed into linear ones. The chapter concludes with stated problems which may be solved by use of linear equations.

EXERCISE 4.5 Review

1 Find the truth set for the statement "H_2O is the chemical symbol for x."

2 Is $x^2 + 25 = (x + 5)^2$ an identity or a conditional equation?

Are the statements in Probs. 3 and 4 true or false?

3 $\dfrac{49}{98} = \dfrac{4}{8}$

4 $\dfrac{2x^2 - x + 4}{2x^2 - x + 8} = \dfrac{1}{2}$

5 Show that 5 is a solution of $x/5 + 1 = x - 3$.

6 Are the equations $6x - 1 = 2x + 3$ and $x = 1$ equivalent?

7 Are the equations $2x^2 - 1 = 3x$ and $4x^2 - 6x = 2$ equivalent?

Solve the equations in each of Probs. 8 to 20.

8 $2x + 5 = 4x - 1$

9 $\frac{2}{9}(4x + 7) = x + 1$

10 $\frac{x + 3}{5} + \frac{3x - 5}{8} = 4$

11 $\frac{5x}{4} + \frac{2}{3} = \frac{4x}{3} - \frac{1}{3}$

12 $\frac{3x}{5} - \frac{2}{7} = \frac{3x}{7} + \frac{2}{35}$

13 $bx - a = x - \frac{a}{b}$

14 $|3x + 1| = 4$

15 $\frac{2}{3x - 5} = \frac{1}{x}$

16 $\frac{x - 4}{x + 1} = \frac{x - 3}{x + 7}$

17 $\frac{2x + 7}{9x - 1} = \frac{4x + 5}{18x - 20}$

18 $\frac{3}{2x + 1} + \frac{4}{x - 2} = \frac{10x + 5}{2x^2 - 3x - 2}$

19 $\frac{1}{x - 2} + \frac{2}{x} = \frac{3}{x - 1}$

20 $\frac{1}{x + 6} - \frac{1}{x + 4} = \frac{1}{x + 5} - \frac{1}{x + 3}$

21 Show that there is no solution to

$$\frac{3}{3x + 1} - \frac{2}{2x + 1} = \frac{7}{6x^2 + 5x + 1}$$

22 If Alfred has x dimes and Ben has three more than twice as many dimes as Alfred has, how many dimes does Ben have?

23 If Abigail was x years old 3 years ago and is now twice as old as Jessica, what is Jessica's age now?

24 The denominator of a fraction is 3 more than the numerator. If the numerator and denominator are each increased by 1, the value of the fraction is $\frac{1}{2}$. What is the original fraction?

25 A chain with 24 links is cut into three pieces. Find the number of links in the shorter piece if it has $\frac{2}{3}$ as many links as each of the other two pieces.

26 A man takes twice as long as a machine to do a certain job. Together they can do the job in 6 hours. How long would it take the man by himself?

27 A chemist wants to make 6 liters of a 10 percent acid solution by mixing a 7 percent and a 12 percent solution. How many liters of each should she use?

28 Solve $s = (a - rl)/(1 - r)$ for r.

29 Show that $a/b = c/d$ if and only if $(a + c)/(b + d) = c/d$.

30 Prove (4.1).

31 Prove (4.2).

5
exponents, roots, and radicals

In Eq. (1.20) we defined a positive integral exponent and the number a^0, $a \neq 0$; furthermore, we derived the laws for the product and the quotient of two positive integral powers of the same number and used these definitions and laws to a limited extent. In fields in which mathematics is used, a broader concept of an exponent is needed. Consequently, in this chapter we shall extend the definition of an exponent so as to include negative and rational values. These extensions will be made so that the laws developed in Chap. 1 shall hold. Furthermore, we shall develop laws for a power of a product and of a quotient and explain how to use the old and new concepts and laws in more complicated situations than occurred in Chap. 1.

5.1 LAWS OF POSITIVE INTEGRAL EXPONENTS

We shall begin by repeating the definition of a positive integral exponent and the laws of exponents developed in Chap. 1. If n is a positive integer, then the product $a \cdot a \cdot a \cdots$ to n a's is called the *nth power of a* and is written as a^n. The number a is the *base* and n is the *exponent*. A symbolic form of the definition and the laws developed in Chap. 1 are listed here for the convenience of the reader.

nth Power
Base
Exponent

Definition	$a^n = a \cdot a \cdot a \cdots$ to n a's	(5.1)
Multiplication	$a^m a^n = a^{m+n}$ m and n positive integers	(5.2)

Division $\dfrac{a^m}{a^n} = a^{m-n}$ $\begin{cases} m \text{ and } n \text{ positive integers} \\ m > n \\ a \neq 0 \end{cases}$ (5.3)

Zero $a^0 = 1$ $a \neq 0$ (5.4)

We shall now develop three more laws for positive integral

exponents. If we apply the definition of a positive integral exponent to the number $(a^m)^n$, we have

$$(a^m)^n = a^m \cdot a^m \cdot a^m \cdots \text{ to } n \text{ factors}$$
$$= a^{m+m+m\cdots\text{ to } n \text{ terms}} \qquad \text{by (5.2)}$$
$$= a^{nm}$$

Consequently, we have

Power of a Power $\qquad (a^m)^n = a^{nm}$ $\qquad\qquad\qquad\qquad\qquad$ **(5.5)**

If we apply the definition to $(ab)^n$, we obtain

$$(ab)^n = ab \cdot ab \cdot ab \cdots \text{ to } n \text{ factors}$$
$$= (a \cdot a \cdot a \cdots \text{ to } n \text{ factors})(b \cdot b \cdot b \cdots \text{ to } n \text{ factors})$$
$$\text{by commutative axiom for multiplication}$$

$$= a^n b^n$$

Therefore,

Power of a Product $\qquad (ab)^n = a^n b^n$ $\qquad\qquad\qquad\qquad\qquad$ **(5.6)**

We can show similarly that

Power of a Quotient $\qquad \left(\dfrac{a}{b}\right)^n = \dfrac{a^n}{b^n} \qquad b \neq 0$ $\qquad\qquad\qquad\qquad\qquad$ **(5.7)**

We shall now illustrate the use of laws (5.1) to (5.7) by several examples.

Example 1 $\qquad\qquad a^5 = a \cdot a \cdot a \cdot a \cdot a$ $\qquad\qquad\qquad$ by (5.1)

Example 2 $\qquad\qquad a^7 a^2 = a^{7+2} = a^9$ $\qquad\qquad\qquad$ by (5.2)

Example 3 $\qquad\qquad \dfrac{a^6 a^5}{a^4} = \dfrac{a^{6+5}}{a^4} = \dfrac{a^{11}}{a^4} = a^{11-4} = a^7$ $\qquad$ by (5.2) and (5.3)

Example 4 $\qquad\qquad (a^5)^3 = a^{(3)(5)} = a^{15}$ $\qquad\qquad\qquad$ by (5.5)

Example 5 $\qquad\qquad (x^2 y^3)^4 = (x^2)^4 (y^3)^4 = x^8 y^{12}$ $\qquad$ by (5.6) and (5.5)

Example 6 $\qquad\qquad \left(\dfrac{x^3}{z^4}\right)^2 = \dfrac{(x^3)^2}{(z^4)^2} = \dfrac{x^6}{z^8}$ $\qquad\qquad$ by (5.7) and (5.5)

Example 7 $\quad (3x^2 y^3)(7x^4 y^5) = 3 \cdot 7 x^2 x^4 y^3 y^5$ $\qquad$ by commutative axiom
$$\qquad\qquad\qquad\qquad = 21 x^6 y^8 \qquad\qquad\qquad\qquad \text{by (5.2)}$$

Example 8 $\qquad\qquad \dfrac{30 x^5 y^7}{15 x^2 y^3} = \left(\dfrac{30}{15}\right)\left(\dfrac{x^5}{x^2}\right)\left(\dfrac{y^7}{y^3}\right)$
$$\qquad\qquad\qquad\qquad = 2 x^3 y^4 \qquad\qquad\qquad\qquad \text{by (5.3)}$$

Simplifying We say that an expression that includes positive integral exponents is *simplified* if all combinations are made that can be made by use of (5.1) to (5.7). The procedures are illustrated in the following example.

Example 9 Simplify

$$\left(\frac{3x^2y^3}{z^4}\right)^3\left(\frac{2y^2z^7}{x^3}\right)^2$$

Solution In order to simplify, we shall raise each product and quotient to the indicated power by use of (5.5), (5.6), and (5.7), then apply (5.2), and finally reduce to lowest terms by use of (5.3).

$$\left(\frac{3x^2y^3}{z^4}\right)^3\left(\frac{2y^2z^7}{x^3}\right)^2 = \frac{3^3(x^2)^3(y^3)^3}{(z^4)^3}\frac{2^2(y^2)^2(z^7)^2}{(x^3)^2} \quad \text{by (5.6) and (5.7)}$$

$$= \frac{27x^6y^9}{z^{12}}\frac{4y^4z^{14}}{x^6} \quad \text{by (5.5)}$$

$$= 27(4)\left(\frac{x^6}{x^6}\right)(y^9y^4)\left(\frac{z^{14}}{z^{12}}\right) \quad \text{by commutative axiom}$$

$$= 108x^{6-6}y^{9+4}z^{14-12} \quad \text{by (5.2) and (5.3)}$$

$$= 108x^0y^{13}z^2$$

$$= 108y^{13}z^2 \quad \text{since by (5.4), } x^0 = 1$$

If we interchange the members of (5.6) and of (5.7), we get

$$a^nb^n = (ab)^n \tag{5.6'}$$

$$\frac{a^n}{b^n} = \left(\frac{a}{b}\right)^n \tag{5.7'}$$

This justifies the practice of multiplying or dividing the two bases and then raising the result to the common power.

Example 10 Simplify

$$\left(\frac{3a^3b^5}{13c^4}\right)^2\left(\frac{26c^5}{9a^2b^3}\right)^2$$

Solution $$\left(\frac{3a^3b^5}{13c^4}\right)^2\left(\frac{26c^5}{9a^2b^3}\right)^2 = \left[\left(\frac{3a^3b^5}{13c^4}\right)\left(\frac{26c^5}{9a^2b^3}\right)\right]^2 \quad \text{by (5.6')}$$

$$= \left(\frac{3\cdot 26}{13\cdot 9}\frac{a^3}{a^2}\frac{b^5}{b^3}\frac{c^5}{c^4}\right)^2 \quad \text{by commutative axiom}$$

$$= \left(\frac{2}{3}ab^2c\right)^2 \quad \text{by (5.3)}$$

$$= \frac{4}{9}a^2b^4c^2 \quad \text{by (5.6), (5.7), and (5.5)}$$

Example 11   Simplify

$$\left(\frac{9x^{3a-4}}{3x^{2a-1}}\right)^3 \left(\frac{y^b}{x^{a-1}}\right)^2$$

Solution $\left(\dfrac{9x^{3a-4}}{3x^{2a-1}}\right)^3 \left(\dfrac{y^b}{x^{a-1}}\right)^2 = (3x^{3a-4-(2a-1)})^3 \left(\dfrac{y^b}{x^{a-1}}\right)^2$ **by (5.3)**

$$= (3x^{a-3})^3 \left(\frac{y^b}{x^{a-1}}\right)^2$$

$$= 27x^{3a-9}\,\frac{y^{2b}}{x^{2a-2}} \qquad \text{\textbf{by (5.5) and (5.7)}}$$

$$= 27x^{a-7}y^{2b} \qquad \text{\textbf{by (5.3)}}$$

Reminder 1   In applying (5.5), be sure to remember that the exponent inside the parentheses is multiplied by the one outside and not raised to the power indicated by it.   Thus,

$$(a^3)^2 = a^{(3)(2)} = a^6 \qquad \text{not} \qquad a^{3^2} = a^9$$

Reminder 2   If x^a is to be multiplied by y^b, we can only indicate the product by writing $x^a y^b$ unless $x = y$ or $a = b$.   If the bases are equal, we *use the common base* and add the exponents.   If the exponents are equal, we *use the common exponent* and multiply the bases.   Thus $a^2 a^3 = a^5$ and $a^2 b^2 = (ab)^2$, as is seen from (5.2) and (5.6′).

EXERCISE 5.1   Positive Integral Exponents

Perform the indicated operations in Probs. 1 to 44.

1 $2^3 2^2$	**2** $3^2 3^4$	**3** $5^2 5$	**4** $4^2 4^3$
5 $a^5 a^3$	**6** $a^2 a^0$	**7** $a^4 a$	**8** $a^3 a^2$
9 $2^5/2^2$	**10** $3^4/3$	**11** $5^3/5^2$	**12** $4^4/4^3$
13 b^9/b^6	**14** b^8/b^3	**15** b^7/b^2	**16** b^9/b^6
17 $(3^2)^3$	**18** $(3^3)^2$	**19** $(4^2)^1$	**20** $(3^5)^0$
21 $(a^3)^2$	**22** $(a^4)^2$	**23** $(a^2)^4$	**24** $(a^1)^6$
25 $(3a^2)(2a^3)$	**26** $(2a^5)(5a^2)$	**27** $(4a^3)(3a^4)$	**28** $(2a^2)(3a^3)$
29 $8a^4/2a^2$	**30** $9a^5/3a^2$	**31** $10a^7/2a^3$	**32** $12a^6/3a^3$
33 $(2a^3)^2$	**34** $(3a^2)^3$	**35** $(5a^7)^0$	**36** $(4a^3)^1$
37 $(a^2 b^3)^2$	**38** $(a^3 b^2)^3$	**39** $(a^4 b)^3$	**40** $(a^0 b^3)^3$
41 $\left(\dfrac{a^2}{b^3}\right)^3$	**42** $\left(\dfrac{a^4}{b}\right)^2$	**43** $\left(\dfrac{a^0}{b}\right)^5$	**44** $\left(\dfrac{a^5}{b^2}\right)^0$

Perform the indicated operations in the following problems and simplify.

45 $\dfrac{28x^3 y^5}{7x^2 y^3}$ 　　　　　　**46** $\dfrac{45x^4 y^6}{9x^3 y^4}$ 　　　　　　**47** $\dfrac{18a^2 b^4}{6a^0 b}$

48 $\dfrac{42a^6b^4}{6a^3b^2}$

49 $(4x^2y^3)(3x^3y^4)$

50 $(3x^4y)(2xy^3)$

51 $(6x^0y^4)(5x^2y)$

52 $(0x^5y^2)(7xy^5)$

53 $\dfrac{2a^3b^2}{3a^2d^3}\dfrac{6c^4d^4}{4bc^2}$

54 $\dfrac{15a^3b^3}{6b^2c^4}\dfrac{18c^5d^7}{5ad^4}$

55 $\dfrac{4x^3y^5}{9y^2z}\dfrac{27z^3w^2}{2wx^2}$

56 $\dfrac{15a^5b^4}{3b^0x^3}\dfrac{7x^5y^7}{5y^5a}$

57 $\left(\dfrac{18a^4b^5}{6a^6b^3}\right)^2$

58 $\left(\dfrac{25a^7b^3}{5a^2b}\right)^3$

59 $\left(\dfrac{14a^6b^3}{7a^5b^4}\right)^4$

60 $\left(\dfrac{35a^9b^3}{21a^7b^5}\right)^4$

61 $\left(\dfrac{8a^2b^3}{3c^2}\right)^2\left(\dfrac{6a^0c}{4b^2}\right)^2$

62 $\left(\dfrac{10x^3y^2}{7z^4}\right)^3\left(\dfrac{21z^4}{20x^4y}\right)^2$

63 $\left(\dfrac{5a^2}{6b^3c}\right)^3\left(\dfrac{18b^4c}{10a^2}\right)^3$

64 $\left(\dfrac{28a^0}{6bc^2}\right)^4\left(\dfrac{3b^2c}{14a^2}\right)^4$

65 $\left[\left(\dfrac{2x^2}{3y}\right)^2\left(\dfrac{6y^2}{4x}\right)^3\right]^2$

66 $\left[\left(\dfrac{3x^4}{7y^5}\right)^4\left(\dfrac{14y^4}{6x^3}\right)^4\right]^3$

67 $\left[\left(\dfrac{4a^5}{3y^3}\right)^4\left(\dfrac{9y^5}{16a^9}\right)^2\right]^5$

68 $\left[\left(\dfrac{3x^4}{4a^3}\right)^4\left(\dfrac{8a^4}{9x^5}\right)^3\right]^4$

69 $a^{3x+2}a^{x-3}$

70 $a^{5x-1}a^{2-4x}$

71 $b^{x+y}b^{2x-y}$

72 $b^{3x+2y}b^{2x-3y}$

73 $\dfrac{a^{4x+5}}{a^{3x+2}}$

74 $\dfrac{a^{3x-1}}{a^{x-3}}$

75 $\dfrac{c^{2-y}}{c^{1+y}}$

76 $\dfrac{c^{x+2y}}{c^{x+y}}$

77 $\left(\dfrac{x^{2a}y^{a-1}}{x^ay^a}\right)^2$

78 $\left(\dfrac{a^{3b+2}c^{d+1}}{a^{2b-1}c^d}\right)^3$

79 $\dfrac{(a^{2+s}b^{s-1})^3}{(a^{s+1}b^{s-3})^2}$

80 $\dfrac{(b^{c+d}a^{c+1})^3}{(b^ca^{c-1})^2}$

5.2 NEGATIVE INTEGRAL EXPONENTS

In this section we shall extend the definition of a^n to include an interpretation of a^{-t}, where $t > 0$ and $a \neq 0$. If we disregard the restriction $m > n$ in law (5.3), we have, since $-t = t - 2t$,

$$a^{-t} = a^{t-2t} = \frac{a^t}{a^{2t}} = \frac{1}{a^t}$$

Therefore, we define a^{-t} as

Negative
Exponent $\qquad a^{-t} = \dfrac{1}{a^t} \qquad a \neq 0$ $\qquad\qquad\qquad$ (5.8)

We shall next prove that laws (5.2) and (5.3) hold for this interpretation of negative exponents. To show that law (5.2) holds, we must prove that $a^{-t}a^{-r} = a^{-t-r}$. By (5.8) we have

$$a^{-t}a^{-r} = \frac{1}{a^t} \cdot \frac{1}{a^r}$$

$$= \frac{1}{a^{t+r}} \qquad \text{by (5.2)}$$

$$= a^{-(t+r)} \qquad \text{by (5.8)}$$

$$= a^{-t-r}$$

To show that (5.3) holds, we must show that $a^{-t}/a^{-r} = a^{-t-(-r)}$. Again using (5.8), we have

$$\frac{a^{-t}}{a^{-r}} = \frac{\dfrac{1}{a^t}}{\dfrac{1}{a^r}}$$

$$= \frac{a^r}{a^t} \qquad \text{multiplying each member of the complex fraction by } a^{t+r}$$

$$= a^{r-t} \qquad \text{by (5.3)}$$

$$= a^{-t+r}$$

$$= a^{-t-(-r)}$$

Since laws (5.4) to (5.7) were derived from laws (5.2) and (5.3), these laws hold also for negative exponents. Therefore, we have removed the restrictions on laws (5.2) to (5.7) that m be greater than n, and the only remaining restriction is that m and n be integers.

It is frequently desirable that a fraction whose numerator or denominator or both include negative exponents should be converted to an equal fraction in which all exponents are positive. We use the fundamental principle of fractions to accomplish this purpose. For example, to convert $a^x b^{-y}/c^z d^{-w}$ to an equal fraction that has no negative exponents, we first notice that $b^{-y} \cdot b^y = b^{-y+y} = b^0 = 1$. Hence, if we multiply the given fraction by b^y/b^y, we obtain

$$\frac{a^x b^{-y}}{c^z d^{-w}} \cdot \frac{b^y}{b^y} = \frac{a^x}{c^z d^{-w} b^y}$$

Similarly, if we multiply the right member by d^w/d^w, we get

$$\frac{a^x d^w}{c^z d^{-w} b^y d^w} = \frac{a^x d^w}{c^z b^y}$$

These two steps can be combined into the single operation of multiplying the given fraction by $b^y d^w / b^y d^w$ and obtaining

$$\frac{a^x b^{-y}}{c^z d^{-w}} = \frac{a^x b^{-y}}{c^z d^{-w}} \cdot \frac{b^y d^w}{b^y d^w} = \frac{a^x d^w}{c^z b^y}$$

The example just given suggests the following procedure by which a fraction, whose numerator or denominator or both are monomials that include negative exponents, may be expressed as an equal fraction with all exponents positive.

Elimination of Negative Exponents   For each negative power of a number that occurs in the numerator or denominator, multiply both numerator and denominator by that number with the numerically equal positive exponent.

Example 1 Convert

$$\frac{a^2b^{-3}c^{-2}}{x^{-1}y^3z^{-3}}$$

into an equal fraction in which all exponents are positive.

Solution   The negative powers of numbers in the numerator and denominator are b^{-3}, c^{-2}, x^{-1}, z^{-3}. Therefore, we multiply each member of the given fraction by $b^3c^2xz^3$ and get

$$\frac{a^2b^{-3}c^{-2}}{x^{-1}y^3z^{-3}} \cdot \frac{b^3c^2xz^3}{b^3c^2xz^3} = \frac{a^2b^{-3+3}c^{-2+2}xz^3}{x^{-1+1}y^3z^{-3+3}b^3c^2} \qquad \text{by commutative axiom and (5.2)}$$

$$= \frac{a^2b^0c^0xz^3}{x^0y^3z^0b^3c^2}$$

$$= \frac{a^2xz^3}{y^3b^3c^2} \qquad \text{since by (5.4), } b^0 = c^0 = z^0 = 1$$

Example 2 Express

$$\frac{2c^{-2}d^{-1}}{3x^{-1}y^3}$$

as an equal fraction having only positive exponents.

Solution $$\frac{2c^{-2}d^{-1}}{3x^{-1}y^3} = \frac{2c^{-2}d^{-1}}{3x^{-1}y^3} \cdot \frac{c^2dx}{c^2dx}$$

$$= \frac{2c^{-2+2}d^{-1+1}x}{3x^{-1+1}y^3c^2d} \qquad \text{by commutative axiom and (5.2)}$$

$$= \frac{2c^0d^0x}{3x^0y^3c^2d}$$

$$= \frac{2x}{3y^3c^2d} \qquad \text{by (5.4)}$$

If the numerator or denominator or both are polynomials and if either or both the numerator and denominator have negative exponents, we use the principle explained above to eliminate the negative exponents. For example, to convert $(x^{-1} + y^{-1})/(x^{-2} - y^{-2})$ to an equal fraction in which all exponents are positive, we notice that x appears with exponents -1 and -2, and y likewise appears with exponents -1 and -2. Therefore, if we multiply the fraction by x^2y^2/x^2y^2, we obtain a fraction in which the exponents of x and of y are positive. The details of the conversion process follow:

$$\frac{x^{-1} + y^{-1}}{x^{-2} - y^{-2}} \cdot \frac{x^2y^2}{x^2y^2} = \frac{x^{-1+2}y^2 + x^2y^{-1+2}}{x^{-2+2}y^2 - x^2y^{-2+2}} = \frac{xy^2 + x^2y}{y^2 - x^2}$$

$$= \frac{xy(y + x)}{(y + x)(y - x)} = \frac{xy}{y - x}$$

Since laws (5.2) to (5.7) now hold for negative as well as positive integral exponents, we can use them for combining powers of the same number regardless of the signs of the exponents. In the following examples we shall make all possible applications of laws (5.2) to (5.7) and shall express the results without zero or negative exponents.

Example 3

$$\frac{12a^{-2}b^3c^{-3}}{4a^3b^{-1}c^{-2}} = \frac{12a^{-2}b^3c^{-3}}{4a^3b^{-1}c^{-2}} \cdot \frac{a^2bc^3}{a^2bc^3}$$ multiplying by a^2bc^3/a^2bc^3

$$= \frac{12a^{-2+2}b^{3+1}c^{-3+3}}{4a^{3+2}b^{-1+1}c^{-2+3}}$$ by (5.2)

$$= \frac{12a^0b^4c^0}{4a^5b^0c}$$

$$= \frac{3b^4}{a^5c}$$ since $a^0 = b^0 = c^0 = 1$ and $\frac{12}{4} = 3$

Example 4

$$\left(\frac{2x^{-3}y^2}{x^4z^3}\right)^{-3} = \frac{2^{-3}x^9y^{-6}}{x^{-12}z^{-9}}$$ by (5.7), (5.6), and (5.5)

$$= \frac{2^{-3}x^9y^{-6}}{x^{-12}z^{-9}} \cdot \frac{2^3y^6x^{12}z^9}{2^3y^6x^{12}z^9}$$ multiplying by $2^3y^6x^{12}z^9/2^3y^6x^{12}z^9$

$$= \frac{2^{-3+3}x^{9+12}y^{-6+6}z^9}{2^3x^{-12+12}y^6z^{-9+9}}$$ by commutative axiom and (5.2)

$$= \frac{2^0x^{21}y^0z^9}{8x^0y^6z^0}$$

$$= \frac{x^{21}z^9}{8y^6}$$ by (5.4)

Example 5

$$\frac{2^{-2} + 2^{-3}}{2^{-4}} = \frac{2^{-2} + 2^{-3}}{2^{-4}} \cdot \frac{2^4}{2^4}$$ multiplying the fraction by $2^4/2^4$

$$= \frac{2^{-2+4} + 2^{-3+4}}{2^{-4+4}}$$ by distributive axiom and (5.2)

$$= \frac{2^2 + 2}{2^0}$$

$$= \frac{4 + 2}{1}$$ by (5.4)

$$= 6$$

Example 6

$$\left(\frac{x^{-1} - y^{-1}}{x^{-1}y^{-1}}\right)^{-2} = \left(\frac{x^{-1} - y^{-1}}{x^{-1}y^{-1}} \cdot \frac{xy}{xy}\right)^{-2}$$ multiplying the fraction inside parentheses by xy/xy, since the negative exponent of x and of y with the greatest absolute value is -1

$$= \left(\frac{x^{-1+1}y - xy^{-1+1}}{x^{-1+1}y^{-1+1}}\right)^{-2}$$ by distributive and commutative axioms and (5.2)

$$= (y - x)^{-2}$$ by (5.4)

$$= \frac{1}{(y - x)^2}$$ by (5.8)

EXERCISE 5.2   Negative Exponents

Find the value of the expression in each of Probs. 1 to 24.

1 2^{-3}	**2** 3^{-2}	**3** 4^{-3}	**4** 2^{-5}
5 $3^{-1}3^{-2}$	**6** $4^{-1}4^{-3}$	**7** $5^{-2}5^3$	**8** $6^{-4}6^2$
9 $\dfrac{4^{-2}}{4^{-3}}$	**10** $\dfrac{5^{-4}}{5^{-2}}$	**11** $\dfrac{3^{-2}}{3^{-5}}$	**12** $\dfrac{4^{-1}}{4^{-4}}$
13 $(3^{-1})^2$	**14** $(3^2)^{-1}$	**15** $(7^0)^{-5}$	**16** $(3^{-2})^{-3}$
17 $(2^{-1}3^2)^{-2}$	**18** $(3^{-1}4)^{-2}$	**19** $(2^{-3}4^{-1})^2$	**20** $(3^{-2}5^2)^2$
21 $\left(\dfrac{2^{-2}}{3^{-1}}\right)^{-1}$	**22** $\left(\dfrac{3^{-2}}{2^3}\right)^{-2}$	**23** $\left(\dfrac{2^{-4}}{3^{-1}}\right)^2$	**24** $\left(\dfrac{2^0}{3^{-2}}\right)^2$

Write each expression in Probs. 25 to 32 without denominators.   Use negative exponents if needed.

25 a^2/b^3	**26** a^3/b^4	**27** a^{-2}/b	**28** a^{-2}/b^{-3}
29 $\dfrac{2^{-1}a^{-2}b^{-3}}{a^{-3}b^{-2}}$	**30** $\dfrac{3^{-2}a^{-1}b^{-3}}{9^{-1}a^2b^0}$	**31** $\dfrac{2r^2s^{-1}v^0}{3^{-1}r^4s^{-2}v^{-3}}$	**32** $\dfrac{2^{-1}x^3y^{-3}z^2}{4^0x^{-1}y^{-1}z^4}$

Make all possible applications of laws (5.2) to (5.8), and express the following without negative exponents.

33 $3a^{-1}a^{-3}$	**34** $2^{-1}a^{-2}a^0$	**35** $2^{-3}x^{-2}x$	**36** $7^{-1}a^{-3}a^{-2}$
37 $\dfrac{x^{-3}}{x^{-4}}$	**38** $\dfrac{x^{-5}}{x^{-1}}$	**39** $\dfrac{a^{-3}}{a^{-2}}$	**40** $\dfrac{a^{-4}}{a^{-5}}$
41 $\dfrac{x^{-2}y^3z}{x^0y^{-1}z^2}$	**42** $\dfrac{x^{-3}y^{-1}z^2}{x^{-2}y^{-3}z^{-1}}$	**43** $\dfrac{p^{-2}d^{-1}q^0}{p^2d^{-2}q^{-1}}$	**44** $\dfrac{s^{-1}a^{-2}m^3}{s^{-2}a^{-1}m^{-1}}$
45 $\dfrac{3^{-1}t^{-1}h^{-2}e}{2^{-4}t^0h^{-1}e^{-2}}$	**46** $\dfrac{4^{-2}h^{-2}a^0m^2}{2^{-3}h^{-3}a^{-1}m^3}$	**47** $\dfrac{2^{-3}e^{-1}a^0t^2}{4^{-1}e^{-3}a^{-1}t^{-1}}$	**48** $\dfrac{8^{-1}s^{-3}a^{-1}d^0}{2^{-4}s^{-1}a^{-3}d^2}$
49 $\left(\dfrac{a^{-2}t^{-1}e^0}{a^2t^{-2}e^2}\right)^{-2}$	**50** $\left(\dfrac{c^{-1}a^2r^{-3}}{c^0a^3r^{-4}}\right)^{-3}$	**51** $\left(\dfrac{h^{-1}a^{-3}t^2}{h^2a^{-5}t}\right)^{-1}$	**52** $\left(\dfrac{t^{-1}a^2b^0}{t^{-2}a^3b^2}\right)^{-4}$
53 $\left(\dfrac{8^{-2}t^{-1}a^0}{4^{-3}t\,a^{-2}}\right)^{-3}$	**54** $\left(\dfrac{3^{-5}m^{-3}a^5}{27^{-2}m^{-2}a^3}\right)^{-2}$	**55** $\left(\dfrac{12^{-2}h^{-4}e^3}{4^{-1}h^{-2}e}\right)^{-1}$	**56** $\left(\dfrac{7^{-2}a^3t^2}{98^{-1}a^2t^{-1}}\right)^{-1}$
57 $x^{-1}-x$	**58** $x^{-2}+x^2$	**59** $a^{-1}b+ab^{-1}$	**60** $a^2b^{-1}-a^{-1}b$
61 $\dfrac{x^{-1}y^{-1}}{x^{-1}-y^{-1}}$	**62** $\dfrac{a^{-2}-b^{-2}}{a^{-1}-b^{-1}}$	**63** $\dfrac{a^{-2}b^{-1}-a^{-1}b^{-2}}{a^{-2}-b^{-2}}$	**64** $\dfrac{x^{-2}-y^{-2}}{x^{-2}y^{-2}}$

65 $2(x+1)^{-3}(x-1)-3(x+1)^{-4}(x-1)^2$

66 $3(x+3)^{-2}(x-2)^2-2(x+3)^{-3}(x-2)^3$

67 $(x+2)^{-4}-4(x+2)^{-5}(x+1)$

68 $(x+3)^{-5}(x+2)^{-2}+5(x+3)^{-6}(x+2)^{-1}$

69 $6(2x-1)^{-1}(3x+2)-2(2x-1)^{-2}(3x+2)^2$

70 $9(2x+3)^{-2}(3x-1)^2-4(2x+3)^{-3}(3x-1)^3$

71 $2(4x-7)^{-3}(2x+5)^{-2}+12(4x-7)^{-4}(2x+5)^{-1}$

72 $12(3x+4)^{-2}(4x-3)^{-4}+6(3x+4)^{-3}(4x-3)^{-3}$

5.3   FRACTIONAL EXPONENTS

We shall now extend the definition of a^n to include situations in which n is a rational fraction. The extension will be made in such manner that laws (5.2) to (5.7) hold. If (5.5) holds for $m = 1/k$ and $n = k$, we then have $(a^{1/k})^k = a^{k/k} = a$. Consequently, we define $a^{1/k}$ to be a number whose kth power is a. Unless we add further restrictions, $a^{1/k}$ may have more than one value. For example, $16^{1/2}$ may be 4 or -4 since $4^2 = 16$ and $(-4)^2 = 16$. We shall remove this ambiguity by the following definitions and discussion.

---

kth Root   The number b is a *kth root* of a if and only if $b^k = a$.

Principal Root   If there is a positive kth root of a, it is called the *principal kth root of a*. If there is no positive kth root of a but there is a negative one, then the negative kth root is called the *principal kth root* of a.

Radical of Order k
Radicand
Index

It is customary to use $\sqrt[k]{a}$ to indicate the principal kth root of a. This symbol is called a *radical* of order k, a is called the *radicand*, and k the *index* of the radical. If the index is not written, it is understood to be 2.

We are now in a position to define $a^{1/k}$. The definition is

$$a^{1/k} = \sqrt[k]{a} \tag{5.9}$$

There is a positive kth root of a if a is positive; hence $a^{1/k}$ is positive for $a > 0$. For example, $64^{1/3} = 4$ and $36^{1/2} = 6$. It can be proved that if a is negative and k is odd, then there is a negative kth root of a. For example, $(-32)^{1/5} = -2$. Finally, there is no real kth root of a if a is negative and k is even, since an even power of any nonzero real number is positive. Thus there is no real square root of -4 and no real fourth root of -81. In this chapter *we shall deal only with real numbers; hence we shall not consider even roots of negative numbers.*

If we replace a by b^j in (5.9), we find that $(b^j)^{1/k} = \sqrt[k]{b^j}$. Consequently $(b^j)^{1/k}$ is a kth root of b^j. Now if we raise $(b^{1/k})^j$ to the kth power, we have

$$[(b^{1/k})^j]^k = (b^{1/k})^{jk} \qquad \text{by (5.5)}$$
$$= (b^{1/k})^{kj} \qquad \text{by commutative axiom}$$
$$= [(b^{1/k})^k]^j$$
$$= b^j \qquad \text{by definition of } b^{1/k}$$

Therefore $(b^j)^{1/k}$ and $(b^{1/k})^j$ are both kth roots of b^j, and it can be proved† that, except for the excluded case, the two roots have the same sign and therefore are equal. Consequently, for $\sqrt[k]{b}$ real, we have

Rational Exponents $$b^{j/k} = (b^j)^{1/k} = \sqrt[k]{b^j} = (b^{1/k})^j = (\sqrt[k]{b})^j \qquad \sqrt[k]{b} \text{ real} \qquad (5.10)$$

By use of (5.10) and arguments similar to the one just given, we can prove that laws (5.2) to (5.7) hold for fractional exponents as defined by (5.9). We shall now work several problems that illustrate the procedures to be followed in dealing with fractional exponents.

Example 1 Evaluate $9^{1/2}$, $27^{2/3}$, and $(-128)^{5/7}$.

Solution
$$9^{1/2} = 3 \qquad \text{by (5.9)}$$
$$27^{2/3} = (\sqrt[3]{27})^2 \qquad \text{by (5.10)}$$
$$= 3^2 = 9$$
$$(-128)^{5/7} = (\sqrt[7]{-128})^5 \qquad \text{by (5.10)}$$
$$= (-2)^5 = -32$$

Example 2 Express $2a^{1/3}b^{2/3}$ and $3a^{2/5}/b^{3/5}$ in radical form.

Solution
$$2a^{1/3}b^{2/3} = 2(ab^2)^{1/3} \qquad \text{by (5.6)}$$
$$= 2\sqrt[3]{ab^2} \qquad \text{by (5.9)}$$
$$\frac{3a^{2/5}}{b^{3/5}} = 3\left(\frac{a^2}{b^3}\right)^{1/5} \qquad \text{by (5.7)}$$
$$= 3\sqrt[5]{\frac{a^2}{b^3}} \qquad \text{by (5.9)}$$

Example 3 Find the product of $3a^{2/5}$ and $2a^{1/3}$, and find the quotient of $8x^{2/3}y^{5/6}$ and $2x^{1/2}y^{1/3}$.

Solution
$$(3a^{2/5})(2a^{1/3}) = 6a^{2/5 + 1/3}$$
$$= 6a^{(6+5)/15}$$
$$= 6a^{11/15}$$
$$\frac{8x^{2/3}y^{5/6}}{2x^{1/2}y^{1/3}} = \frac{8}{2}x^{2/3 - 1/2}y^{5/6 - 1/3}$$
$$= 4x^{(4-3)/6}y^{(5-2)/6}$$
$$= 4x^{1/6}y^{3/6} = 4x^{1/6}y^{1/2}$$

†See P.K. Rees, F. W. Sparks and C. S. Rees, "Algebra and Trigonometry," 3d ed., p. 114, McGraw-Hill Book Company, New York, 1975.

Example 4   Use the laws of exponents to make all possible combinations and express the result without zero or negative exponents in

$$\left(\frac{16x^{-1}y^{5/6}z^{3/4}}{9x^3y^{1/3}z^{1/2}}\right)^{1/2}$$

Solution $\left(\dfrac{16x^{-1}y^{5/6}z^{3/4}}{9x^3y^{1/3}z^{1/2}}\right)^{1/2} = \left(\dfrac{16}{9}\dfrac{x^{-1}}{x^3}\dfrac{y^{5/6}}{y^{1/3}}\dfrac{z^{3/4}}{z^{1/2}}\dfrac{x}{x}\right)^{1/2}$ multiplying by x/x

$$= \left(\frac{16}{9}\frac{x^{-1+1}}{x^{3+1}}y^{5/6-1/3}z^{3/4-1/2}\right)^{1/2}$$ by (5.2) and (5.3)

$$= \left(\frac{16}{9}\frac{x^0y^{1/2}z^{1/4}}{x^4}\right)^{1/2}$$

$$= \frac{4}{3}\frac{y^{1/4}z^{1/8}}{x^2}$$ by (5.4), (5.6), and (5.7)

EXERCISE 5.3   Radical and Exponential Expressions

Express the number in each of Probs. 1 to 20 without exponents or radicals.

1 $(9)^{1/2}$ 2 $64^{1/3}$ 3 $(-32)^{1/5}$ 4 $(-128)^{1/7}$ 5 $8^{2/3}$

6 $64^{5/6}$ 7 $81^{3/4}$ 8 $4^{3/2}$ 9 $\sqrt{25^3}$ 10 $\sqrt[4]{16^3}$

11 $\sqrt[3]{27^2}$ 12 $\sqrt[5]{32^3}$ 13 $8^{-2/3}$ 14 $25^{-3/2}$ 15 $32^{-2/5}$

16 $81^{-3/4}$ 17 $\sqrt{.09}$ 18 $\sqrt[3]{.064}$ 19 $\sqrt[5]{.000032}$ 20 $\sqrt[3]{.001}$

Make all possible combinations by use of the laws of exponents and express each final form without zero or negative exponents and without fractional exponents in the denominator.

21 $(2x^{1/3}y^2)(3x^{2/3}y^{-1})$ 22 $(4x^{3/4}y^{-2})(3x^{1/4}y^3)$

23 $(5x^{2/5}y^{-2})(2x^{3/5}y^2)$ 24 $(3x^{4/7}y^{-3})(2x^{3/7}y^3)$

25 $(8^{2/3}a^{2/3}b^{4/3})(4^{3/2}a^{1/3}b^{1/2})$ 26 $(4^{1/2}a^{3/5}b^{2/3})(27^{2/3}a^{1/3}b^{1/5})$

27 $(4^{3/2}a^{1/4}b^{2/5})(25^{1/2}a^{2/5}b^{3/4})$ 28 $(9^{1/2}a^{1/3}b^{2/5})(8^{2/3}a^{3/5}b^{1/3})$

29 $\dfrac{27^{2/3}x^{3/5}y^{1/4}}{4^{3/2}x^{1/3}y^{1/5}}$ 30 $\dfrac{16^{3/4}x^{1/5}y^{3/4}}{8^{2/3}x^{2/3}y^{1/3}}$

31 $\dfrac{9^{1/2}x^{2/7}y^{3/2}}{27^{2/3}x^{1/3}y^{3/5}}$ 32 $\dfrac{32^{3/5}x^{2/9}y^{2/3}}{81^{3/4}x^{1/5}y^{5/7}}$

33 $\dfrac{(4^{3/2}x^{1/3}y^{2/5})(9^{1/2}x^{2/3}y^{3/5})}{16^{3/4}x^{1/5}y^{1/4}}$ 34 $\dfrac{(27^{2/3}x^{2/5}y^{3/7})(8^{1/3}x^{3/5}y^{4/7})}{81^{3/4}x^{1/3}y^{3/5}}$

35 $\dfrac{32^{3/5}x^{4/7}y^{2/3}}{(9^{1/2}x^{1/3}y^{2/5})(4^{3/2}x^{1/2}y^{1/4})}$ 36 $\dfrac{8^{4/3}x^{1/4}y^{3/5}}{(16^{3/4}x^{1/2}y^{2/3})(4^{3/2}x^{2/3}y^{3/4})}$

37 $(2x^{1/3}y^{1/5})^2$ 38 $(3x^{-1/2}y^{1/4})^4$

39 $(5^{1/2}x^{2/5}y^{-1/3})^4$ 40 $(4x^{2/5}y^{-1/4})^5$

41 $(3a^{-2}b^{2/5})^{1/2}(9^{-1/2}a^{-1}b^{2/5})^{-1}$ 42 $(32x^{-2/3}y^{5/6})^{6/5}(2x^{1/2}y^{-1})^{-2}$

43 $(2^{-1}a^{1/3}b^{-1/3})^{-3}(2a^{2/5}b^{2/3})^{1/2}$ 44 $(16x^{4/7}y^{8/9})^{-1/4}(2x^{-1}y^{1/4})^{-4}$

45 $\left(\dfrac{27x^{-3/4}y^0}{8x^{1/4}y^{-1/4}}\right)^{-4/3}$

46 $\left(\dfrac{243x^{1/3}y^{-2/3}}{32x^{-1/2}y}\right)^{-2/5}$

47 $\left(\dfrac{5^{4/3}x^{5/6}y^{-1/3}}{16^{2/3}x^{1/2}y^{5/9}}\right)^{3/4}$

48 $\left(\dfrac{16x^{-2/5}y^0}{9x^{2/5}y^{-2/5}}\right)^{-5/2}$

49 $\left(\dfrac{a^{x+y}}{a^{2y}}\right)^{1/(x-y)}$

50 $\left(\dfrac{a^{-x-y}}{a^{x-2y}}\right)^{x/(2x-y)}$

51 $\left(\dfrac{a^{x-y}}{a^x}\right)^{(y-x)/y}$

52 $\left(\dfrac{a^{x+3y}}{a^{2x+y}}\right)^{y/(x-2y)}$

53 $(x^{1/2}-y^{1/2})^2$

54 $(x^{1/2}+y^{1/2})(x^{1/2}-y^{1/2})$

55 $(x^{1/3}-y^{1/3})(x^{2/3}+x^{1/3}y^{1/3}+y^{2/3})$

56 $(x^{1/3}+y^{1/3})(x^{2/3}-x^{1/3}y^{1/3}+y^{2/3})$

57 $\left(\dfrac{a^{-1}-b^{-1}}{a^{-1}b^{-1}}\right)^{-1}$

58 $\left(\dfrac{a^{-1}+b^{-1}}{a^{-2}-b^{-2}}\right)^{-1}$

59 $\left(\dfrac{a^{-2}+b^{-2}}{a^{-2}b^{-2}}\right)^{-2}$

60 $\left(\dfrac{a^{-2}-b^{-2}}{a^{-1}-b^{-1}}\right)^{-2}$

61 $(3x-2)(x+1)^{-1/2}+6(x+1)^{1/2}$

62 $(2x+5)(x-2)^{-1/3}+3(x-2)^{2/3}$

63 $4(x+1)(2x-3)^{-3/5}+5(2x-3)^{2/5}$

64 $(x+3)(3x+1)^{-2/3}+(3x+1)^{1/3}$

65 $(2x-1)^{1/2}(x-2)^{-1/3}+3(x-2)^{2/3}(2x-1)^{-1/2}$

66 $(3x+4)^{1/3}(4x-3)^{-3/4}+(3x+4)^{-2/3}(4x-3)^{1/4}$

67 $(5x-3)^{2/5}(2x-5)^{-1/2}+2(5x-3)^{-3/5}(2x-5)^{1/2}$

68 $2(4x+3)^{1/4}(3x+2)^{-1/3}+(3x+2)^{2/3}(4x+3)^{-3/4}$

5.4 LAWS OF RADICALS

In this section we shall develop and apply three laws of radicals. Since (5.6) and (5.7) are valid for $n = 1/k$, we have

$$(ab)^{1/k} = a^{1/k}b^{1/k} \quad \text{and} \quad \left(\frac{a}{b}\right)^{1/k} = \frac{a^{1/k}}{b^{1/k}}$$

Now making use of the relation between fractional exponents and radicals, we have

Root of a Product $\quad \sqrt[k]{ab} = \sqrt[k]{a}\,\sqrt[k]{b}$ (5.11)

and

Root of a Quotient $\quad \sqrt[k]{\dfrac{a}{b}} = \dfrac{\sqrt[k]{a}}{\sqrt[k]{b}}$ (5.12)

If we replace m by $1/j$ and n by $1/k$ in (5.5), we have $(a^{1/j})^{1/k} = a^{1/jk}$; *hence, in terms of radicals, we get*

Root of a Root $\quad \sqrt[k]{\sqrt[j]{a}} = \sqrt[kj]{a}$ (5.13)

We can use (5.11) to remove rational factors from the radicand, to multiply radicals of the same order, and to insert factors into the radicand, as is seen in the following examples:

Example 1 $\sqrt[3]{108} = \sqrt[3]{(27)(4)}$ expressing 108 as the product of the perfect cube 27 and 4

$= \sqrt[3]{27}\,\sqrt[3]{4}$ by use of (5.11)

$= 3\sqrt[3]{4}$

Example 2 $\sqrt{5}\,\sqrt{7} = \sqrt{(5)(7)} = \sqrt{35}$ by use of (5.11) from right to left

Example 3 $2\sqrt[4]{3} = \sqrt[4]{2^4}\,\sqrt[4]{3}$ since $2 = \sqrt[4]{2^4}$

$= \sqrt[4]{(2^4)(3)} = \sqrt[4]{48}$ by use of (5.11) from right to left

Example 4 $\sqrt[5]{32a^6b^{13}} = \sqrt[5]{2^5a^5a(b^2)^5b^3}$ factoring into fifth powers as far as possible

$= 2ab^2\sqrt[5]{ab^3}$ removing the fifth powers from the radicand

Example 5 $\sqrt[3]{5ab}\,\sqrt[3]{2ac} = \sqrt[3]{10a^2bc}$ by use of (5.11)

Example 6 $a\sqrt[4]{2ab^2} = \sqrt[4]{a^4}\,\sqrt[4]{2ab^2}$ since $a = \sqrt[4]{a^4}$

$= \sqrt[4]{2a^5b^2}$ by (5.11) from right to left

Example 7 $2\sqrt{3} \cdot 3\sqrt{6} = 6\sqrt{18}$ by (5.11) from right to left

$= 6\sqrt{9 \cdot 2}$ factoring 8

$= 18\sqrt{2}$ by (5.11) from left to right

Law (5.12) is used in obtaining the quotient of two radicals of the same order and in rationalizing monomial denominators, as is illustrated now.

Example 8 $\dfrac{\sqrt[3]{250a^7b^5}}{\sqrt[3]{2ab^3}} = \sqrt[3]{\dfrac{250a^7b^5}{2ab^3}}$ by use of (5.12) from right to left

$= \sqrt[3]{125a^6b^2}$

$= \sqrt[3]{5^3(a^2)^3b^2}$ since $125 = 5^3$ and $a^6 = (a^2)^3$

$= 5a^2\sqrt[3]{b^2}$ since $\sqrt[3]{5^3(a^2)^3} = 5a^2$

Rationalizing a Denominator   We often need to convert a radical with a fractional radicand to an equal fraction with no radicals in the denominator. If this is done, we say the denominator has been *rationalized*. In order to rationalize a monomial denominator, we must multiply the denominator of a radical of order n by the necessary factors so as to make it an nth power and, of course, multiply the numerator by the same factors. Thus, in order to rationalize the denominator of

$$\sqrt[4]{\frac{2a}{3a^3b^2}}$$

we multiply denominator and numerator by 3^3ab^2, since that makes the denominator a fourth power. Hence, we have

$$\sqrt[4]{\frac{2a}{3a^3b^2}} = \sqrt[4]{\frac{2a}{3a^3b^2} \frac{3^3ab^2}{3^3ab^2}}$$

$$= \frac{\sqrt[4]{54a^2b^2}}{3ab} \qquad \text{since } \sqrt[4]{3^4a^4b^4} = 3ab$$

Furthermore, in order to rationalize the denominator of $\sqrt{2x/3y}$, we multiply the numerator and denominator by $3y$ and have

$$\sqrt{\frac{2x}{3y}} = \sqrt{\frac{2x}{3y} \frac{3y}{3y}}$$

$$= \frac{\sqrt{6xy}}{3y} \qquad \text{if } y > 0$$

$$= -\frac{\sqrt{6xy}}{3y} \qquad \text{if } y < 0$$

These can be combined into

$$\sqrt{\frac{2x}{3y}} = \frac{\sqrt{6xy}}{3|y|}$$

5.5 CHANGING THE ORDER OF A RADICAL

Decreasing the Index   If the index of a radical can be factored and if the radicand can be expressed as a power with the exponent of the power equal to one of the factors of the index, then we can decrease the order of the radical. Thus,

$$\sqrt[6]{8x^3} = \sqrt{\sqrt[3]{(2x)^3}} = \sqrt{2x}$$

Another example is

$$\sqrt[15]{a^{10}y^{25}} = \sqrt[3]{\sqrt[5]{(a^2y^5)^5}} = \sqrt[3]{a^2y^5} = \sqrt[3]{a^2y^3y^2} = y\sqrt[3]{a^2y^2}$$

Alternative Procedure   An alternative procedure for obtaining the same result is to change from radical form to fractional-exponent form, reduce the fractional exponents to lowest terms, and then express the result in radical form. If this procedure is used in connection with the last example, we have

$$\sqrt[15]{a^{10}y^{25}} = (a^{10}y^{25})^{1/15} = a^{10/15}y^{25/15}$$

$$= a^{2/3}y^{5/3} \qquad \text{since } \tfrac{10}{15} = \tfrac{2}{3} \text{ and } \tfrac{25}{15} = \tfrac{5}{3}$$

$$= \sqrt[3]{a^2y^5} = y\sqrt[3]{a^2y^2}$$

Simplify Radicals   In the following exercise, we shall use the word *simplify* to indicate that all possible combinations that can be made by use of the laws of exponents are to be made, the result is to be expressed without zero or negative exponents, all monomial denominators are to be rationalized, and all possible rational factors are to be removed from the radicand.

EXERCISE 5.4   Simplifying Radicals

Simplify the following radicals.

1 $\sqrt{121}$ **2** $\sqrt{36}$ **3** $\sqrt{64}$ **4** $\sqrt{169}$

5 $\sqrt[3]{64}$ **6** $\sqrt[4]{81}$ **7** $\sqrt[6]{64}$ **8** $\sqrt[5]{1024}$

9 $\sqrt{18}$ **10** $\sqrt{75}$ **11** $\sqrt[3]{108}$ **12** $\sqrt[4]{162}$

13 $\sqrt{12}\sqrt{27}$ **14** $\sqrt{50}\sqrt{98}$ **15** $\sqrt[3]{800}\sqrt[3]{270}$ **16** $\sqrt[3]{28}\sqrt[3]{98}$

17 $\sqrt{20}\sqrt{12}$ **18** $\sqrt{63}\sqrt{32}$ **19** $\sqrt{28}\sqrt{21}$ **20** $\sqrt{54}\sqrt{50}$

21 $\dfrac{\sqrt{27}}{\sqrt{12}}$ **22** $\dfrac{\sqrt{50}}{\sqrt{98}}$ **23** $\dfrac{\sqrt[3]{40}}{\sqrt[3]{135}}$ **24** $\dfrac{\sqrt[4]{144}}{\sqrt[4]{243}}$

25 $\dfrac{\sqrt[3]{24}}{\sqrt[3]{576}}$ **26** $\dfrac{\sqrt[3]{16}}{\sqrt[3]{54}}$ **27** $\dfrac{\sqrt[3]{192}}{\sqrt[3]{108}}$ **28** $\dfrac{\sqrt[5]{486}}{\sqrt[5]{64}}$

29 $\sqrt{18x^2y^3}$ **30** $\sqrt{75x^5y^4}$ **31** $\sqrt[3]{108x^3y^5}$ **32** $\sqrt[3]{192x^7y^5}$

33 $\sqrt[4]{32x^4y^7}$ **34** $\sqrt[4]{162x^5y^6}$ **35** $\sqrt[5]{96x^6y^9}$ **36** $\sqrt[6]{1458x^8y^7}$

37 $\dfrac{\sqrt{12a^3b^5}}{\sqrt{75ab^2}}$ **38** $\dfrac{\sqrt{108a^3b^2}}{\sqrt{12ab^3}}$ **39** $\dfrac{\sqrt{27a^5b^4}}{\sqrt{48a^3b}}$ **40** $\dfrac{\sqrt{40a^2b^5}}{\sqrt{45ab^3}}$

41 $\sqrt{8x^{-1}y^3}\sqrt{75x^2y^{-1}}$ **42** $\sqrt{18x^3y^{-3}}\sqrt{50x^2y^5}$ **43** $\sqrt{32x^{-5}y}\sqrt{12xy^2}$

44 $\sqrt{28x^5y^{-7}}\sqrt{63xy^4}$ **45** $\dfrac{\sqrt{3x^{-4}y}}{\sqrt{48x^{-1}y^3}}$ **46** $\dfrac{\sqrt{7x^3y^{-5}}}{\sqrt{63x^{-3}y^3}}$

47 $\dfrac{\sqrt{150x^{-2}y^{-3}}}{\sqrt{54x^{-1}y^{-2}}}$ **48** $\dfrac{\sqrt{147x^{-1}y^3}}{\sqrt{50x^{-2}y^{-2}}}$ **49** $\dfrac{\sqrt{5xy^{-1}}\sqrt{2x^{-3}y}}{\sqrt{40x^{-3}y^{-2}}}$

50 $\dfrac{\sqrt{50x^{-2}y}\sqrt{98xy^{-3}}}{\sqrt{3xy^{-1}}}$ **51** $\dfrac{\sqrt[3]{54x^{-1}y^5}}{\sqrt[3]{8x^2y^{-1}}\sqrt[3]{250x^{-5}y^3}}$ **52** $\dfrac{\sqrt[4]{162x^3y^9}}{\sqrt[4]{32x^{-1}y^2}\sqrt[4]{625x^{-3}y^{-4}}}$

Reduce the order of each of the following radicals and simplify.

53 $\sqrt[4]{25}$ **54** $\sqrt[4]{36}$ **55** $\sqrt[6]{125}$ **56** $\sqrt[6]{49}$ **57** $\sqrt[6]{64}$

58 $\sqrt[8]{64}$ **59** $\sqrt[12]{64}$ **60** $\sqrt[15]{64}$ **61** $\sqrt[4]{9x^6y^2}$ **62** $\sqrt[4]{25x^2y^6}$

63 $\sqrt[6]{27x^3y^6}$ **64** $\sqrt[6]{216x^6y^3}$ **65** $\sqrt[8]{64x^4y^2}$ **66** $\sqrt[9]{64x^3y^6}$ **67** $\sqrt[10]{32x^5y^{15}}$

68 $\sqrt[12]{64x^3y^6}$ **69** $\dfrac{\sqrt[6]{8a^3y^9}}{\sqrt[4]{4a^6y^2}}$ **70** $\dfrac{\sqrt[6]{4x^6y^2}}{\sqrt[4]{8x^9y^3}}$ **71** $\dfrac{\sqrt[12]{125x^6y^3}}{\sqrt[8]{49x^4y^2}}$ **72** $\dfrac{\sqrt[15]{27x^9y^6}}{\sqrt[20]{256x^8y^{16}}}$

5.6   RATIONALIZING BINOMIAL DENOMINATORS

The product of the sum and the difference of the same two numbers is the square of the first minus the square of the second;

hence, the product of $\sqrt{a} + \sqrt{b}$ and $\sqrt{a} - \sqrt{b}$ is $(\sqrt{a})^2 - (\sqrt{b})^2 = a - b$. Consequently, if the denominator of a fraction is a binomial that contains second-order radicals, we can rationalize it by multiplying each member of the fraction by the binomial obtained by changing the sign between the terms in the denominator.

Example 1   Rationalize the denominator in

$$\frac{2 + \sqrt{5}}{3 - \sqrt{5}}$$

Solution $\dfrac{2 + \sqrt{5}}{3 - \sqrt{5}} = \dfrac{2 + \sqrt{5}}{3 - \sqrt{5}} \dfrac{3 + \sqrt{5}}{3 + \sqrt{5}} = \dfrac{6 + 5\sqrt{5} + 5}{9 - 5} = \dfrac{11 + 5\sqrt{5}}{4}$

Example 2   Rationalize the denominator in

$$\frac{5 + \sqrt{7}}{\sqrt{2} - \sqrt{3}}$$

Solution $\dfrac{5 + \sqrt{7}}{\sqrt{2} - \sqrt{3}} = \dfrac{5 + \sqrt{7}}{\sqrt{2} - \sqrt{3}} \cdot \dfrac{\sqrt{2} + \sqrt{3}}{\sqrt{2} + \sqrt{3}} = \dfrac{5\sqrt{2} + 5\sqrt{3} + \sqrt{14} + \sqrt{21}}{-1}$

Example 3   Rationalize the denominator in

$$\frac{\sqrt{3} + \sqrt{6}}{\sqrt{10} - \sqrt{3}}$$

Solution $\dfrac{\sqrt{3} + \sqrt{6}}{\sqrt{10} - \sqrt{3}} \cdot \dfrac{\sqrt{10} + \sqrt{3}}{\sqrt{10} + \sqrt{3}} = \dfrac{\sqrt{30} + 3 + \sqrt{60} + \sqrt{18}}{7}$

$\qquad\qquad = \dfrac{\sqrt{30} + 3 + 2\sqrt{15} + 3\sqrt{2}}{7}$

Example 4   Rationalize the denominator in

$$\frac{3\sqrt{x} - \sqrt{y}}{\sqrt{x} - 2\sqrt{y}}$$

Solution $\dfrac{3\sqrt{x} - \sqrt{y}}{\sqrt{x} - 2\sqrt{y}} = \dfrac{3\sqrt{x} - \sqrt{y}}{\sqrt{x} - 2\sqrt{y}} \dfrac{\sqrt{x} + 2\sqrt{y}}{\sqrt{x} + 2\sqrt{y}} = \dfrac{3x + 5\sqrt{xy} - 2y}{x - 4y}$

5.7   ADDITION OF RADICALS

Two or more terms that have a common factor can be combined into a single term by use of the distributive axiom.   Thus,

$$2\sqrt{5} - 3\sqrt{5} + 5\sqrt{5} = (2 - 3 + 5)\,\sqrt{5} \qquad \textbf{by distributive axiom}$$
$$= 4\sqrt{5}$$

This example illustrates the procedure used in adding radicals. It is advisable to simplify each radical that is to be added, since any common factor can then be detected.   For example,

$$\sqrt{2} + \sqrt{128} - \sqrt{18} = \sqrt{2} + \sqrt{(64)(2)} - \sqrt{(9)(2)}$$
$$= (1 + 8 - 3)\sqrt{2} \qquad \textbf{by distributive axiom}$$
$$= 6\sqrt{2}$$

It may happen that the terms to be added do not contain a common factor but are such that they can be grouped so that there is a common factor in each group, as illustrated below.

$$\sqrt{8} + \sqrt[3]{16} + \sqrt[3]{54} - \sqrt{72} - \sqrt[3]{24}$$
$$= 2\sqrt{2} + 2\sqrt[3]{2} + 3\sqrt[3]{2} - 6\sqrt{2} - 2\sqrt[3]{3}$$
$$= (2 - 6)\sqrt{2} + (2 + 3)\sqrt[3]{2} - 2\sqrt[3]{3} \qquad \textbf{by distributive axiom}$$
$$= -4\sqrt{2} + 5\sqrt[3]{2} - 2\sqrt[3]{3}$$
$$\sqrt{2x^3y} + \sqrt[4]{4x^2y^6} + \sqrt[3]{16x^4y} = x\sqrt{2xy} + \sqrt{\sqrt{4x^2y^6}} + 2x\sqrt[3]{2xy}$$
$$= x\sqrt{2xy} + \sqrt{2xy^3} + 2x\sqrt[3]{2xy}$$
$$= x\sqrt{2xy} + y\sqrt{2xy} + 2x\sqrt[3]{2xy}$$
$$= (x + y)\sqrt{2xy} + 2x\sqrt[3]{2xy}$$
$$\textbf{by distributive axiom}$$

In adding radicals, we must bear in mind that two radicals can be added if and only if they have the same index and the same radicand.

EXERCISE 5.5   Operations on Radicals

Find the following products.

1 $(\sqrt{2} + \sqrt{3})(\sqrt{2} - \sqrt{3})$

2 $(\sqrt{7} - \sqrt{3})(\sqrt{7} + \sqrt{3})$

3 $(\sqrt{5} + \sqrt{2})(\sqrt{5} - \sqrt{2})$

4 $(\sqrt{6} - \sqrt{5})(\sqrt{6} + \sqrt{5})$

5 $(\sqrt{2} - 2\sqrt{3})(3\sqrt{2} + \sqrt{3})$

6 $(\sqrt{5} + 2\sqrt{2})(3\sqrt{5} - \sqrt{2})$

7 $(\sqrt{7} + 3\sqrt{5})(2\sqrt{7} - \sqrt{5})$

8 $(2\sqrt{3} - \sqrt{5})(\sqrt{3} - 3\sqrt{5})$

9 $(6 + \sqrt{5})(6 - \sqrt{5})$

10 $(\sqrt{6} + 3)(\sqrt{6} - 3)$

11 $(5 + 2\sqrt{3})(5 - 3\sqrt{3})$

12 $(7 + 2\sqrt{3})(7 - 5\sqrt{3})$

13 $(\sqrt{2} + \sqrt{3} + \sqrt{5})(\sqrt{2} + \sqrt{3} - \sqrt{5})$

14 $(\sqrt{6} - \sqrt{3} - \sqrt{2})(\sqrt{6} - \sqrt{3} + \sqrt{2})$

15 $(\sqrt{7} - \sqrt{5} + \sqrt{3})(\sqrt{7} + \sqrt{5} - \sqrt{3})$

16 $(\sqrt{2} - \sqrt{7} - \sqrt{3})(\sqrt{2} + \sqrt{7} + \sqrt{3})$

Rationalize the denominator in each of Probs. 17 to 28.

17 $\dfrac{\sqrt{3}+2}{\sqrt{3}-2}$ 18 $\dfrac{\sqrt{7}-3}{\sqrt{7}+3}$ 19 $\dfrac{4-\sqrt{3}}{4+\sqrt{3}}$ 20 $\dfrac{5+\sqrt{6}}{5-\sqrt{6}}$

21 $\dfrac{2\sqrt{3}-\sqrt{5}}{\sqrt{3}-2\sqrt{5}}$ 22 $\dfrac{3\sqrt{5}-\sqrt{7}}{\sqrt{5}+2\sqrt{7}}$ 23 $\dfrac{\sqrt{6}+3\sqrt{2}}{\sqrt{6}-\sqrt{2}}$ 24 $\dfrac{2\sqrt{7}+\sqrt{5}}{\sqrt{7}+2\sqrt{5}}$

25 $\dfrac{\sqrt{3}+\sqrt{5}}{\sqrt{15}-\sqrt{3}}$ 26 $\dfrac{\sqrt{2}-\sqrt{3}}{\sqrt{3}-\sqrt{6}}$ 27 $\dfrac{\sqrt{14}-2\sqrt{3}}{\sqrt{7}-\sqrt{2}}$ 28 $\dfrac{\sqrt{10}+2\sqrt{2}}{3\sqrt{2}-\sqrt{5}}$

Simplify the radicals in each of the following problems and, if possible, combine.

29 $\sqrt{3}+\sqrt{12}+\sqrt{27}$ 30 $\sqrt{2}-\sqrt{8}+\sqrt{18}$ 31 $\sqrt{20}-\sqrt{45}+\sqrt{80}$

32 $\sqrt{12}+\sqrt{48}-\sqrt{108}$ 33 $\sqrt[3]{2}+\sqrt[3]{16}-\sqrt[3]{54}$ 34 $\sqrt[3]{3}-\sqrt[3]{24}+\sqrt[3]{81}$

35 $\sqrt[3]{5}+\sqrt[3]{40}+\sqrt[3]{320}$ 36 $\sqrt[3]{4}-\sqrt[3]{32}+\sqrt[3]{108}$ 37 $2\sqrt{3}+5\sqrt[3]{3}+\sqrt{27}$

38 $3\sqrt{2}+4\sqrt[3]{16}+\sqrt{50}$ 39 $2\sqrt{5}+\sqrt[3]{250}+\sqrt[3]{432}$ 40 $\sqrt[3]{9}-\sqrt{27}+\sqrt[3]{72}$

41 $\sqrt[4]{9}+\sqrt{12}+\sqrt[3]{24}$ 42 $\sqrt[6]{8}+\sqrt[3]{54}+\sqrt{50}$ 43 $\sqrt[6]{8}+3\sqrt[4]{4}-\sqrt{8}$

44 $\sqrt{12}+\sqrt[6]{216}+\sqrt[3]{144}$ 45 $\sqrt{8x^2y}+\sqrt{18y^3}-y\sqrt{50y}$

46 $\sqrt{27x^3y}+2x\sqrt{48xy}+\sqrt{12xy^3}$ 47 $\sqrt{20x^3}-\sqrt{45xy^2}-2x\sqrt{45x}$

48 $y\sqrt{48y}+\sqrt{12x^2y}-\sqrt{75y^3}$

49 $4x\sqrt{x^3y^7}+3xy^2\sqrt{x^3y^3}-7y\sqrt{x^5y^5}+xy^{-1}\sqrt{25x^3y^9}$

50 $\sqrt{18x^5y^2}+xy\sqrt{8x^3}-3x^{-1}y^{-2}\sqrt{50x^7y^7}+2xy^{-1}\sqrt{98x^3y^5}$

51 $x\sqrt[3]{24x^4y^2}+2xy\sqrt[3]{81x^4y^5}-6x^3y^2\sqrt[3]{3x^{-2}y^2}-3x^4y\sqrt[3]{192x^{-2}y^5}$

52 $\sqrt[3]{5x^8y^4}-2x\sqrt[3]{40x^5y^4}+3xy^{-1}\sqrt[3]{135x^5y^7}-3x^{-1}y^0\sqrt[3]{320x^{11}y^4}$

53 $5x\sqrt{\dfrac{3y^2}{2}}-3y\sqrt{\dfrac{8x^2}{3}}+2\sqrt{\dfrac{3x^2y^2}{2}}$ 54 $\sqrt{\dfrac{3y}{4x^2}}+\dfrac{2y}{x}\sqrt{\dfrac{3}{4y}}-\dfrac{2}{3x}\sqrt{3y}$

55 $\sqrt{3x}+\sqrt[3]{3x^2}-\sqrt{\dfrac{3}{x^3}}-\sqrt[3]{\dfrac{3}{x^4}}$ 56 $\sqrt[3]{\dfrac{x^4}{3}}-\sqrt{\dfrac{x^3}{3}}-\sqrt[3]{\dfrac{xy^3}{3}}+\dfrac{1}{3}\sqrt{3xy^2}$

5.8 SUMMARY

We begin this chapter by recalling the definition of a^n, for n a positive integer. Then we give the value of a product and a quotient of two positive integral powers of the same base, and the value of the zeroth power of a nonzero number. These are put in the symbolic form

$$a^n = a \cdot a \cdot a \cdots \text{to } n \text{ } a\text{'s} \tag{5.1}$$

$$a^m a^n = a^{m+n} \tag{5.2}$$

$$a^m/a^n = a^{m-n} \tag{5.3}$$

$$a^0 = 1 \qquad a \neq 0 \tag{5.4}$$

We then develop formulas for a power of a power, a power of a product, and a power of a quotient and thereby have

$$(a^m)^n = a^{nm} \tag{5.5}$$
$$(ab)^n = a^n b^n \tag{5.6}$$
$$(a/b)^n = a^n/b^n \tag{5.7}$$

We then define *simplified* as it applies to terms with positive integral exponents in Sec. 5.1. Then, in Sec. 5.2, we give the definition of a negative power of a number in the form

$$a^{-t} = \frac{1}{a^t} \tag{5.8}$$

and simplify expressions that contain integral exponents.

In Sec. 5.3 we define fractional powers and roots and give laws of radicals. In this discussion, we develop

$$a^{1/k} = \sqrt[k]{a} \tag{5.9}$$
$$b^{j/k} = \sqrt[k]{b^j} = (\sqrt[k]{b})^j \qquad \sqrt[k]{b} \text{ real} \tag{5.10}$$
$$\sqrt[k]{ab} = \sqrt[k]{a}\sqrt[k]{b} \tag{5.11}$$
$$\sqrt[k]{\frac{a}{b}} = \frac{\sqrt[k]{a}}{\sqrt[k]{b}} \tag{5.12}$$
$$\sqrt[k]{\sqrt[j]{a}} = \sqrt[kj]{a} \tag{5.13}$$

Finally, we discuss rationalizing binomial denominators and addition of radical expressions.

EXERCISE 5.6   Review

Perform the indicated operations in Probs. 1 to 30, and leave the results in a form without negative exponents and without fractional exponents or radicals in the denominator.

1 $a^6 a^2$ 　　　2 $a^8 a$ 　　　3 $\dfrac{a^6}{a^2}$ 　　　4 $\dfrac{a^8}{a^0}$

5 $(a^3)^2$ 　　6 $(a^0)^7$ 　　7 $(3a^3)(2a^2)$ 　　8 $\dfrac{8a^8}{2a^2}$

9 $(4a^3)^2$ 　10 $\left(\dfrac{a^3}{b^2}\right)^3$ 　11 $\dfrac{72x^4y^5}{9x^0y}$ 　12 $\left(\dfrac{21x^2y^3}{7xy^2}\right)^5$

13 $\left(\dfrac{a^{2b-1}c^{3b+2}}{a^{1+b}c^{2b-3}}\right)^2$ 　14 $\left(\dfrac{a^{x+3}b^{y-1}}{a^3 b^y}\right)^3$ 　15 2^{-4} 　16 3^{-2}

17 $\dfrac{5^{-3}}{5^{-2}}$ 　18 $5^{-1}5^1$ 　19 $\dfrac{2x^2y^{-1}z^0}{5^{-1}xy^{-3}z^{-2}}$ 　20 $\dfrac{b^{-2}a^{-1}m^{-3}}{b^{-5}am^{-1}}$

21 $3(x+1)^{-4}(x-1)^2 + 2(x+1)^{-3}(x-1)$

22 $(x + 3)^{1/2}(2x - 1)^{-1/2} + (x + 3)^{-1/2}(2x - 1)^{1/2}$

23 $(3x^{1/4}y^{2/3})(2x^{1/3}y^{-1/4})$ **24** $(5x^{-1/3}y^{2/5})(2x^{3/7}y^{-1/3})$

25 $(2a^{-4}b^{2/3})^{1/2}(2^{-1/5}a^{-1}b^{2/5})^{5/2}$

26 $\left(\dfrac{a^{x+y}}{a^{x-y}}\right)^{x/2y}$ **27** $\sqrt[4]{162x^5y^9}$

28 $\sqrt{50x^{-1}y^3}$ **29** $(\sqrt{7} - \sqrt{5})(\sqrt{7} + \sqrt{5})$

30 $(2\sqrt{5} - \sqrt{3})(\sqrt{3} - \sqrt{15})$

Rationalize the denominator in Probs. 31 and 32.

31 $\dfrac{2\sqrt{5} - \sqrt{7}}{\sqrt{7} + \sqrt{5}}$ **32** $\dfrac{2\sqrt{2} + \sqrt{5}}{\sqrt{10} - 2\sqrt{2}}$

33 Express the reciprocal of $4 + 3\sqrt{2}$ in the form $a + b\sqrt{2}$.

34 Show that the reciprocal of $a + b\sqrt{2}$ is $(a - b\sqrt{2})/(a^2 - 2b^2)$.

35 Show that the two commutative and associative field postulates, along with the distributive property of fields for $\{a + b\sqrt{2} \,|\, a, b \text{ rational}\}$, are "inherited" from the reals.

36 Show that the field properties other than those mentioned in Prob. 35 hold; hence, that $\{a + b\sqrt{2} \,|\, a, b \text{ rational}\}$ is a field.

6
quadratic equations

In Chap. 4 we considered situations that can be solved by means of linear equations. Many problems in pure mathematics and in applications lead to equations in which the variable enters to a degree greater than 1. In this chapter, we shall study equations in which the square of the variable occurs.

6.1 INTRODUCTORY REMARKS

We shall begin with a definition:

Quadratic
Equation

The equation $f(x) = g(x)$ is a **quadratic equation in x** if (1) one of $f(x)$ and $g(x)$ is a polynomial of second degree in x and the other is a polynomial of first degree in x or is a constant, or (2) both $f(x)$ and $g(x)$ are polynomials of second degree in x with different coefficients of x^2.

Any such equation can be put in the form $ax^2 + bx + c = 0$ where a, b, and c are constants.

The following equations are examples of quadratics.

$$3x^2 + 2x - 1 = 5x + 6$$
$$5x^2 - 7x = 4$$
$$2x^2 + 3 = 8x - 7$$
$$6x^2 + 5x - 4 = 3x^2 - 2x$$
$$2x^2 = 3$$
$$5x^2 = x$$

Before discussing the solution of the general quadratic equation, we shall obtain the solution of $ax^2 - b = 0$. We begin by solving for x^2 and obtaining

$$x^2 = \frac{b}{a}$$

Therefore, taking the square root of each member, we have

$$x = \pm\sqrt{\frac{b}{a}} \qquad \text{since} \qquad \left(\pm\sqrt{\frac{b}{a}}\right)^2 = \frac{b}{a}$$

Hence, the solution set is $\{\sqrt{b/a}, -\sqrt{b/a}\}$.

It can be proved that any quadratic equation is equivalent to one of the form $(x + d)^2 = k$. Hence $x + d = \pm\sqrt{k}$, since $(\pm\sqrt{k})^2 = k$ and $x = -d \pm\sqrt{k}$. Consequently, the solution set is $\{-d - \sqrt{k}, -d + \sqrt{k}\}$, and there are two solutions.

Example To solve $x^2 + 6x + 7 = 0$, we add 2 to both members. Thus, $x^2 + 6x + 9 = 2$ and $(x + 3)^2 = 2$, and it follows that $x + 3 = \pm\sqrt{2}$ or $x = -3 \pm \sqrt{2}$.

6.2 SOLUTION BY FACTORING

If the left member of the quadratic equation $ax^2 + bx + c = 0$ can be factored, the roots can be obtained by setting each factor equal to 0 and solving for x. This procedure is justified since the product of two factors is 0 if either factor is 0.

Example By factoring, solve the equation $2x^2 = x + 6$.

Solution Since neither member of the above equation is 0, we obtain an equivalent equation with one member 0 by adding $-x - 6$ to each member. This operation yields the equation

$$2x^2 - x - 6 = 0$$

Then the method of solving by factoring can be applied. The complete process follows:

$2x^2 = x + 6$	given equation
$2x^2 - x - 6 = 0$	adding $- x - 6$ to each member
$(2x + 3)(x - 2) = 0$	factoring the left member
$2x + 3 = 0$	setting the first factor equal to zero
$2x = -3$	adding -3 to each member
$x = -\frac{3}{2}$	multiplying each member by $\frac{1}{2}$
$x - 2 = 0$	setting the second factor equal to zero
$x = 2$	adding 2 to each member

Hence, the solution set is $\{-\frac{3}{2}, 2\}$.

Verification If $x = -\frac{3}{2}$, we have

$$2(-\tfrac{3}{2})^2 = \tfrac{18}{4} = \tfrac{9}{2} \qquad \text{for the left member}$$

and

$$-\tfrac{3}{2} + 6 = \frac{-3 + 12}{2} = \tfrac{9}{2} \qquad \text{for the right member}$$

Hence,

$$2(-\tfrac{3}{2})^2 = -\tfrac{3}{2} + 6$$

Furthermore, if $x = 2$, we have

$$2(2)^2 = 8 \qquad \text{for the left member}$$

$$2 + 6 = 8 \qquad \text{for the right member}$$

Consequently, $2(2)^2 = 2 + 6$.

Note   We wish to impress the reader with the fact that this method is applicable *only when the right member of the equation is 0.* If one of the factors of the left member is 0, their product is 0, regardless of the value of the other factor. However, if the right member of the equation is not 0, as in

$$(x - 1)(x + 2) = 6$$

we cannot arbitrarily assign a value to either factor without at the same time fixing the value of the other. For example, if in the preceding example we let $x - 1 = 3$, then surely $x + 2 = 2$ if their product is 6. Obviously, these two conditions cannot be satisfied by the same value of x.

EXERCISE 6.1   Solution by Factoring

1 $x^2 - 9 = 0$	2 $x^2 - 16 = 0$	3 $x^2 = 36$
4 $x^2 - 25 = 0$	5 $49x^2 - 1 = 0$	6 $64x^2 = 1$
7 $100x^2 - 1 = 0$	8 $81x^2 - 1 = 0$	9 $4x^2 = 9$
10 $9x^2 - 16 = 0$	11 $16x^2 - 25 = 0$	12 $25x^2 - 36 = 0$
13 $x^2 - 5x + 6 = 0$	14 $x^2 - 7x + 12 = 0$	15 $x^2 - 6x + 5 = 0$
16 $x^2 - 9x + 14 = 0$	17 $x^2 + 4x + 3 = 0$	18 $x^2 + 5x + 6 = 0$
19 $x^2 + 5x + 4 = 0$	20 $x^2 + 9x + 14 = 0$	21 $x^2 + x - 6 = 0$
22 $x^2 + 4x - 5 = 0$	23 $x^2 - 6x + 8 = 0$	24 $x^2 - 9x + 18 = 0$
25 $6x^2 - x - 1 = 0$	26 $8x^2 - 2x + 1 = 0$	27 $6x^2 + x - 1 = 0$
28 $10x^2 + 3x - 1 = 0$	29 $6x^2 - 5x - 6 = 0$	30 $10x^2 + x - 2 = 0$
31 $12x^2 + 13x - 4 = 0$	32 $10x^2 + 11x - 6 = 0$	33 $15x^2 + 16x - 15 = 0$
34 $12x^2 + 17x + 6 = 0$	35 $6x^2 - 7x - 3 = 0$	36 $6x^2 + x - 2 = 0$
37 $10x^2 + x - 2 = 0$	38 $10x^2 + 13x - 3 = 0$	39 $12x^2 - 25x + 12 = 0$
40 $6x^2 + 11x - 10 = 0$	41 $16x^2 - 8x - 3 = 0$	42 $15x^2 + 34x + 15 = 0$
43 $9x^2 - 9x - 10 = 0$	44 $12x^2 - x - 6 = 0$	45 $14x^2 + 17x - 6 = 0$
46 $12x^2 + 20x - 25 = 0$	47 $12x^2 + 11x - 5 = 0$	48 $6x^2 + 17x - 14 = 0$

49 $6x^2 + dx - 12d^2 = 0$ **50** $8x^2 - 2dx - 3d^2 = 0$ **51** $2x^2 + dx - 6d^2 = 0$

52 $3x^2 + 8dx - 3d^2 = 0$ **53** $6a^2x^2 - 11ax - 10 = 0$ **54** $6a^2x^2 - ax - 12 = 0$

55 $10a^2x^2 + 11ax - 6 = 0$ **56** $12a^2x^2 + ax - 6 = 0$

6.3 COMPLEX NUMBERS

We stated in Sec. 6.1 that every quadratic equation is equivalent to an equation of the form $(x + d)^2 = k$ and that the roots are $x = -d \pm\sqrt{k}$. If k is a negative number, then $\sqrt{k}$ is not a real number, since the square of any nonzero real number is a positive number. Consequently, we must either ignore such numbers as the square root of a negative number or else define an extension of the real number system. We shall choose the latter course.

If $i^2 = -1$, then $i = \sqrt{-1}$. Thus, if $n > 0$, we define $\sqrt{-n}$ to be $\sqrt{-1}\sqrt{n}$ and let i represent $\sqrt{-1}$. Therefore,

$$\sqrt{-n} = i\sqrt{n} \qquad n > 0 \tag{6.1}$$

Such numbers were not understood by early mathematicians, who called them *imaginary*. In the eighteenth century, however, Gauss and Argand devised a geometrical representation of imaginary numbers, and these numbers have become very real in mathematics, physics, and electrical engineering.

Complex Number | **A number of the form $a + bi$, a and b real, is called a *complex number*.** $\tag{6.2}$

Thus, $3 + 5i$ is a complex number.

Pure Imaginary | If $a = 0$ and $b \neq 0$, then $a + bi$ becomes bi and is called a *pure imaginary* number. Hence, $2i$ is a complex number which is pure imaginary. If $b = 0$ and $a \neq 0$, then the complex number reduces to the real number a. Therefore, real numbers and pure imaginary numbers are subsets of the set of complex numbers $a + bi$.

Example | $\sqrt{-7} = i\sqrt{7}$, $\sqrt{-8} = i\sqrt{8} = 2i\sqrt{2}$, and $\sqrt{-9} = i\sqrt{9} = 3i$.

It is interesting to note that $i^2 = -1$, $i^3 = i^2i = -i$, and $i^4 = (i^2)^2 = (-1)^2 = 1$. Thus, $i^5 = i$ and $i^6 = i^2$.

6.4 SOLUTION BY COMPLETING THE SQUARE

As pointed out in Sec. 6.1, the solution of $(x + d)^2 = k$ is $-d \pm\sqrt{k}$ and is readily obtained. We can thus find the solution of the general quadratic $ax^2 + bx + c = 0$ if we can express it as the square of a binomial equal to a constant. To do this, we add $-c$ to each member and then divide through by a. Thus we get

$$x^2 + \frac{bx}{a} = -\frac{c}{a} \tag{1}$$

From our work on special products and factoring, we know that a quadratic trinomial with leading coefficient 1 is a perfect square provided the constant term is the square of half the coefficient of x. If we apply this to Eq. (1), we see we must add $[\frac{1}{2}(b/a)]^2 = b^2/4a^2$ to each member to make the left member a perfect square.

Example 1   Solve $x^2 + 6x = 27$ by completing the square.

Solution   Since the coefficient of x^2 is given as 1 and the constant term is already on the right, we begin by adding the square of half the coefficient of x to each member of the given equation. Thus, we get

$$x^2 + 6x + [\tfrac{1}{2}(6)]^2 = 27 + 3^2$$
$$x^2 + 6x + 3^2 = 36 \qquad \text{factoring}$$
$$(x + 3)^2 = 6^2$$
$$x + 3 = \pm 6 \qquad \text{taking the square roots}$$
$$x = -3 \pm 6$$
$$= 3, -9 \qquad \text{since } -3 + 6 = 3;\ -3 - 6 = -9$$

These possible solutions can be checked by replacing x in the given equation by each of them in turn. Thus, for $x = 3$, the left member becomes $3^2 + 6(3) = 27$; hence, 3 is a root. Similarly, $(-9)^2 + 6(-9) = 81 - 54 = 27$, and -9 is also a root.

Example 2   By the method of completing the square, solve the equation

$$3x - 9 = -x^2 \tag{1}$$

Solution   Since our first objective is to convert Eq. (1) into an equivalent equation in which the left member is the square of a binomial of the type $x + d$ and the right member is a constant, we must first convert (1) to an equivalent equation in which the first term is x^2, the second term involves x, and the right member is a constant. Therefore, we add $x^2 + 9$ to each member and get

$$3x - 9 + x^2 + 9 = -x^2 + x^2 + 9$$

or

$$x^2 + 3x = 9$$

The next step is to complete the square by adding $[\frac{1}{2}(3)]^2 = \frac{9}{4}$ to each member. Thus, we get

$$x^2 + 3x + \tfrac{9}{4} = 9 + \tfrac{9}{4}$$

or

$$(x + \tfrac{3}{2})^2 = \tfrac{45}{4}$$ **since** $x^2 + 3x + \tfrac{9}{4} = (x + \tfrac{3}{2})^2$ **and** $9 + \tfrac{9}{4} = (36 + 9)/4 = \tfrac{45}{4}$

Now we equate the principal† square root of the left member to both square roots of the right and get

$$x + \tfrac{3}{2} = \pm \sqrt{\tfrac{45}{4}}$$
$$= \pm \frac{3\sqrt{5}}{2}$$ **since** $\sqrt{\tfrac{45}{4}} = \sqrt{(\tfrac{9}{4})(5)} = \frac{3\sqrt{5}}{2}$

Finally, we solve the two linear equations and get

$$x = \frac{-3}{2} \pm \frac{3\sqrt{5}}{2}$$
$$= \tfrac{3}{2}(-1 \pm \sqrt{5})$$

Hence, the solution set is

$$\{\tfrac{3}{2}(-1 + \sqrt{5}), \tfrac{3}{2}(-1 - \sqrt{5})\}$$

We next check these roots by substituting them for x in each member of Eq. (1). For the left member, we get

$$3[\tfrac{3}{2}(-1 \pm \sqrt{5})] - 9 = -\frac{9}{2} \pm \frac{9\sqrt{5}}{2} - \frac{18}{2} = -\frac{27}{2} \pm \frac{9\sqrt{5}}{2}$$

and for the right,

$$-[\tfrac{3}{2}(-1 \pm \sqrt{5})]^2 = -\tfrac{9}{4}(1 \mp 2\sqrt{5} + 5) = -\tfrac{9}{4}(6 \mp 2\sqrt{5})$$
$$= -\frac{27}{2} \pm \frac{9\sqrt{5}}{2}$$

Example 3   By the method of completing the square, solve the equation

$$8x - 2 = 3x^2 \qquad\qquad (1)$$

Solution   Again the first objective is to convert Eq. (1) to an equivalent equation in which the left member is the square of a binomial of the type $x + d$. Consequently, we must first convert (1) to an equivalent equation in which the first term of the left member is x^2, the second term involves x, and the right member is a con-

---

†The question arises here: Why not use both square roots of the left member? The answer is that such procedure will yield two linear equations that are equivalent to those obtained by use of the principal square root, since $-(x + \tfrac{3}{2}) = \pm 3\sqrt{5}/2$ is equivalent to $x + \tfrac{3}{2} = \mp 3\sqrt{5}/2$.

stant. Therefore, we add $-3x^2 + 2$ to each member and divide the resulting equation by -3. Thus, we obtain

$-3x^2 + 8x = 2$ adding $-3x^2 + 2$ to each member of Eq. (1)

$x^2 - \dfrac{8x}{3} = -\dfrac{2}{3}$ dividing each member by -3

We now add the square of one-half the coefficient of x to each member and complete the process as follows:

$$x^2 - \frac{8x}{3} + \frac{16}{9} = -\frac{2}{3} + \frac{16}{9} \qquad \text{adding } [\tfrac{1}{2}(-\tfrac{8}{3})]^2 = \tfrac{16}{9} \text{ to each member}$$

$$= \frac{-6 + 16}{9}$$

$$= \frac{10}{9}$$

Now we express

$$x^2 - \frac{8x}{3} + \frac{16}{9} \qquad \text{as} \qquad (x - \tfrac{4}{3})^2$$

and have

$$(x - \tfrac{4}{3})^2 = \tfrac{10}{9}$$

$$x - \tfrac{4}{3} = \pm\sqrt{\tfrac{10}{9}}$$

$$x = \frac{4}{3} \pm \frac{\sqrt{10}}{3} \qquad \text{since } \sqrt{10/9} = \sqrt{10}/3$$

$$x = \tfrac{1}{3}(4 \pm \sqrt{10})$$

Consequently, the solution set is

$$\{\tfrac{1}{3}(4 + \sqrt{10}), \tfrac{1}{3}(4 - \sqrt{10})\}$$

We check the solution set by substituting $\tfrac{1}{3}(4 \pm \sqrt{10})$ in each member of Eq. (1). The substitution is as follows:

Left member

$8[\tfrac{1}{3}(4 \pm \sqrt{10})] - 2$

$= \tfrac{32}{3} \pm \tfrac{8}{3}\sqrt{10} - 2$

$= \dfrac{32 \pm 8\sqrt{10} - 6}{3}$

$= \dfrac{26 \pm 8\sqrt{10}}{3}$

Right member

$3[\tfrac{1}{3}(4 \pm \sqrt{10})]^2$

$= 3[\tfrac{1}{9}(16 \pm 8\sqrt{10} + 10)]$

$= \tfrac{1}{3}(26 \pm 8\sqrt{10})$

$= \dfrac{26 \pm 8\sqrt{10}}{3}$

Consequently, the two members of the equation are equal if $x = \tfrac{1}{3}(4 \pm \sqrt{10})$.

Example 4   Solve the equation $2x^2 + 9 = 3x$.

Solution

$2x^2 + 9 = 3x$	given equation copied
$2x^2 - 3x = -9$	adding $-3x - 9$ to each member
$x^2 - \dfrac{3x}{2} = -\dfrac{9}{2}$	dividing each member by the coefficient of x^2
$x^2 - \dfrac{3x}{2} + \dfrac{9}{16} = -\dfrac{9}{2} + \dfrac{9}{16}$	adding $[\frac{1}{2}(\frac{3}{2})]^2$ to each member
$\phantom{x^2 - \dfrac{3x}{2} + \dfrac{9}{16}} = \dfrac{-72 + 9}{16}$	
$\phantom{x^2 - \dfrac{3x}{2} + \dfrac{9}{16}} = -\frac{63}{16}$	simplifying right member
$(x - \frac{3}{4})^2 = -\frac{63}{16}$	expressing left member as $(x - \frac{3}{4})^2$
$x - \frac{3}{4} = \pm \sqrt{-\frac{63}{16}}$	equating the square roots of the members
$\phantom{x - \frac{3}{4}} = \pm \dfrac{3\sqrt{-7}}{4}$	since $\sqrt{-63} = \sqrt{9(-7)} = 3\sqrt{-7}$ and $\sqrt{16} = 4$
$\phantom{x - \frac{3}{4}} = \pm \dfrac{3i\sqrt{7}}{4}$	since $\sqrt{-7} = i\sqrt{7}$
$x = \dfrac{3}{4} \pm \dfrac{3i\sqrt{7}}{4}$	solving for x
$ = \frac{3}{4}(1 \pm i\sqrt{7})$	

Therefore, the solution set is

$$\{\tfrac{3}{4}(1 + i\sqrt{7}), \tfrac{3}{4}(1 - i\sqrt{7})\}$$

To check the solution, we substitute $\frac{3}{4}(1 \pm i\sqrt{7})$ for x in each member of the original equation and get the following results:

Left member

$$2[\tfrac{3}{4}(1 \pm i\sqrt{7})]^2 + 9$$
$$= 2[\tfrac{9}{16}(1 \pm 2i\sqrt{7} + 7i^2)] + 9$$
$$= \tfrac{9}{8}(1 \pm 2i\sqrt{7} - 7) + 9$$
$$= \tfrac{9}{8}(-6 \pm 2i\sqrt{7}) + 9$$
$$= -\frac{54}{8} \pm \frac{18i\sqrt{7}}{8} + 9$$
$$= \frac{18}{8} \pm \frac{18i\sqrt{7}}{8}$$
$$= \frac{9}{4} \pm \frac{9i\sqrt{7}}{4}$$

Right member

$$3[\tfrac{3}{4}(1 \pm i\sqrt{7})]$$
$$= \tfrac{9}{4}(1 \pm i\sqrt{7})$$
$$= \frac{9}{4} \pm \frac{9i\sqrt{7}}{4}$$

EXERCISE 6.2   Solution by Completing the Square

Solve the equation in each of Probs. 1 to 52.

1 $x^2 - 5x + 6 = 0$ **2** $x^2 - 5x + 4 = 0$ **3** $x^2 - 7x + 12 = 0$

4 $x^2 - 7x + 10 = 0$ **5** $x^2 + 5x + 6 = 0$ **6** $x^2 + 6x + 5 = 0$

7 $x^2 + 4x + 3 = 0$ **8** $x^2 + 7x + 12 = 0$ **9** $x^2 - x - 2 = 0$

10 $x^2 - x - 6 = 0$ **11** $x^2 - 3x - 10 = 0$ **12** $x^2 - x - 12 = 0$

13 $2x^2 - 3x - 2 = 0$ **14** $2x^2 + x - 3 = 0$ **15** $3x^2 - x - 4 = 0$

16 $4x^2 + 5x - 6 = 0$ **17** $4x^2 - 5x + 1 = 0$ **18** $3x^2 - 10x - 8 = 0$

19 $3x^2 + 2x - 8 = 0$ **20** $4x^2 - 7x - 15 = 0$ **21** $6x^2 - 7x + 2 = 0$

22 $6x^2 + 11x + 3 = 0$ **23** $12x^2 + 7x - 12 = 0$ **24** $8x^2 - 14x - 15 = 0$

25 $10x^2 + 21x - 10 = 0$ **26** $8x^2 + 10x - 7 = 0$ **27** $15x^2 + x - 6 = 0$

28 $6x^2 - 19x + 15 = 0$ **29** $x^2 - 4x + 1 = 0$ **30** $x^2 - 6x + 7 = 0$

31 $x^2 - 10x + 22 = 0$ **32** $x^2 + 2x - 1 = 0$ **33** $2x^2 - 2x - 1 = 0$

34 $9x^2 + 12x + 2 = 0$ **35** $25x^2 - 10x - 2 = 0$ **36** $4x^2 - 6x + 1 = 0$

37 $x^2 - 2x + 2 = 0$ **38** $x^2 - 4x + 5 = 0$ **39** $x^2 - 6x + 13 = 0$

40 $x^2 - 8x + 25 = 0$ **41** $2x^2 - 2x + 1 = 0$ **42** $4x^2 - 8x + 5 = 0$

43 $9x^2 - 24x + 20 = 0$ **44** $8x^2 - 4x + 5 = 0$ **45** $x^2 + 4ax - 5a^2 = 0$

46 $x^2 + mx - 6m^2 = 0$ **47** $x^2 + (b - 2c)x - 2bc = 0$

48 $x^2 - (b + 3c)x + 3bc = 0$ **49** $a^2x^2 - abx - 2b^2 = 0$

50 $b^2x^2 + abx - 6a^2 = 0$ **51** $bx^2 - 3bx + a + 2b = ax$

52 $3abx^2 + (6b^2 - a^2)x = 2ab$

6.5   SOLUTION BY FORMULA

If we solve the equation

$$ax^2 + bx + c = 0 \qquad (6.3)$$

for x, we get a formula that can be used for obtaining the roots of any quadratic equation. We shall derive the formula by solving (6.3) by completing the square, and then we shall explain its use.

$$ax^2 + bx + c = 0 \qquad \text{Eq. (6.3) copied}$$

$$ax^2 + bx = -c \qquad \text{adding } -c \text{ to both members}$$

$$x^2 + \frac{bx}{a} = -\frac{c}{a} \qquad \text{dividing each member by } a \neq 0$$

$$x^2 + \frac{bx}{a} + \left(\frac{b}{2a}\right)^2 = -\frac{c}{a} + \frac{b^2}{4a^2} \qquad \text{adding } [\tfrac{1}{2}(b/a)]^2 \text{ to each member}$$

$$\left(x + \frac{b}{2a}\right)^2 = \frac{b^2 - 4ac}{4a^2}$$

$$x + \frac{b}{2a} = \pm\frac{\sqrt{b^2 - 4ac}}{2a} \dagger$$

$$x = -\frac{b}{2a} \pm \frac{\sqrt{b^2 - 4ac}}{2a}$$

Since the two denominators in the right member are the same, we can write

$$x = \frac{-b \pm \sqrt{b^2 - 4ac}}{2a} \tag{6.4}$$

Quadratic Formula Equation (6.4) is known as the *quadratic formula*. It expresses the roots of Eq. (6.3) in terms of the constant term and the coefficients of x^2 and x. By properly matching the coefficients and the constant term in any quadratic equation with those in (6.3), we can determine the values of a, b, and c, and then we can use the formula to get the roots of the equation. We shall illustrate the use of (6.4) with two examples.

Example 1 Solve the equation $3x^2 - 5x + 2 = 0$ by the quadratic formula.

Solution To solve $3x^2 - 5x + 2 = 0$ with the quadratic formula, we compare the equation with (6.3) and see that $a = 3$, $b = -5$, and $c = 2$. Hence, if we substitute these values in (6.4), we get

$$x = \frac{-(-5) \pm \sqrt{(-5)^2 - 4(3)(2)}}{2(3)}$$

$$= \frac{5 \pm \sqrt{25 - 24}}{6}$$

$$= \frac{5 \pm 1}{6} = 1, \tfrac{2}{3}$$

Hence, the solutions are $\tfrac{2}{3}$ and 1. The reader should verify that this is true.

If the terms in a given equation do not occur in the same order

---

†Here we have tacitly assumed that a is positive. If, however, a is negative, then $\sqrt{4a^2} = -2a$, since by definition the principal square root of a number is positive. Then we have

$$x + \frac{b}{2a} = \pm\sqrt{\frac{b^2 - 4ac}{4a^2}} = \pm\frac{\sqrt{b^2 - 4ac}}{\sqrt{4a^2}} = \mp\frac{\sqrt{b^2 - 4ac}}{2a}$$

Hence,

$$x = \frac{-b \mp \sqrt{b^2 - 4ac}}{2a}$$

This equation yields the same values of x as (6.4), but the values are in reverse order.

as those in (6.3), we convert the equation to an equivalent equation in which the terms have the desired order before applying the quadratic formula.

Example 2 By the quadratic formula, solve the equation $4x^2 = 8x - 7$.

Solution The first step in solving the given equation is to convert it to an equivalent equation in which the terms are in the same order as in (6.3). For this purpose we add $-8x + 7$ to each member and obtain $4x^2 - 8x + 7 = 0$. By comparing this equation with (6.3), we see that $a = 4$, $b = -8$, and $c = 7$. If we substitute these values in (6.4), we get

$$x = \frac{-(-8) \pm \sqrt{(-8)^2 - 4(4)(7)}}{2(4)}$$

$$= \frac{8 \pm \sqrt{64 - 112}}{8}$$

$$= \frac{8 \pm \sqrt{-48}}{8}$$

$$= \frac{8 \pm 4i\sqrt{3}}{8}$$

$$= \tfrac{1}{2}(2 \pm i\sqrt{3})$$

Therefore, the solutions are $\tfrac{1}{2}(2 + i\sqrt{3})$ and $\tfrac{1}{2}(2 - i\sqrt{3})$. The reader should verify that these two numbers are roots of the given equation.

EXERCISE 6.3 Solution by Formula

Solve each of the following equations by use of the formula.

1 $x^2 + 2x - 3 = 0$	2 $x^2 - 3x - 4 = 0$	3 $x^2 + 2x - 8 = 0$
4 $x^2 + x - 6 = 0$	5 $2x^2 + x - 3 = 0$	6 $3x^2 + 4x - 4 = 0$
7 $3x^2 + 8x - 3 = 0$	8 $2x^2 + x - 1 = 0$	9 $6x^2 - 11x + 3 = 0$
10 $12x^2 + 5x - 2 = 0$	11 $10x^2 + 9x - 7 = 0$	12 $12x^2 + x - 20 = 0$
13 $x^2 - 2x - 1 = 0$	14 $x^2 - 4x + 1 = 0$	15 $x^2 + 6x + 4 = 0$
16 $x^2 + 4x - 2 = 0$	17 $x^2 - 2x - 2 = 0$	18 $x^2 + 4x + 2 = 0$
19 $x^2 - 6x + 2 = 0$	20 $x^2 - 10x + 22 = 0$	21 $9x^2 - 12x + 1 = 0$
22 $4x^2 - 12x + 7 = 0$	23 $4x^2 + 6x + 1 = 0$	24 $9x^2 + 24x + 10 = 0$
25 $x^2 - 2x + 2 = 0$	26 $x^2 - 4x + 5 = 0$	27 $x^2 - 6x + 13 = 0$
28 $x^2 - 4x + 13 = 0$	29 $x^2 + 6x + 13 = 0$	30 $x^2 + 4x + 13 = 0$
31 $x^2 + 8x + 20 = 0$	32 $x^2 + 6x + 18 = 0$	33 $x^2 - 2x + 4 = 0$
34 $x^2 + 4x + 6 = 0$	35 $x^2 + 6x + 14 = 0$	36 $x^2 - 8x + 19 = 0$

37 $4x^2 + 12x + 11 = 0$ **38** $9x^2 + 12x + 7 = 0$ **39** $9x^2 - 6x + 7 = 0$

40 $25x^2 + 10x + 6 = 0$ **41** $x^2 - 2ax + a^2 = b^2$ **42** $x^2 + 4ax + 4a^2 = b^2$

43 $x^2 - 2rx + r^2 = 4s^2$ **44** $x^2 + 6rx + 9r^2 = s^2$

45 $3x^2 + 3cd + 6d^2 = cx + 11dx$ **46** $6x^2 + a^2 = 5ax + 2bx + 4b^2$

47 $5d^2x^2 - 3bdx - 2b^2 = 0$ **48** $6x^2 + 3x - a - 2ax = 0$

49 $(2m^2 + 3mn + n^2)x^2 + (m - n)x - 6 = 0$ **50** $(3m^2 - 4mn + n^2)x^2 - 2mx - 1 = 0$

51 $(2a^2 - 3ab - 2b^2)x^2 + (a + 8b)x - 6 = 0$

52 $(a + 3b)x^2 + (3 - a^2 - 6ab - 9b^2)x - 3(a + 3b) = 0$

6.6 EQUATIONS IN QUADRATIC FORM

In a quadratic equation, the unknown need not be x. In fact, it may be any quantity, but it must occur to the second power and may occur also to the first power. Thus,

$$a[f(x)]^2 + b[f(x)] + c = 0 \qquad a \neq 0 \tag{1}$$

is a quadratic equation with $f(x)$ as the unknown. We can solve for $f(x)$ by either of the methods studied and can then solve for x if $f(x)$ is a linear or quadratic expression.

Example 1 Solve $4(x^2 - x)^2 - 11(x^2 - x) + 6 = 0$ for x.

Solution We begin by solving for $x^2 - x$, since the given equation is a quadratic if we think of $x^2 - x$ as the unknown. Since $a = 4$, $b = -11$, and $c = 6$, the quadratic formula gives

$$x^2 - x = \frac{11 \pm \sqrt{(-11)^2 - 4(4)(6)}}{2(4)} = \frac{11 \pm \sqrt{25}}{8}$$

$$= \frac{11 \pm 5}{8} = 2, \frac{3}{4}$$

We now complete the solution of the given equation for x by setting $x^2 - x$ equal to each of these values and solving for x. Thus,

$$x^2 - x = 2 \qquad \text{and} \qquad x^2 - x = \tfrac{3}{4}$$
$$x^2 - x - 2 = 0 \qquad \text{and} \qquad 4x^2 - 4x - 3 = 0$$
$$(x - 2)(x + 1) = 0 \qquad \text{and} \qquad (2x - 3)(2x + 1) = 0$$
$$x = 2, -1 \qquad \text{and} \qquad x = \tfrac{3}{2}, -\tfrac{1}{2}$$

Therefore, the values of x which satisfy the given equation are 2, -1, $\tfrac{3}{2}$, and $-\tfrac{1}{2}$. This statement can be verified by substituting these values for x in the given equation.

Example 2 Solve $3x^4 = 2x^2 + 1$ for x.

Solution　This is a quadratic equation with x^2 as the unknown, and the standard form of the equation is $3(x^2)^2 - 2(x^2) - 1 = 0$. Hence,

$$x^2 = \frac{2 \pm \sqrt{(-2)^2 - 4(3)(-1)}}{2(3)} = \frac{2 \pm \sqrt{16}}{6}$$

$$= \frac{2 \pm 4}{6} = 1, -\tfrac{1}{3}$$

We now solve

$$x^2 = 1 \qquad \text{and} \qquad x^2 = -\tfrac{1}{3}$$
$$x = \pm 1 \qquad \text{and} \qquad x = \pm i\sqrt{3}/3$$

Therefore, the solutions for x are 1, -1, $i\sqrt{3}/3$, and $-i\sqrt{3}/3$. They can be verified in the usual manner.

Example 3　Solve $\left(\dfrac{x+7}{2x-1}\right)^2 - 4\left(\dfrac{x+7}{2x-1}\right) + 3 = 0$ for x.

Solution　If we think of $(x+7)/(2x-1)$ as the unknown, the given equation is a quadratic. Now, by use of the factoring method, we get

$$\left(\frac{x+7}{2x-1} - 1\right)\left(\frac{x+7}{2x-1} - 3\right) = 0$$

$$\frac{x+7}{2x-1} = 1, 3$$

Hence

$$x + 7 = 2x - 1 \qquad \text{and} \qquad x + 7 = 3(2x - 1)$$
$$8 = x \qquad\qquad \text{and} \qquad\qquad 10 = 5x$$

Therefore, the solutions of the given equation are $x = 8, 2$.

EXERCISE 6.4　Equations in Quadratic Form

Solve each of the following equations.

1　$x^4 - 10x^2 + 9 = 0$　　　2　$x^4 - 3x^2 - 4 = 0$　　　3　$x^4 - 16 = 0$

4　$x^4 - 8x^2 - 9 = 0$　　　5　$x^6 - 9x^3 + 8 = 0$　　　6　$x^6 + 26x^3 - 27 = 0$

7　$x^6 + 28x^3 + 27 = 0$　　8　$x^6 + 9x^3 + 8 = 0$　　　9　$x^8 - 17x^4 + 16 = 0$

10　$x^8 - 82x^4 + 81 = 0$　　11　$16x^8 - 65x^4 + 4 = 0$　　12　$81x^8 - 325x^4 + 4 = 0$

13　$x^{-2} - 5x^{-1} + 6 = 0$　　14　$x^{-2} + 5x^{-1} - 6 = 0$　　15　$3x^{-2} + 13x^{-1} + 4 = 0$

16　$2x^{-2} + x^{-1} - 6 = 0$　　17　$x^{-6} + 7x^{-3} - 8 = 0$　　18　$x^{-6} - 26x^{-3} - 27 = 0$

19　$8x^{-6} - 65x^{-3} + 8 = 0$　　20　$27x^{-6} + 26x^{-3} - 1 = 0$　　21　$x^{-4} - 3x^{-2} - 4 = 0$

22　$x^{-4} - 8x^{-2} - 9 = 0$　　23　$4x^{-4} - 37x^{-2} + 9 = 0$　　24　$16x^{-4} - 65x^{-2} + 4 = 0$

25　$(x^2 + 1)^2 - 7(x^2 + 1) + 10 = 0$　　　　26　$(x^2 + 2)^2 - (x^2 + 2) - 6 = 0$

27　$(x^2 + 4)^2 + 2(x^2 + 4) - 15 = 0$　　　　28　$(x^2 - 2)^2 - 9(x^2 - 2) + 14 = 0$

29 $(2x^2 + 5x)^2 + 2(2x^2 + 5x) - 3 = 0$ **30** $(3x^2 + x)^2 - 4(3x^2 + x) + 3 = 0$

31 $(5x^2 + 3x)^2 - 3(5x^2 + 3x) + 2 = 0$ **32** $(2x^2 + 3x)^2 - 7(2x^2 + 3x) + 10 = 0$

33 $(2x^2 - 5x - 1)^2 + 2(2x^2 - 5x - 1) - 8 = 0$

34 $(x^2 + 4x + 2)^2 - 4(x^2 + 4x + 2) - 5 = 0$

35 $(x^2 + 3x + 1)^2 - 2(x^2 + 3x + 1) - 3 = 0$

36 $(2x^2 - 3x + 2)^2 - 4(2x^2 - 3x + 2) - 21 = 0$

37 $\left(\dfrac{x^2 + 3x + 1}{x + 1}\right)^2 + 2\dfrac{x^2 + 3x + 1}{x + 1} - 3 = 0$

38 $\left(\dfrac{x^2 + 2x - 1}{x + 2}\right)^2 - 5\dfrac{x^2 + 2x - 1}{x + 2} + 6 = 0$

39 $\left(\dfrac{2x^2 - 3x + 2}{2x + 1}\right)^2 - \dfrac{2x^2 - 3x + 2}{2x + 1} - 2 = 0$

40 $\left(\dfrac{3x^2 - 4x + 1}{x + 3}\right)^2 - 2\dfrac{3x^2 - 4x + 1}{x + 3} - 3 = 0$

41 $2\dfrac{x + 2}{2x + 1} - 5 + 3\dfrac{2x + 1}{x + 2} = 0$ *Hint:* Multiply through by $(x + 2)/(2x + 1)$.

42 $3\dfrac{x + 3}{2x - 1} + 2 - 5\dfrac{2x - 1}{x + 3} = 0$ **43** $\dfrac{2x - 3}{x + 1} - 2 - 3\dfrac{x + 1}{2x - 3} = 0$

44 $5\dfrac{2x - 5}{x + 2} + 3 - 2\dfrac{x + 2}{2x - 5} = 0$

45 $2x + 3 - 4\sqrt{2x + 3} + 3 = 0$ *Hint:* Use $\sqrt{2x + 3}$ as the variable.

46 $x - 2 - 5\sqrt{x - 2} + 6 = 0$ **47** $3x - 1 - 3\sqrt{3x - 1} + 2 = 0$

48 $2x + 5 + 6\sqrt{2x + 5} + 8 = 0$

49 $x + 5 - 3\sqrt{x + 3} = 0$ *Hint:* Subtract 2 from each number.

50 $2x + 1 - 6\sqrt{2x + 3} = -7$ **51** $3x - 4\sqrt{3x - 4} = 9$

52 $4x - 4 - \sqrt{4x - 1} = 3$

6.7 RADICAL EQUATIONS

Radical Equation An equation in which either or both members contain a radical that has the unknown in the radicand is called a *radical equation*.

Thus, $\sqrt{x + 7} - 3 = \sqrt{2x - 17}$ is a radical equation. If the radicals are of second order, we can often solve the equation. The method depends on the following theorem:

Any root of a given equation is also a root of the equation obtained by equating the squares of the members of the given equation.

Proof If r is a root of $f(x) = g(x)$, then $f(r) = g(r)$, and it follows that $f(r) - g(r) = 0$. If we multiply each member of this equation by $f(r) + g(r)$ and equate the products, we have

$$[f(r) - g(r)][f(r) + g(r)] = 0[f(r) + g(r)]$$

Now, performing the indicated multiplications, we get $f^2(r) - g^2(r) = 0$, and adding $g^2(r)$ to each member gives $f^2(r) = g^2(r)$. This completes the proof, since replacing x by r in $f^2(x) = g^2(x)$ gives $f^2(r) = g^2(r)$.

The converse of this theorem is not true, as is seen by noting that -2 is a solution of $x^2 = 4$ but not of $x = 2$.

As an example, any solution of $\sqrt{x + 7} - 3 = \sqrt{2x - 17}$ is also a solution of $(\sqrt{x + 7} - 3)^2 = (\sqrt{2x - 17})^2$.

If an equation contains three or fewer radicals and all are of order 2, the steps for solving it are:

Solving a **1** Obtain an equation that is equivalent to the given equation
Radical and that has one radical and no other term in one member.
Equation This process is called *isolating a radical*.
2 Equate the squares of the members of the equation obtained in step 1.
3 If the equation obtained in step 2 contains one or more radicals, repeat the process until an equation free of radicals is obtained. The equation obtained is called the *rationalized equation*.
4 Solve the rationalized equation.
5 Substitute the roots obtained in step 4 in the original equation in order to determine which of these numbers satisfy the original equation.

Note Step 5 is an essential step in solving a radical equation, since the rationalized equation may have roots that are not roots of the given equation.

We shall illustrate the process with four examples.

Example 1 Solve the equation

$$x = \sqrt{5x - 1} - 1 \tag{1}$$

Solution

$$x = \sqrt{5x - 1} - 1 \qquad \text{given Eq. (1)}$$

$$x + 1 = \sqrt{5x - 1} \qquad \text{isolating radical by adding 1 to each member of Eq. (1)}$$

$$x^2 + 2x + 1 = 5x - 1 \qquad \text{equating squares of the members}$$

$$x^2 + 2x + 1 - 5x + 1 = 5x - 1 - 5x + 1 \qquad \text{adding } -5x + 1 \text{ to each member}$$

$$x^2 - 3x + 2 = 0 \qquad \text{combining similar terms}$$

$$(x - 2)(x - 1) = 0 \qquad \text{factoring left member}$$

$$x = 2 \qquad \text{setting } x - 2 = 0 \text{ and solving}$$

$$x = 1 \qquad \text{setting } x - 1 = 0 \text{ and solving}$$

We check these roots by substituting each of them for x in each member of Eq. (1) and obtaining

Value of x	Left member	Right member
2	2	$\sqrt{10 - 1} - 1 = 3 - 1 = 2$
1	1	$\sqrt{5 - 1} - 1 = 2 - 1 = 1$

Therefore, since the left and right members of Eq. (1) are equal for $x = 2$ and for $x = 1$, the solution set of Eq. (1) is $\{2, 1\}$.

Example 2 Solve the equation

$$\sqrt{5x - 11} - \sqrt{x - 3} = 4 \tag{1}$$

Solution

$$\sqrt{5x - 11} - \sqrt{x - 3} = 4 \qquad \text{given Eq. (1)}$$

$$\sqrt{5x - 11} = \sqrt{x - 3} + 4 \qquad \text{isolating } \sqrt{5x - 11}$$

$$5x - 11 = x - 3 + 8\sqrt{x - 3} + 16 \qquad \text{equating squares of the members}$$

$$5x - 11 - x + 3 - 16 = 8\sqrt{x - 3} \qquad \text{isolating } 8\sqrt{x - 3}$$

$$4x - 24 = 8\sqrt{x - 3} \qquad \text{combining similar terms}$$

$$x - 6 = 2\sqrt{x - 3} \qquad \text{dividing each member by 4}$$

$$x^2 - 12x + 36 = 4(x - 3) \qquad \text{equating squares of the members}$$

$$x^2 - 12x + 36 = 4x - 12 \qquad \text{by distributive axiom}$$

$$x^2 - 16x + 48 = 0 \qquad \text{adding } -4x + 12 \text{ to each member}$$

$$(x - 12)(x - 4) = 0 \qquad \text{factoring left member}$$

$$x = 12 \qquad \text{setting } x - 12 = 0 \text{ and solving}$$

$$x = 4 \qquad \text{setting } x - 4 = 0 \text{ and solving}$$

Check

$$\sqrt{60 - 11} - \sqrt{12 - 3} = \sqrt{49} - \sqrt{9} \qquad \text{substituting 12 for } x \text{ in left member of Eq. (1)}$$

$$= 7 - 3$$

$$= 4$$

Therefore, since the right member of Eq. (1) is also 4, then 12 is a root.

$$\sqrt{20 - 11} - \sqrt{4 - 3} = \sqrt{9} - \sqrt{1} \qquad \text{substituting 4 for } x \text{ in left member of Eq. (1)}$$

$$= 3 - 1 = 2$$

Since the right member of Eq. (1) is not 2, then 4 is not a root. Therefore, the solution set of Eq. (1) is {12}.

Example 3   Solve the equation

$$\sqrt{x + 1} + \sqrt{2x + 3} - \sqrt{8x + 1} = 0 \tag{1}$$

Solution

$$\sqrt{x + 1} + \sqrt{2x + 3} - \sqrt{8x + 1} = 0 \qquad \text{given Eq. (1)}$$

$$\sqrt{x + 1} + \sqrt{2x + 3} = \sqrt{8x + 1} \qquad \begin{array}{l}\text{isolating}\\ \sqrt{8x + 1}\end{array}$$

$$x + 1 + 2\sqrt{(x + 1)(2x + 3)} + 2x + 3 = 8x + 1 \qquad \begin{array}{l}\text{equating the}\\ \text{squares of the}\\ \text{members}\end{array}$$

$$2\sqrt{(x + 1)(2x + 3)} = 8x + 1 - x - 1 - 2x - 3$$
$$\begin{array}{l}\text{isolating}\\ 2\sqrt{(x + 1)(2x + 3)}\end{array}$$

$$2\sqrt{2x^2 + 5x + 3} = 5x - 3 \qquad \begin{array}{l}\text{simplifying}\\ \text{radicand and}\\ \text{combining}\\ \text{similar terms}\end{array}$$

$$4(2x^2 + 5x + 3) = 25x^2 - 30x + 9$$
$$\begin{array}{l}\text{equating squares}\\ \text{of the members}\end{array}$$

$$8x^2 + 20x + 12 = 25x^2 - 30x + 9$$
$$\begin{array}{l}\text{by distributive}\\ \text{axiom}\end{array}$$

$$17x^2 - 50x - 3 = 0 \qquad \begin{array}{l}\text{adding } -25x^2 +\\ 30x - 9 \text{ to}\\ \text{each member}\\ \text{and dividing}\\ \text{by } -1\end{array}$$

$$x = \frac{50 \pm \sqrt{2500 + 204}}{34}$$
$$\begin{array}{l}\text{by quadratic}\\ \text{formula}\end{array}$$

$$= \frac{50 \pm \sqrt{2704}}{34}$$

$$= \frac{50 \pm 52}{34}$$

$$= \frac{102}{34} \quad \text{and} \quad \frac{-2}{34}$$

$$= 3 \quad \text{and} \quad \frac{-1}{17}$$

It can be shown by substitution in (1) that 3 in a root and $-\frac{1}{17}$ is not.

Example 4   Solve the equation

$$\sqrt{x+3} - \sqrt{x-2} = 5 \tag{1}$$

Solution
$$\sqrt{x+3} - \sqrt{x-2} = 5 \qquad \text{given Eq. (1)}$$
$$\sqrt{x+3} = \sqrt{x-2} + 5 \qquad \text{isolating } \sqrt{x+3}$$
$$x+3 = x-2+10\sqrt{x-2}+25 \qquad \text{equating squares of the members}$$
$$-10\sqrt{x-2} = 20 \qquad \text{isolating } -10\sqrt{x-2} \text{ and combining terms}$$
$$\sqrt{x-2} = -2 \qquad \text{dividing by } -10$$
$$x-2 = 4 \qquad \text{equating squares of the members}$$
$$x = 6 \qquad \text{adding 2 to each member}$$

Check
$$\sqrt{6+3} - \sqrt{6-2} = \sqrt{9} - \sqrt{4} \qquad \text{replacing } x \text{ by 6 in left member of Eq. (1)}$$
$$= 3 - 2 = 1$$

Consequently, 6 is not a root of Eq. (1) since the right member of Eq. (1) is 5. Furthermore, we conclude that Eq. (1) has no roots since the set of roots of the rationalized equation includes the set of roots of the given equation.

EXERCISE 6.5   Radical Equations

Solve the following equations.

1   $\sqrt{2x-1} = 3$

2   $\sqrt{3x+1} = 4$

3   $\sqrt{5x-1} = -2$

4   $\sqrt{3x+1} = 5$

5   $\sqrt{2x+3} = \sqrt{3x}$

6   $\sqrt{x+5} = \sqrt{2x-2}$

7   $\sqrt{x+3} = \sqrt{5x-1}$

8   $\sqrt{x+4} = \sqrt{2x-1}$

9   $\sqrt{4x+5} = \sqrt{6x-5}$

10   $\sqrt{7x+11} = \sqrt{1-3x}$

11   $\sqrt{3x-4} = \sqrt{2x+1}$

12   $\sqrt{2x+9} = \sqrt{4x-3}$

13   $\sqrt{x+3} = x+3$

14   $\sqrt{2x+5} = 2x-1$

15   $\sqrt{5x+1} = x+1$

16   $\sqrt{2x+1} = x-1$

17   $\sqrt{4x-7}/\sqrt{x-3} = 3$

18   $\sqrt{3-x}/\sqrt{5x+9} = 1$

19   $\sqrt{2-x}/\sqrt{x+3} = 2$

20   $\sqrt{3x-2}/\sqrt{x-1} = 2$

21   $\sqrt{x^2+x+2} = x+1$

22   $\sqrt{x^2-3x-9} = x+6$

23   $\sqrt{x^2+x-2} = x$

24   $\sqrt{x^2-6x+2} = x+4$

25   $\sqrt{x^2+7x-4} = \sqrt{10x+6}$

26   $\sqrt{x^2+3x-7} = \sqrt{4x+5}$

27   $\sqrt{2x^2+3x+2} = \sqrt{5x+6}$

28   $\sqrt{x^2+4x+7} = \sqrt{1-x}$

29   $\sqrt{x^2+3x+13} = 2x-7$

30   $\sqrt{3x^2-8x+4} = x+2$

31   $\sqrt{5x^2-4x+3} = x+1$

32   $\sqrt{x^2-2x+6} = 2x-3$

33 $\sqrt{2x + 7} + 1 = \sqrt{3x + 9}$

34 $\sqrt{3x + 1} = 2 + \sqrt{2x - 6}$

35 $\sqrt{2x - 1} - 1 = \sqrt{x - 1}$

36 $\sqrt{2x + 3} - 1 = \sqrt{x + 1}$

37 $\sqrt{4x - 3} + \sqrt{2x - 2} = \sqrt{7x + 4}$

38 $\sqrt{5x - 4} - \sqrt{2x + 1} = \sqrt{2x - 7}$

39 $\sqrt{5x - 1} - \sqrt{3x - 2} = \sqrt{x - 1}$

40 $\sqrt{2x + 3} + \sqrt{x - 2} = \sqrt{5x + 1}$

41 $\sqrt{5x + 1} - \sqrt{2x} = \sqrt{3x + 1}$

42 $\sqrt{7x + 2} - \sqrt{3x - 2} = \sqrt{6x - 8}$

43 $\sqrt{x + 3} + \sqrt{2x - 1} = \sqrt{7x + 2}$

44 $\sqrt{3x + 1} - \sqrt{2x - 1} = \sqrt{x - 4}$

45 $\sqrt{x^2 + 3x - 1} - 1 = \sqrt{x^2 + x - 2}$

46 $\sqrt{x^2 - x - 4} - 2 = \sqrt{x^2 - 4x - 1}$

47 $\sqrt{x^2 + 3x - 3} - \sqrt{x^2 - 3x + 5} = 2$

48 $\sqrt{x^2 - x - 2} - \sqrt{x^2 - 2x - 2} = 1$

6.8   SUM, PRODUCT, AND TYPE OF ROOTS

We shall consider the general quadratic $ax^2 + bx + c = 0$ and find the sum, product, and type of the roots.   If we let

Sum of the Roots
$$r = \frac{-b + \sqrt{b^2 - 4ac}}{2a} \quad \text{and} \quad s = \frac{-b - \sqrt{b^2 - 4ac}}{2a}$$

we find that the sum of the roots is

$$r + s = \frac{-b + \sqrt{b^2 - 4ac}}{2a} + \frac{-b - \sqrt{b^2 - 4ac}}{2a}$$

Hence

$$r + s = \frac{-b}{a} \tag{6.5}$$

Their product is

Product of the Roots
$$rs = \frac{-b + \sqrt{b^2 - 4ac}}{2a} \ \frac{-b - \sqrt{b^2 - 4ac}}{2a}$$

$$= \frac{b^2 - (b^2 - 4ac)}{4a^2} = \frac{c}{a}$$

Thus

$$rs = \frac{c}{a} \tag{6.6}$$

Example 1   Without solving, find the sum and the product of the roots of $2x^2 + 5x - 8 = 0$.

Solution   Since $a = 2$, $b = 5$, and $c = -8$, we have $r + s = -b/a = -5/2$ and $rs = c/a = -8/2 = -4$.

We shall now prove the following theorem.

Theorem If r and s are the roots of the equation $ax^2 + bx + c = 0$, for $a \neq 0$, then

$$ax^2 + bx + c = a(x - r)(x - s) \tag{6.7}$$

Proof We can write

$$ax^2 + bx + c = a\left(x^2 + \frac{b}{a}x + \frac{c}{a}\right)$$

by factoring out a. By use of (6.5) and (6.6), we can replace b/a by $-(r + s)$, and c/a by rs. Thus, we have

$$ax^2 + bx + c = a[x^2 - (r + s)x + rs]$$

$$ax^2 + bx + c = a(x - r)(x - s) \tag{6.8}$$

as we wished to prove.

Example 2 Form the quadratic equation that has 3 and $\frac{1}{2}$ as roots. By use of (6.8), the equation is $a(x - 3)(x - \frac{1}{2}) = 0$, where a may be any constant. If we choose $a = 2$, we get $2(x - 3)(x - \frac{1}{2}) = (x - 3)$ $(2x - 1) = 2x^2 - 7x + 3 = 0$ as the desired equation.

We shall now assume that the coefficients a, b, and c in the equation $ax^2 + bx + c = 0$ are rational numbers and determine the type of roots for several ranges of values of the discriminant $D = b^2 - 4ac$.

1 If $D = 0$, then $r = -b/2a$ and $s = -b/2a$; hence, the roots are rational and equal.
2 If $D < 0$, then $\sqrt{D} = \sqrt{b^2 - 4ac}$ is an imaginary number, and so the roots are conjugate complex numbers.
3 If $D > 0$, then $\sqrt{D}$ is real and not zero. However, it may be a perfect square or it may not be a perfect square.
   a If $D > 0$ and D is the square of a rational number, the roots are rational and unequal.
   b If $D > 0$ and D is not the square of a rational number, the roots are irrational and unequal.

Example 3 Determine the type of roots of (a) $4x^2 - 12x + 9 = 0$, (b) $4x^2 - 12x + 11 = 0$, (c) $4x^2 - 12x + 5 = 0$, (d) $4x^2 - 12x + 3 = 0$.

Solution (a) $D = (-12)^2 - 4(4)9 = 0$; hence the roots are rational and equal.
(b) $D = (-12)^2 - 4(4)11 = -32 < 0$; hence, the roots are conjugate complex numbers.
(c) $D = (-12)^2 - 4(4)5 = 64 = 8^2$; hence, the roots are rational and unequal.
(d) $(D) = (-12)^2 - 4(4)3 = 96$, not a perfect square; hence, the roots are irrational and unequal.

EXERCISE 6.6   Sum, Product, and Type of Roots

Without solving, determine the sum, product, and type of roots of each equation given below.

1 $x^2 - 4x + 4 = 0$	2 $x^2 - 6x + 9 = 0$	3 $x^2 + 2x + 1 = 0$
4 $x^2 + 10x + 25 = 0$	5 $9x^2 - 12x + 4 = 0$	6 $16x^2 + 24x + 9 = 0$
7 $4x^2 + 20x + 25 = 0$	8 $25x^2 - 40x + 16 = 0$	9 $x^2 + 4 = 0$
10 $x^2 + 9 = 0$	11 $x^2 + 16 = 0$	12 $x^2 + 1 = 0$
13 $9x^2 - 12x + 5 = 0$	14 $2x^2 - 6x + 5 = 0$	15 $16x^2 + 16x + 13 = 0$
16 $9x^2 + 30x + 29 = 0$	17 $2x^2 - 7x + 6 = 0$	18 $3x^2 - 5x + 2 = 0$
19 $3x^2 + 5x - 12 = 0$	20 $5x^2 + 7x - 6 = 0$	21 $6x^2 - 7x + 2 = 0$
22 $15x^2 - 11x + 2 = 0$	23 $15x^2 - x - 6 = 0$	24 $8x^2 - 22x - 21 = 0$
25 $x^2 - 2x - 1 = 0$	26 $x^2 - 4x + 1 = 0$	27 $x^2 + 6x + 6 = 0$
28 $x^2 + 10x + 23 = 0$	29 $9x^2 - 12x + 2 = 0$	30 $2x^2 - 6x + 2 = 0$
31 $5x^2 + 2x - 1 = 0$	32 $16x^2 + 16x - 1 = 0$	

Form a quadratic equation that has the pair of numbers given in each of Probs. 33 to 48 as roots.

33 $2, 2$	34 $3, 3$	35 $-5, -5$	36 $-4, -4$
37 $\frac{2}{3}, -\frac{1}{2}$	38 $\frac{3}{4}, \frac{2}{5}$	39 $\frac{3}{5}, -\frac{1}{2}$	40 $-\frac{2}{7}, -\frac{7}{2}$
41 $1 \pm 2i$	42 $2 \pm 3i$	43 $\dfrac{3 \pm \sqrt{2}i}{2}$	44 $\dfrac{5 \pm \sqrt{3}i}{3}$
45 $\dfrac{1 \pm \sqrt{5}}{2}$	46 $\dfrac{3 \pm \sqrt{2}}{2}$	47 $\dfrac{-3 \pm \sqrt{5}}{7}$	48 $\dfrac{-2 \pm \sqrt{3}}{5}$

6.9   PROBLEMS THAT LEAD TO QUADRATIC EQUATIONS

Many stated problems, especially those which deal with products or quotients involving the unknown, lead to quadratic equations. The method of obtaining the equation for solving such problems is the same as that in Sec. 4.9, and the reader should review that section at this point. It should be noted here that often a problem which can be solved by the use of a quadratic equation has only one solution, while the equation has two solutions. In such cases the root which does not satisfy the conditions of the problem is discarded.

Example 1   A rectangular building whose depth is twice its frontage is divided into two parts by a partition that is 30 feet from, and parallel to, the front wall. If the rear portion of the building contains 3500 square feet, find the dimensions of the building.

Solution   Let

x = the frontage of the building in feet

Then

$2x =$ the depth in feet

Also,

$2x - 30 =$ the length of the rear portion in feet

since partition is 30 feet from front wall

and

$x =$ the width of the rear portion

Consequently,

$x(2x - 30) =$ the area of the rear portion in square feet

since area of a rectangle is equal to the product of length and width

Therefore,

$x(2x - 30) = 3500$ **since area of the rear portion is 3500 square feet**

We solve this equation as follows:

$2x^2 - 30x - 3500 = 0$ **performing indicated operations and adding**
−3500 to each member

$(x - 50)(2x + 70) = 0$ **factoring left member**

$x = 50$ **setting x − 50 = 0 and solving**

$x = -35$ **setting 2x − 70 = 0 and solving**

Since, however, no dimension of the building can be negative, we discard −35 and have

$x = 50$ feet frontage

$2x = 100$ feet depth

Example 2 The periods of time required by two painters to paint a square yard of floor differ by 1 minute. Together they can paint 27 square yards in 1 hour. How long does it take each to paint 1 square yard?

Solution Let

$x =$ number of minutes required by faster painter to paint 1 square yard

Then

$x + 1 =$ number of minutes required by other painter

Consequently,

$\dfrac{1}{x} =$ fraction of a square yard the first paints in 1 minute

and

$$\frac{1}{x+1} = \text{fraction of a square yard the other paints in 1 minute}$$

Hence,

$$\frac{1}{x} + \frac{1}{x+1} = \text{fraction of a square yard painted by both in 1 minute}$$

However, since together they painted 27 square yards in 60 minutes, they covered $\frac{27}{60} = \frac{9}{20}$ square yard in 1 minute. Therefore,

$$\frac{1}{x} + \frac{1}{x+1} = \frac{9}{20}$$

Solving this equation, we have

$20(x + 1) + 20x = 9x(x + 1)$ **multiplying each member by 20x(x + 1)**

$20x + 20 + 20x = 9x^2 + 9x$ **by distributive axiom**

$-9x^2 + 31x + 20 = 0$ **adding −9x² − 9x to each member**

$$x = \frac{-31 \pm \sqrt{(31)^2 - 4(-9)(20)}}{2(-9)}$$

by quadratic formula

$$= \frac{-31 \pm \sqrt{961 + 720}}{-18}$$

$$= \frac{-31 \pm \sqrt{1681}}{-18}$$

$$= \frac{-31 \pm 41}{-18}$$

$$= -\tfrac{5}{9}, 4$$

We discard $-\frac{5}{9}$, since a negative time has no meaning in this problem. Hence,

$$x = 4 \quad \text{and} \quad x + 1 = 5$$

Thus, the painters require 4 and 5 minutes, respectively, to paint 1 square yard.

EXERCISE 6.7   Problems Solvable by Quadratic Equations

1   Find two consecutive positive integers whose product exceeds their sum by 19.

2   Find a number that is 56 less than its square.

3   Find two consecutive even integers whose product is 30 more than half the square of the larger.

4   Two numbers differ by 15, and their product is four times their sum. Find the numbers.

5 Find two integers that differ by 5, and the sum of whose squares is 25 times the smaller.

6 The sum of a number and its reciprocal is $\frac{13}{6}$. Find the number.

7 Two numbers differ by 5, and their reciprocals differ by $\frac{1}{10}$. Find the numbers.

8 The sum of two numbers is 17, and the sum of their squares is 169. Find the numbers.

9 The number of square yards in the area of a square exceeds the number of linear yards in the perimeter by 2300. Find the dimensions of the square.

10 The length and width of a rectangle differ by 5 feet, and the length of the diagonal is 25 feet. Find the dimensions.

11 A circular flower garden is surrounded by a walk 2 feet wide. If the combined areas of the walk and garden are 1.44 times the area of the garden, find the radius of the latter.

12 The length of a rectangular fish pond exceeds the width by 4 feet, and the pond is surrounded by a walk 4 feet wide. If the combined areas of the walk and the pond are 672 square feet, find the dimensions of the pond.

13 Airfield A is 660 miles due north of B. A plane flew from A to B and returned in $6\frac{2}{3}$ hours of flying time. If the wind was blowing from the south at 20 miles per hour during both flights, find the airspeed of the plane.

14 Airfield B is 480 miles north of field A, and field C is 400 miles east of B. A pilot flew from A to B, delayed an hour, and then flew to C. If the wind blew from the north at 20 miles per hour during the flight from A to B, but changed to the west at 20 miles per hour during the delay, and the entire trip required 6 hours, find the airspeed of the plane.

15 A highway patrolman left his headquarters and cruised at a constant speed for 28 miles, and then was notified of an accident. He drove to the scene of the accident 8 miles away at a speed that was 45 miles per hour faster than his cruising speed. If he had been on duty 54 minutes when he reached the accident, find his cruising speed.

16 A rancher drove 100 miles to a city to accept delivery of a new car, and returned in the new car. His average speed to the city was 10 miles per hour more than his returning speed, and the entire trip required $3\frac{2}{3}$ hours of driving time. Find the speed on each part of the trip.

17 A carpenter and her helper can build a garage in $2\frac{2}{3}$ days. Each of them working alone built similar garages, and the helper required 2 more days than the carpenter. How many days were required for each construction?

18 Two brothers washed the walls of their room in 3 hours. How long will it take each boy working alone to wash the walls of a similar room if the older boy can do the job in $2\frac{1}{2}$ hours less time than the younger?

19 A farmer and his helper, each driving a tractor, ploughed a tract of land in 6 days. Two years before, the helper ploughed the tract alone with the smaller tractor. The next year, the farmer ploughed the tract in 5 days less time with the larger tractor. How many days were required on each of the 2 years?

20 The outlet pipe drains a full irrigation reservoir in 2 hours less time than it takes the inlet pipe to fill it.   One day, at the start of an irrigation job when the reservoir was full, the farmer opened both pipes, and the reservoir was empty at the end of 24 hours.   In how many hours can the intake pipe refill the reservoir if the outlet is closed?

21   A square piece of lineoleum was 4 feet too short to cover the floor of a rectangular room with an area of 192 square feet.   Find the dimensions of a supplementary piece necessary to complete the coverage.

22   A clothesline 35 feet long is stretched diagonally between the corners of a rectangular service yard.   If the yard is enclosed by 98 feet of fencing, find its dimensions.

23   The outside dimensions of a framed picture are 20 by 18 inches.   Find the width of the frame if its area is one-fourth of the area enclosed by it.

24   A woman built a garage door whose width exceeded its height by 11 feet, and in the construction she used 252 linear feet of lumber 6 inches wide.   Find the dimensions of the door.

25   A builder used 54 linear feet of lumber to make the form for a concrete sidewalk 4 inches thick.   If the walk contained 24 cubic feet of concrete, find its dimensions.

26   A grocer sold two lots of eggs for $18 and $26, respectively, and there were 10 dozen more eggs in the latter than in the former.   If the price per dozen of the second lot was 5 cents more than that of the first, and the price of each was more than 50 cents, how many dozen were in each lot?

27   The brothers in a family bought a used car for $1200 and shared equally in the cost. After 6 months one of the brothers left home and sold his share to the others for $240. If each of the remaining brothers' shares in this purchase was $220 less than his share in the original price, find the number of brothers in the family.

28   The driver for a car pool estimated that the cost of driving her car to a plant was $6, which was divided equally among the passengers, including herself.   When two additional people joined the pool, the cost per person per trip was reduced by 50 cents.   How many people, including the driver, were in the original pool?

29   Someone bought two apartment buildings for $48,000 apiece.   There were 14 apartments in all, and the cost per apartment in one building was $2000 more than in the other.   How many apartments were in each building?.

30   The expense of the annual club party is shared equally by the members attending. In two consecutive years the expense was $100 and $105, respectively, and the share of each member was 25 cents less the second year.   How many members attended each year if the attendance the second year was 10 more than the first?

31   A swimming pool that holds 1800 cubic feet of water can be drained at a rate 15 cubic feet per minute faster than it can be filled.   If it takes 20 minutes longer to fill it than to drain it, find the drainage rate.

32   A contractor who owned two power shovels with different capacities agreed to do three excavation jobs, each of which required moving the same number of cubic feet of soil.   He started two of the jobs at the same time, using the larger shovel on the first and the smaller on the second.   When the first job was finished, he started the third job with the larger machine.   When the second job was finished 7 days later, he moved the smaller machine to the third job, and with the two machines operating, the excavation was finished in 8 days. How many days were required for the first excavation?

6.10 SUMMARY

We define the quadratic equation as $ax^2 + bx + c = 0$ where a, b, and c are constants. We then discuss solution by factoring, completing the square, and formula. The formula is $(-b \pm \sqrt{b^2 - 4ac})/2a$. By use of this, we find that the sum and product of the roots are $-b/a$ and c/a, respectively. We also find that the equation with roots r and s is $a(x - r)(x - s) = 0$. We further find that if the coefficients are rational the roots are conjugate complex numbers for $D < 0$, rational and equal for $D = 0$, and real unequal for $D > 0$. We also find that the roots are rational and unequal if D is the square of a rational number. The solution of radical equations, equations in quadratic form, and problems that lead to quadratic equations are also discussed.

EXERCISE 6.8 Review

Solve the equation in each of Probs. 1 to 6 by factoring.

1 $x^2 - 7x + 10 = 0$ **2** $x^2 + 2x - 3 = 0$ **3** $2x^2 - 11x + 12 = 0$

4 $3x^2 + 5x - 2 = 0$ **5** $6x^2 + 17x + 12 = 0$ **6** $12x^2 + x - 6 = 0$

Solve the equation in each of Probs. 7 to 12 by completing the square.

7 $2x^2 - x - 6 = 0$ **8** $6x^2 - 7x - 20 = 0$ **9** $x^2 - 2x - 2 = 0$

10 $4x^2 + 12x + 7 = 0$ **11** $4x^2 - 12x + 13 = 0$ **12** $9x^2 - 18x + 25 = 0$

Solve the equation in each of Probs. 13 to 20 by use of the quadratic formula.

13 $3x^2 - 7x - 6 = 0$ **14** $6x^2 + x - 15 = 0$ **15** $4x^2 - 12x + 3 = 0$

16 $25x^2 + 40x + 9 = 0$ **17** $9x^2 - 6x + 10 = 0$ **18** $16x^2 + 24x + 13 = 0$

19 $a^2x^2 - abx - 6b^2 = 0$ **20** $2x^2 + (a - b)x = a^2 - 2ab + b^2$

Solve the equation in each of Probs. 21 to 26 for x.

21 $4x^4 + 13x^2 - 12 = 0$ **22** $(2x^2 - 5x)^2 - 9(2x^2 - 5x) - 36 = 0$

23 $\left(\dfrac{x + 2}{2x - 1}\right)^2 + 2\dfrac{x + 2}{2x - 1} - 3 = 0$ **24** $\sqrt{2x + 3} = \sqrt{x + 1} + 1$

25 $\sqrt{x - 1} + \sqrt{3x + 1} = \sqrt{5x - 1}$ **26** $\sqrt{x^2 + x + 3} - \sqrt{x^2 - x - 1} = 2$

Find the sum, product, and type of roots of the equation in each of Probs. 27 to 30 without solving.

27 $8x^2 - 10x - 3 = 0$ **28** $4x^2 - 20x + 21 = 0$

29 $9x^2 + 12x + 7 = 0$ **30** $16x^2 - 8x - 5 = 0$

31 Two painters can paint the outside of a house in 6 days if they work together. How long will it take each working alone to do the job if one can do it in 5 days less than the other?

7
relations, functions, and graphs

We discussed the concept of one-to-one correspondence between the elements of two sets in Chap. 1. In this chapter, we shall consider a type of correspondence which leads to the concept of a relation, and a special case of it which leads to the important concept of a function. We shall make use of the terms *variable* and *constant*; hence, we begin this chapter with a review of the definitions of those terms.

7.1 RELATIONS

Variable
Constant

Domain

Range

If a symbol may take on the value of any element of a set, we say that the symbol is a *variable*. If there is only one element in the set, the symbol is a *constant*. If the symbol A represents a person's age, it may be any real number from zero to the age at death; if it represents the person's age to the nearest year, then A may be any nonnegative integer from 0 to the age at death; finally, if A is the age at death, then A is a constant. The set of values that a variable may take on its called the *domain* of the variable.

If two variables are so related that one or more values of the second are determined when a value in its domain is assigned to the first, we say there is a relation between them. The set of values taken on by the second variable is called its *range*.

Relation Any set of ordered pairs of numbers is called a *relation*.

Consequently, $\{(2, 1), (4, 3), (6, 7), (6, 9)\}$ is a relation, and so is $\{(2, 1), (4, 3), (6, 7), (8, 9)\}$.

Domain
Range

In a relation, the set of first numbers in the ordered pairs is the *domain* and the set of second numbers is called the *range*, in keeping with the definitions of those terms given above.

The relation between two variables may be determined by

various means, including an equation which involves the two variables.

Example 1   If a relation between x and y is defined by $y = r(x) = x^2 - 1$, find the value of y for $x = 0, 1, 3$.

Solution   All we need to do is replace x by 0, 1, and then 3 in the defining equation. If this is done, we have $y = r(0) = 0^2 - 1 = -1$ for $x = 0$, $y = r(1) = 1^2 - 1 = 0$ for $x = 1$; and $y = r(3) = 3^2 - 1 = 8$ for $x = 3$.

Example 2   If $y^2 = 25 - x^2$ defines the relation between x and y, find the values of y that correspond to $x = 0, 3, 4, 5$.

Solution   In order to find the desired values of y, we need only replace x in the defining equation by each given value of x, and solve the resulting equation for y. If this is done, we obtain the following table of corresponding values of x and y:

x	0	3	4	5
y	± 5	± 4	± 3	0

7.2   FUNCTIONS

If we refer to Example 2 of Sec. 7.1, we see that there are two values of y for some values of x. In Example 1, however, there is exactly one value of y for each value of x.

Function   If a relation is such that there is exactly one value of y for each value of x, we say that y is a *function* of x.

Thus, the equation $y = 5x - 3$ defines y as a function of x.

As a result of the definition of a function, we see that we may think of an ordered number pair as an element or building block of the function. The first number in the ordered pair is an element x of the domain D, and the second number is the corresponding element y of the range R. Quite often, the definition of a function is given as:

Function   If D is a set of numbers, if there exists a rule such that for each element x of D exactly one number y is determined, and if R is the set of all numbers y, then the set of ordered pairs $\{(x, y) \mid x \in D \text{ and } y \in R\}$ is a function with domain D and range R.

The essential difference between a relation and a function is that for a function there can be only one value of the second number for a given first number, whereas for a relation there can be several values of the second number for a given value of the first. Thus $y^2 = x$ is a relation but not a function, while $y = \sqrt{x}$ and $y = -\sqrt{x}$ are both functions.

The rule that establishes the correspondence that determines the function is often an equation which may be indicated by $y = f(x)$.

Example 1 Find the set of ordered pairs that make up the function determined by the equation $y = f(x) = x^2 - x + 1$ for $D = \{x \mid x$ is an integer and $-1 \leq x \leq 3\}$.

Solution The only numbers in the domain are $x = -1, 0, 1, 2,$ and $3,$ and we must find the value of y that corresponds to each of them by use of the defining equation $y = f(x) = x^2 - x + 1$. If $x = -1,$ then $y = f(-1) = (-1)^2 - (-1) + 1 = 3$. Therefore, $(-1, 3)$ is an element of the function. Similarly, $y = f(0) = 0^2 - 0 + 1 = 1,$ and $(0, 1)$ is another element of the function. We find in like manner that $f(1) = 1, f(2) = 3,$ and $f(3) = 7$. Consequently, the function is

$$\{(-1, 3), (0, 1), (1, 1), (2, 3), (3, 7)\}$$

Example 2 If $D = \{1, 3, 6, 10\}$, find the ordered pairs that make up $\{[x, f(x)] \mid f(x) = x^2 - 7x - 2 \text{ and } x \in D\}$.

Solution We must find the value of $f(x)$ which corresponds to each element x of D by substituting for x in $f(x)$. Thus,

$$f(1) = 1^2 - 7(1) - 2 = 1 - 7 - 2 = -8$$
$$f(3) = 3^2 - 7(3) - 2 = 9 - 21 - 2 = -14$$
$$f(6) = 6^2 - 7(6) - 2 = 36 - 42 - 2 = -8$$
$$f(10) = 10^2 - 7(10) - 2 = 100 - 70 - 2 = 28$$

Consequently, the function is $\{(1, -8), (3, -14), (6, -8), (10, 28)\}$.

Independent Variable
Dependent Variable
In the function $\{(x, y) \mid y = f(x)\}$ defined by the equation $y = f(x)$, x is often called the *independent variable*, and y is the *dependent variable*. Thus, the independent variable in a function is the variable whose replacement set is the set of first numbers in the ordered pairs that make up the function. The dependent variable is the one whose replacement set is the set of second numbers in the ordered pairs. If the domain is not specified, it is understood to be the set R of real numbers.

EXERCISE 7.1 Relations and Functions

1 Is $\{(1, 7), (2, 8), (3, 6), (3, 6)\}$ a function? Why?

2 Is $\{(7, 1), (8, 2), (6, 3), (9, 3)\}$ a function? Why?

3 Is $\{(0, 1), (1, 2), (2, 3), (3, 0)\}$ a function? Why?

4 Is $\{(3, 2), (3, -1), (5, .99), (10, 1.01)\}$ a function? Why?

5 Find the set of ordered pairs obtained by pairing the nth element of $\{2, 4, 8, 16\}$ with the $(n + 1)$st element of $\{1, 3, 7, 13, 21\}$.

6 Find the set of ordered pairs obtained by pairing the nth element of $\{b, a, s, e\}$ with the $(4 - n)$th element of $\{b, a, l, l\}$ for $n = 1, 2, 3$. Is this a function? Why?

7 Find the set of ordered pairs obtained by pairing the nth element of $\{s, h, e, e, r\}$ with the nth element of $\{c, l, i, f, f\}$. Why is this not a function?

8 Find the set of ordered pairs obtained by pairing the nth element of $\{m, a, t, h\}$ with the $(4 - n)$th element of $\{w, h, i, z\}$, for $n = 1, 2, 3$. Is this a function?

9 Is the set or ordered pairs $\{(x, x^2) \mid x \in \{0, 1, 2, 3\}\}$ a function? Why?

10 Is the set of ordered pairs $\{(x, \sqrt{x}) \mid x \in \{0, 1, 4, 9\}\}$ a function? Why?

11 Is the set of ordered pairs $\{(x, y) \mid y^2 = x, x \in \{0, 1, 4, 9\}\}$ a function? Why?

12 Is the set of ordered pairs $\{(x, y) \mid y^2 = x^2, x \in \{1, 4, 9, 16\}\}$ a function? Why?

13 If $f(x) = 2x - 3$, find $f(0)$, $f(2)$, and $f(5)$.

14 If $f(x) = 6 - 5x$, find $f(1)$, $f(3)$, and $f(7)$.

15 Find $f(-2)$, $f(0)$, and $f(3)$ if $f(x) = x^2 + x - 2$.

16 Find $f(-3)$, $f(1)$, and $f(4)$ if $f(x) = 2x^2 - 7x + 3$.

In Probs. 17 to 20, find the range for the given function and domain.

17 $f = \{(x, 2x + 1)\}$; $D = \{1, 3, 5, 7\}$

18 $f = \{(x, 3x - 5)\}$; $D = \{0, 1, 2, 4\}$

19 $f = \{(x, \sqrt{9 - x})\}$; $D = \{0, 2, 5, 9\}$

20 $f = \{(x, \sqrt{37 - x^2})\}$; $D = \{1, 2\sqrt{3}, \sqrt{21}, 2\sqrt{7}, \sqrt{37}\}$

Write out the ordered pairs determined in each of Probs. 21 to 28, and state whether it is a function.

21 $\{(x, 3x - 2) \mid x \text{ is an integer and } -1 \leq x < 3\}$

22 $\{(x, 2x + 1) \mid x \text{ is an odd integer and } -6 < x < 0\}$

23 $\{(x, x^2 - 3) \mid x \text{ is an integer and } -2 \leq x \leq 2\}$

24 $\{(x, 2x^2 - 3x + 4) \mid x \text{ is divisible by 3 and } -4 \leq x < 9\}$

25 $\{(x, y) \mid y = x^2 - x + 4, D = \{-1, 0, 2, 4\}\}$

26 $\{(x, y) \mid y = x^3 - x^2 + 5, D = \{-1, 0, 1, 2\}\}$

27 $\{(x, y) \mid y = 5x + 4, D = \{-3, -1, 0, 1, 2\}\}$

28 $\{(x, y) \mid y = 3x + 5, D = \{-2, -1, 0, 3\}\}$

29 Find $\dfrac{f(x + h) - f(x)}{h}$ if $f(x) = 3x + 1$.

30 Find $\dfrac{f(x + h) - f(x)}{h}$ if $f(x) = 2x - 5$.

31 If $f(x) = x^2 - 2x + 3$, find $\dfrac{f(x + h) - f(x)}{h}$.

32 If $f(x) = x^2 + 3x - 1$, find $\dfrac{f(x + h) - f(x)}{h}$.

In Probs. 33 to 36, find $f \cup g$ and $f \cap g$.

33 $f = \{(x, x - 2)\}; g = \{(x, 2x + 1)\}; D = \{-3, -1, 3, 5\}$

34 $f = \{(x, x + 1)\}; g = \{(x, 2x - 3)\}; D = \{-4, -2, 1, 4\}$

35 $f = \{(x, 3x^2 + 2x - 3)\}; g = \{(x, 2x^2 + x - 1)\}; D = \{-2, -1, 0, 1\}$

36 $f = \{(x, 2x^2 + 3x - 1)\}; g = \{(x, x^2 + 2x - 2)\}; D = \{-2, 0, 1, 3\}$

7.3 THE RECTANGULAR COORDINATE SYSTEM

In this section we introduce a device for associating an ordered pair of numbers with a point in a plane. Invented by the French mathematician and philosopher Rene Descartes (1596–1650), it is called the *rectangular* or *cartesian coordinate system.*

In order to set up this system, we construct two perpendicular number lines in the plane and choose a suitable scale on each. For convenience these lines are horizontal and vertical, and the unit length on each is the same (see Fig. 7.1), although neither restriction is necessary. The two lines are called the *coordinate axes*, the horizontal line being the X axis and the vertical line the Y axis. The intersection of the two lines is the *origin*, designated by the letter O. The coordinate axes separate the plane into four sections called *quadrants*. These quadrants are numbered I, II, III, and IV counterclockwise, as indicated in Fig. 7.1*a*.

Axes

Origin

Quadrant

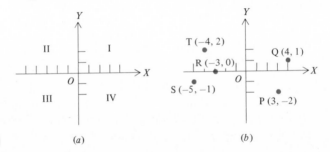

FIGURE 7.1 (a) (b)

Cartesian Plane   A plane in which the coordinate axes have been constructed is called a *cartesian plane*.

Directed Distance   Next, it is agreed that horizontal distances measured to the right from the Y axis are positive, and horizontal distances measured to the left are negative. Similarly, vertical distances measured upward from the X axis are positive, and vertical distances measured downward are negative. These distances, because of their signs, are called *directed distances*. Finally, we agree that the first number in an ordered pair of numbers represents the directed distance from the Y axis to a point, and the second number in the pair represents the directed distance from the X axis to the point. It follows then that an ordered pair of numbers uniquely determines the position of a point in the plane. For example, (4, 1) determines the point Q in Fig. 7.1b that is 4 units to the right of the Y axis and 1 unit above the X axis. Similarly, the ordered pair $(-5, -1)$ determines the point S in Fig. 7.1b that is 5 units to the left of the Y axis and 1 unit below the X axis. Conversely, each point in the plane determines a unique ordered pair of numbers. For example, the point P in Fig. 7.1b is 3 units to the right of the Y axis and 2 units below the X axis, and so P determines the ordered pair $(3, -2)$.

Abscissa

Ordinate   The two numbers in an ordered pair that is associated with a point in the cartesian plane are called the *coordinates* of the point. The first number is called the *abscissa* of the point, and it is the directed distance from the Y axis to the point. The second number in the pair is the *ordinate* of the point, and it represents the directed distance from the X axis to the point.

Plotting   The procedure for locating a point in the plane by means of its coordinates is called *plotting* the point. The notation $P(a, b)$ means that P is the point whose coordinates are (a, b). In order to plot the point $T(-4, 2)$ we count 4 units to the left of the origin on the X axis and then upward 2 units and thus arrive at the point. Similarly, the point $R(-3, 0)$ is 3 units to the left of the origin and on the X axis. The general point and its coordinates are written $P(x, y)$.

7.4   THE GRAPH OF A FUNCTION AND OF A RELATION

By use of the rectangular coordinate system, we can obtain a geometric representation, or a geometric "picture" of a function. For this purpose, we require that each ordered pair of numbers (x, y) of a function be the coordinates of a point in the cartesian plane with x as the abscissa and y as the ordinate. We then define the graph of a function as follows.

Graph of
a Function
The *graph of a function* is the totality of points (x, y) whose coordinates constitute the set of ordered pairs of the function, with x a number in the domain D and y the corresponding number in the range R.

The graphs of most functions we shall discuss in this chapter are smooth† continuous curves. When we say that the graph of a function is a curve, we mean that the point determined by each ordered pair of numbers in the function is on the curve and that the coordinates of each point on the curve are an ordered pair of numbers in the function.

Example 1
We shall illustrate the procedure for obtaining the graph of a function by explaining the steps in the construction of the graph of the function defined by $y = x^2 - 3x - 1$ for $-2 \leq x \leq 5$. Note that this function is

$$\{(x, y) \mid y = x^2 - 3x - 1\} \text{ and } x \text{ belongs to } D = \{x \mid -2 \leq x \leq 5\}\}$$

The first step is to assign each of the integers in D to x and then calculate each corresponding value of y by using the defining equation $y = x^2 - 3x - 1$. Before doing this, however, it is advisable to make a table like the one below in which to record the corresponding values.

Now we shall calculate the corresponding value of y when x is assigned the numbers $-2, -1, 0, 1, 2, 3, 4,$ and 5.

$$y = \begin{cases} (-2)^2 - 3(-2) - 1 = 9 & \text{for } x = -2 \\ (-1)^2 - 3(-1) - 1 = 3 & \text{for } x = -1 \\ 0^2 - 3(0) - 1 = -1 & \text{for } x = 0 \end{cases}$$

Continuing, we obtain the additional ordered pairs $(1, -3)$, $(2, -3)$, $(3, -1)$, $(4, 3)$, and $(5, 9)$ and enter the results in the table:

x	-2	-1	0	1	2	3	4	5
y	9	3	-1	-3	-3	-1	3	9

Now we plot the points (x, y) thus determined, as shown in Fig. 7.2. Because there is an ordered pair in the function for every intermediate real value of x, we connect the plotted points with a smooth curve. This is the portion of the graph defined by $y = x^2 - 3x - 1$ over the specified domain.

---

†At present we are not in a position to give a rigorous definition of a "smooth continuous curve." For our purposes, however, the following *description* will suffice. A smooth continuous curve contains no breaks or gaps, and there are no sudden or abrupt changes in its direction.

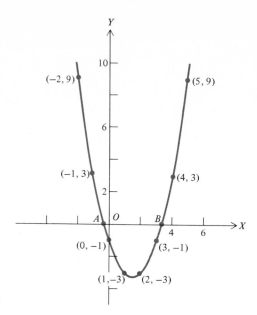

FIGURE 7.2

We now investigate the nature of the graph as x increases through values greater than 5 and decreases through values less than -2. For this purpose we first replace x in $y = x^2 - 3x - 1$ by $5 + h$ and get $y = h^2 + 7h + 9$. From the latter equation we see that as h increases from zero, y increases from 9. Hence the graph extends upward and to the right from the point (5, 9).

Similarly, if we replace x by $-2 - k$, we get $y = k^2 + 7k + 9$, and we therefore conclude that as k increases from zero, y increases from 9. Hence the graph extends upward and to the left from the point $(-2, 9)$.

Zero   A *zero* of a function defined by $y = f(x)$ is a value of the independent variable x for which $y = 0$. Hence the zeros of a function are the abscissas of the points where the graph crosses the X axis. The zeros of many classes of functions can be obtained by algebraic methods, but we must depend upon graphical methods for others. The zeros of the function $\{(x, y) \mid y = x^2 - 3x - 1\}$ are the abscissas of the points A and B in Fig. 7.2 and are approximately -0.3 and 3.3.

Roots   The zeros of a function are called the *roots* of the equation which defines the function.

If the domain of a function is not specified, it is assumed to be the real number system. In such cases, to get a set of ordered pairs for constructing the graph, it is usually advisable to start by assigning consecutive small integers to x, and to continue the process until a sufficient number of points are obtained to determine the nature of the graph. At times the points obtained by assigning consecutive integers to x are too far apart to enable one to sketch the curve. For example, in the function defined by $y = 4x^2 - 1$, if we assign -1, 0, and 1 to x, we get the pairs $(-1, 3)$, $(0, -1)$, and $(1, 3)$. These pairs determine the points A, B, and C in Fig. 7.3, and it is evident that these points alone do not show the nature of the graph. We can get two more points by assigning $-\frac{1}{2}$ and $\frac{1}{2}$ to x and obtaining the pairs $(-\frac{1}{2}, 0)$ and $(\frac{1}{2}, 0)$. When these points are plotted, we get the points D and E, and then the curve in Fig. 7.3 can be sketched.

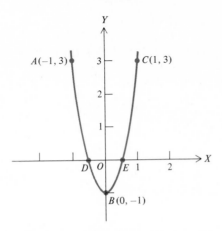

FIGURE 7.3

The graph of a function is cut in only one place by a line parallel to the Y axis, whereas the graph of a relation that is not a function is cut in more than one place by one or more lines that are parallel to the Y axis.

Example 2   Sketch the graph of $y^2 = x + 4$.

Solution   We begin by assigning values to x and finding each corresponding value of y. We must not assign a value less than -4 to x if y is to be real. If we assign the integers from -4 to 5 to x, the values of y are as follows.

x	-4	-3	-2	-1	0	1	2	3	4	5
y^2	0	1	2	3	4	5	6	7	8	9
y	0	± 1	$\pm\sqrt{2}$	$\pm\sqrt{3}$	± 2	$\pm\sqrt{5}$	$\pm\sqrt{6}$	$\pm\sqrt{7}$	$\pm 2\sqrt{2}$	± 3

We now locate the points (x, y) determined by the table and draw a smooth curve through them. We thus have the graph shown in Fig. 7.4.

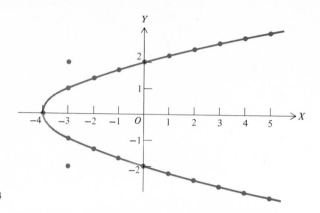

FIGURE 7.4

Example 3 Sketch the graph of the relation determined by $4x^2 + y^2 = 16$.

Solution We make a table of corresponding values of x and y as usual, locate the points determined, draw a smooth curve through them, and thereby get the graph shown in Fig. 7.5. It is cut in two places by any vertical line between $x = -2$ and $x = 2$.

x	-2	-1	0	1	2
y^2	0	12	16	12	0
y	0	$\pm 2\sqrt{3}$	± 4	$\pm 2\sqrt{3}$	0

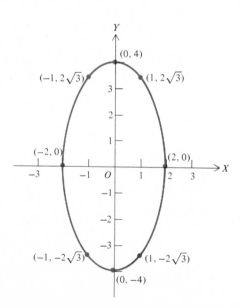

FIGURE 7.5

7.5 LINEAR FUNCTIONS

It is proved in analytic geometry that if the domain is the set of real numbers, the graph of the function $\{(x, y) \mid y = ax + b\}$ is a

Linear Function straight line. Such a function is called a *linear function*, and its graph is completely determined by two points. If the graph does not pass through the origin or is not parallel to either axis, the points where the graph crosses the axes are readily determined by assigning 0 to x in the equation $y = ax + b$ and obtaining the point $(0, b)$, and then assigning 0 to y and solving for x and getting $(-b/a, 0)$. These points determine the graph, but it is advisable to calculate a third point as a check.

Intercepts The abscissa of the point where the line crosses the X axis is called the X *intercept*, and the ordinate of the point where it crosses the Y axis is the Y *intercept*. The graph of $\{(x, y) \mid y = ax\}$ passes through the origin since $(0, 0)$ satisfies the equation $y = ax$. Hence, we must assign a number other than zero to x to get a second point to determine the graph.

The graph of $\{(x, y) \mid x = a\}$ is the line parallel to the Y axis at the directed distance of a units from it. The graph of $\{(x, y) \mid y = b\}$ is the line parallel to the X axis and at the directed distance of b units from it. These graphs are shown in Fig. 7.6a.

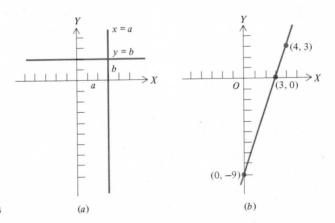

FIGURE 7.6 (*a*) (*b*)

Example Construct the graph of the function defined by $y = 3x - 9$.

Solution We find the intercepts by assigning 0 to x and solving for y and by assigning 0 to y and solving for x. We find a third point by assigning 4 to x and solving for y. Thus we get the following table of corresponding numbers:

x	0	3	4
y	-9	0	3

We now plot the points determined by these pairs of numbers, draw a straight line through them, and obtain the graph in Fig. 7.6b.

EXERCISE 7.2   Graphs

1   Plot the points determined by the following ordered pairs of numbers:

$(3, 2)$, $(0, 6)$, $(-2, 4)$, $(-7, 0)$, $(4, 0)$, $(5, -4)$, $(-6, 0)$, $(-3, -5)$.

2   Describe the line on which each of the following sets of points is located:
   (a) The two coordinates of each point are equal but of opposite sign.
   (b) The two coordinates are numerically equal, with the first positive and the second negative.
   (c) The abscissa of each point is zero.
   (d) The ordinate of each point is 3.

3   Describe the line on which each of the following sets of points is located:
   (a) The two coordinates of each point are equal.
   (b) The two coordinates of each point are equal and nonnegative.
   (c) The abscissa of each point is 4.
   (d) The ordinate of each point is -2.

4   If n is a negative number, give the quadrant in which each of the following points is located: $(n, 2)$, $(n, -3)$, $(-n, -1)$, $(-n, 6)$, $(3, n)$, $(-2, n)$, $(-n, -n)$, $(n, -n)$.

Construct the graph of the function or relation defined by the equation in each of Probs. 5 to 24.

5   $y = 2x - 1$ 6 $y = 3x + 4$ 7 $y = 5x - 3$

8   $y = 3x + 6$ 9 $y = -x + 2$ 10 $y = -3x - 1$

11   $y = -4x - 3$ 12 $y = -2x + 3$ 13 $y = x^2$

14   $y = -x^2$ 15 $y = -2x^2$ 16 $y = 4x^2$

17   $y = 4x^2 + 4x + 2$ 18 $y = -2x^2 + 3x - 1$ 19 $y = -3x^2 + 2x - 4$

20   $y = 5x^2 + 7x - 4$ 21 $y^2 = 16 - x^2$; $D = \{x \mid -4 \le x \le 4\}$

22   $y^2 = 9 - x^2$; $D = \{x \mid -3 \le x \le 3\}$

23   $y^2 = 25 - x^2$; $D = \{x \mid 0 \le x \le 5\}$

24   $y^2 = 36 - x^2$; $D = \{x \mid -6 \le x \le 0\}$

In each of Probs. 25 to 28, find both intercepts and then sketch the curve.

25   $\{(x, y) \mid y = 3x + 6\}$; $D = \{x \mid -4 \le x \le 1\}$

26   $\{(x, y) \mid y = 2x - 5\}$; $D = \{x \mid 1 \le x \le 5\}$

27   $\{(x, y) \mid y = -x - 4\}$; $D = \{x \mid -6 \le x \le 1\}$

28   $\{(x, y) \mid y = -2x + 7\}$; $D = \{x \mid 0 \le x \le 4\}$

7.6   SOME SPECIAL FUNCTIONS

In this section, we shall discuss some functions whose graphs are not continuous curves, and some which are. We begin by considering

$$f = \{(x, y) \mid y = x - 2, x \geq 1\} \cup \{(x, y) \mid y = -x + 1, x < 1\}$$

The graph is defined by the equations

$$y = x - 2 \qquad \text{for } x \geq 1 \tag{1}$$
$$y = -x + 1 \qquad \text{for } x < 1 \tag{2}$$

The graph of $y = x - 2$ is a straight line with x intercept 2 and y intercept -2, but we are interested only in the part for which $x \geq 1$; hence, the part of the graph in which we are interested is the ray that begins at $(1, -1)$ and extends upward and to the right as shown in Fig. 7.7. The graph of $y = -x + 1$ is also a line, and we are interested only in the part for which $x < 1$; hence, the ray that we want begins at $(1, 0)$ but does not include that point and extends to the left and upward as shown in Fig. 7.7.

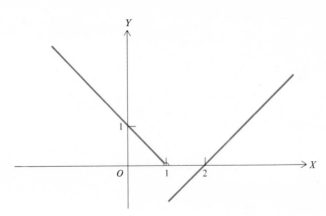

FIGURE 7.7

Bracket Function   Another interesting function is the bracket function. It is important in the theory of numbers and is represented by $\{(x, y) \mid y = [x]\}$. By $[x]$, we mean the largest integer than is smaller than or equal to the number x. Thus, $[\frac{1}{2}] = 0$ since zero is the largest integer that is smaller than or equal to $\frac{1}{2}$; furthermore, $[1.8] = 1$; $[4.01] = 4$; and $[n] = n$ if n is a positive integer. Values of $y = [x]$ for several values of x are shown below:

$$y = \begin{cases} 0 & \text{for } 0 \leq x < 1 \\ 1 & \text{for } 1 \leq x < 2 \\ 2 & \text{for } 2 \leq x < 3 \\ \cdots\cdots\cdots\cdots\cdots \\ n - 1 & \text{for } n - 1 \leq x < n, \ n \text{ an integer} \end{cases}$$

Example 1   Sketch the graph of the bracket function $y = [x]$.

Solution   The values of y for a variety of values of x are shown just above this example. We note that y has the same value for a variation of 1 in x, beginning with each integer. The graph is shown in Fig. 7.8.

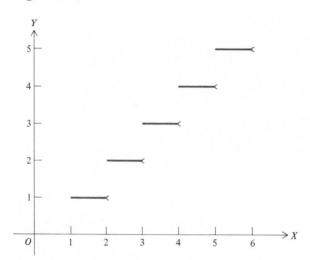

FIGURE 7.8

Example 2   Sketch the graph of $y = |x - 1|$.

Solution   If we make a table in the usual manner, beginning with $x = -3$ and continuing to $x = 4$, we obtain

x	-3	-2	-1	0	1	2	3	4
$x - 1$	-4	-3	-2	-1	0	1	2	3
y	4	3	2	1	0	1	2	3

Now, locating the points determined by the table and drawing a curve through them, we get the curve shown in Fig. 7.9.

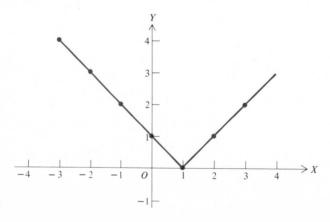

FIGURE 7.9

7.7   THE CUBIC AND QUARTIC

We shall now consider the cubic equation $y = ax^3 + bx^2 + cx + d$ and the quartic $y = ax^4 + bx^3 + cx^2 + dx + e$. There are at most three values of x for a given value of y in the cubic, and at most four values of x for a given value of y in the quartic.

Example 1   Sketch the graph of $y = 6x^3 - 11x^2 - 4x + 4$ for x between -1.5 and 2.5.

Solution   If we assign the endpoints and the integers in the designated interval to x and determine each value of y, we get the following table.

x	-1.5	-1	0	1	2	2.5
y	-35	-9	4	-5	0	19

We get the curve shown in Fig. 7.10 by locating these points and drawing a smooth curve through them.

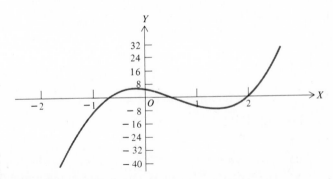

FIGURE 7.10

Example 2   Sketch the graph of $y = 2x^4 + 3x^3 - 4x^2 - 3x + 2$ from $x = -2.5$ to $x = 1.5$.

Solution   If we assign the endpoints and the integers in the given interval to x and calculate each corresponding value of y, we get the following table.

x	-2.5	-2	-1	0	1	1.5
y	15.75	0	0	2	0	8.75

Now, locating the points and drawing a smooth curve through them, we obtain the curve shown in Fig. 7.11.

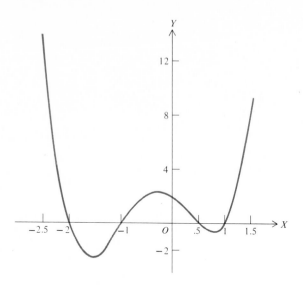

FIGURE 7.11

7.8 THE INVERSE OF A FUNCTION

We considered relations in Sec. 7.1 and functions in Sec. 7.2. We shall now consider the two relations:

$$f = \{(1, 2), (3, 4), (5, 6)\} \quad \text{and} \quad g = \{(1, 2), (3, 4), (5, 4)\}.$$

The domain of each is $\{1, 3, 5\}$, but the range of f is $\{2, 4, 6\}$ whereas that of g is $\{2, 4\}$.

If we interchange the first and second elements of f, we obtain

$$F = \{(2, 1), (4, 3), (6, 5)\}$$

This is a function with domain $\{2, 4, 6\}$ and range $\{1, 3, 5\}$. If, however, we interchange first and second elements in g, we get $\{(2, 1), (4, 3), (4, 5)\}$; this is a relation but not a function, since two pairs have the same first element 4 and different second elements 3 and 5.

Inverse
Relation

Inverse
Function

The set of ordered pairs obtained by interchanging the elements in each ordered pair of a function is called the *inverse relation*. Furthermore, if the inverse relation is a function, it is called the *inverse function.*

If the function is designated by f, we use f^{-1} to indicate the inverse function. Therefore, if $f = \{(1, 2), (3, 4), (5, 6)\}$, then $f^{-1} = \{(2, 1), (4, 3), (6, 5)\}$.

It follows from the definition of a function that if a function f is such that no two of its ordered pairs with different first elements have the same second element, the inverse function f^{-1} exists.

If the function is expressed as a set of ordered pairs, then the inverse function is obtained by interchanging the first and second elements of each ordered pair in f.

If the function is designated by

$$f = \{(x, y) \mid y = f(x)\}$$

Finding f^{-1} then $y = f(x)$ defines the function, and $x = f(y)$ defines the inverse function since it has the effect of interchanging elements of each ordered pair. If $x = f(y)$ can be solved for y, the solution is ordinarily put in the form $y = f^{-1}(x)$, the inverse relation is designated by

$$f^{-1} = \{(x, y) \mid y = f^{-1}(x)\}$$

and it is the inverse function if $y = f^{-1}(x)$ is a function.

Example 1 Find the inverse of $f = \{(x, y) \mid y = 3x - 6\}$ and sketch the graphs of the function and its inverse.

Solution Since $y = 3x - 6$ defines the function, it follows that $x = 3y - 6$ defines the inverse relation. Now, solving for y gives $y = \frac{1}{3}x + 2$. Consequently,

$$f^{-1} = \{(x, y) \mid y = \tfrac{1}{3}x + 2\}$$

is the inverse, and it is a function since there is only one y for each x. Both graphs are shown in Fig. 7.12.

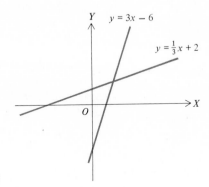

FIGURE 7.12

Example 2 If $g = \{(x, y) \mid y = g(x) = \sqrt{x^2 + 9}\}$, $x \geq 0$, find g^{-1}.

Solution The function g is defined by the equation $y = \sqrt{x^2 + 9}$; hence, the inverse is defined by $x = \sqrt{y^2 + 9}$ and $x \geq 3$. Now, solving this equation for y gives

$$x^2 = y^2 + 9 \qquad \text{squaring each member}$$
$$y^2 = x^2 - 9 \qquad \text{solving for } y^2$$
$$y = \sqrt{x^2 - 9} \qquad \text{for } x \geq 3$$

Hence, $g^{-1} = \{(x, y) \mid y = g^{-1}(x) = \sqrt{x^2 - 9}\}$ is the inverse, and it is a function. Notice that the domain of g^{-1} is $x \geq 3$, and its range is $y \geq 0$; notice further that the range of g is $y \geq 3$, and its domain is $x \geq 0$, the same as the range of g^{-1}.

The graphs of the function and its inverse are shown in Fig. 7.13. Each is the reflection of the other in the line $y = x$.

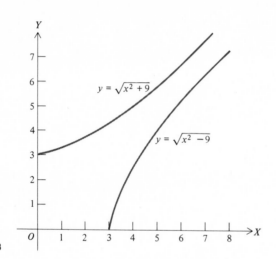

FIGURE 7.13

EXERCISE 7.3   Special Functions and Inverses

Construct the graphs of the functions defined by the equations in Probs. 1 to 28.

1 $y = \begin{cases} x & \text{for } x \geq 1 \\ x - 1 & \text{for } x < 1 \end{cases}$

2 $y = \begin{cases} x + 2 & \text{for } x > 0 \\ x + 1 & \text{for } x \leq 0 \end{cases}$

3 $y = \begin{cases} 2x - 3 & \text{for } x > -1 \\ 3x - 2 & \text{for } x \leq -1 \end{cases}$

4 $y = \begin{cases} 2x - 3 & \text{for } x \geq 2 \\ 3x - 2 & \text{for } x < 2 \end{cases}$

5 $y = \begin{cases} x^2 & \text{for } x \leq 0 \\ -x & \text{for } x > 0 \end{cases}$

6 $y = \begin{cases} \sqrt{x^2 - 9} & \text{for } x \geq 3 \\ x - 3 & \text{for } x < 3 \end{cases}$

7 $y = \begin{cases} 2x - 5 & \text{for } x > 4 \\ \sqrt{16 - x^2} & \text{for } 0 \leq x \leq 4 \end{cases}$

8 $y = \begin{cases} x^2 - x & \text{for } x \geq 1 \\ -x + 1 & \text{for } x < 1 \end{cases}$

9 $y = |x|$

10 $y = |-x|$

11 $y = |x| + 1$

12 $y = |x| + x$

13 $y = |x - 1| + |x + 2|$

14 $y = |x - 1| - |x - 2|$

15 $y = |x + 3| + |x + 1|$

16 $y = |x - 2| - |x + 1|$

17 $y = [x]$ for $0 \leq x \leq 4$

18 $y = [x + 2]$ for $-1 \leq x \leq 3$

19 $y = [x] + 2$ for $-1 \leq x \leq 3$.

20 $y = |[x]|$ for $-2 \leq x \leq 2$

21 $y = 3x^3 + 5x^2 + 4x - 4$

22 $y = 3x^3 - x^2 - 8x - 4$

23 $y = 2x^3 + 5x^2 + x - 2$

24 $y = 2x^3 - 5x^2 + x + 2$

25 $y = 2x^4 - x^3 - 8x^2 + x + 6$

26 $y = 2x^4 + 5x^3 - 2x^2 - 10x - 4$

27 $y = 2x^4 + x^3 - 4x^2 + x - 6$

28 $y = 3x^4 - x^3 + 4x^2 - 2x - 4$

Find the equation which defines the inverse of the function defined by the equation in each of Probs. 29 to 40. Solve the equation of the inverse for y, and find the domain of the inverse.

29 $y = x + 1, x \geq 2$

30 $y = -x + 3, x \geq 3$

31 $y = 2x - 5, x \geq 2.5$

32 $y = 3x + 9, x \leq 1$

33 $y = \dfrac{x + 1}{x - 1}, x \geq 2$

34 $y = \dfrac{2x - 3}{x + 2}, x \geq 0$

35 $y = \dfrac{x + 3}{x}, x > 0$

36 $y = \dfrac{x}{x - 4}, x < 4$

37 $y = 2 + \sqrt{x - 1}, x \geq 1$

38 $y = 5 - \sqrt{x + 3}, x \geq -3$

39 $y = 3 - \sqrt{2x + 5}, x \geq -1$

40 $y = 1 + \sqrt{3x - 2}, x \geq 1$

Find the equation which defines the inverse of the relation defined by the equation in each of Probs. 41 to 44. State whether the inverse is a function or a relation, and give its domain.

41 $y^2 = x^2 - 4, 2 \leq x \leq 5$

42 $y^2 = 4 - x^2, -2 \leq x \leq 2$

43 $y^2 = 4x - 12, x \geq 3$

44 $y^2 = -3x + 6, x \leq 2$

In each of Probs. 45 to 48, show that the domain and range of f are the range and domain, respectively, of f^{-1}.

45 $f = \{(x, y) \mid y = \sqrt{x + 2}, x = -2, 2, 7, 14\}$

46 $f = \{(x, y) \mid y = \sqrt{x^2 - 1}, x = 1, \sqrt{2}, \sqrt{5}, \sqrt{10}\}$

47 $f = \{(x, y) \mid y = 2x + 3, -1 \leq x \leq 3\}$

48 $f = \{(x, y) \mid y = 3x - 1, 1 \leq x \leq 5\}$

In each of Probs. 49 to 52, show that $f[f^{-1}(x)] = f^{-1}[f(x)] = x$.

49 $f = \{(x, y) \mid y = 2x - 6\}$

50 $f = \{(x, y) \mid y = 3x + 1\}$

51 $f^{-1} = \{(x, y) \mid y = \sqrt{x}, x \geq 0\}$

52 $f^{-1} = \{(x, y) \mid y = \sqrt{x - 1}, x \geq 1\}$

7.9 SUMMARY

The first section of this chapter gives the definitions of variable, constant, and relation. In the next section we give the definition of a function as a special case of a relation and then as a set of ordered pairs. We then discuss the rectangular coordinate system, the procedure for sketching the graph of a function, the graphs of some special functions, and, finally, the inverse of a function.

EXERCISE 7.4   Review

1 Is $\{(2, 1), (4, 2), (6, 3), (5, 5)\}$ a function? Why?

2 Is $\{(2, 1), (4, 2), (6, 3), (5, 3)\}$ a function? Why?

3 Find the set of ordered pairs obtained by associating with each odd number on the face of a clock, the even integer that is one larger than it.

4 Is $\{(x, x^2) \,|\, x \in \{1, 2, 3, 5\}\}$ a function? Why?

5 Is $\{(x^2, x) \,|\, x \in \{1, 2, 3, 5\}\}$ a function? Why?

6 Is $\{(x, y) \,|\, y^2 = x, x = 1, 2, 3, 5\}$ a function? Why?

7 Find the range of $f = \{(x, 2x - 1)\}$, $D = \{-1, 0, 1, 2\}$.

8 If $f(x) = 3x^2 - 2x + 5$, find $f(2 + h) - f(2)$.

9 If $f = \{(x, 3x + 5)\}$ and $g = \{(x, x + 1)\}$, $D = \{-3, -2, -1, 0\}$, find $f \cap g$

Sketch the graph of the function defined by the equation in each of Probs. 10 to 22.

10 $\{(x, y) \,|\, y = 2x - 3, -2 \le x \le 5\}$

11 $\{(x, y) \,|\, y = 2x^2 + x + 1, -3 \le x \le 2\}$

12 $\{(x, y) \,|\, y = \sqrt{49 - x^2}, 0 \le x \le 7\}$

13 $y = |x| - x$

14 $y = [x] + 3$

15 $y = \begin{cases} x + 2 & \text{for } x \ge 1 \\ x - 1 & \text{for } x < 1 \end{cases}$

16 $y = \begin{cases} 3x + 1 & x > 3 \\ \sqrt{9 - x^2} & 0 \le x \le 3 \end{cases}$

17 $y = \begin{cases} x^2 + 4 & \text{for } -2 \le x \le 2 \\ x + 1 & \text{for } x > 2 \end{cases}$

18 $y = \begin{cases} \sqrt{x^2 - 4} & \text{for } 2 \le x < 4 \\ x - 2 & \text{for } x < 2 \end{cases}$

19 $y = |x - 3| + |x + 1|$

20 $y = |x + 2| - |x - 1|$

21 $y = 2x^3 - 5x^2 + 3x + 2$

22 $y = 2x^4 - 2x^3 - 3x^2 - x - 2$

Find the equation which defines the inverse of the function defined in each of Probs. 23 to 26.   Find the domain of the inverse.

23 $y = 2x - 1, x \ge 0.5$

24 $y = (x + 2)/(x - 2), x > 4$

25 $y = \sqrt{4 - x^2}, -2 \le x < 0$

26 $y = 2 + \sqrt{x - 3}, x \ge 3$

27 Show that the domain and range of $\{(x, y) \,|\, y = x^2 - 1\}$, $D = \{0, 1, 2, 4\}$ are the range and domain, respectively, of the inverse function.

28 For $f = \{(x, y) \,|\, y = 2x + 5\}$, show that $f^{-1}[f(x)] = f[f^{-1}(x)] = x$.

29 For $f = \{(x, y) \,|\, y = 3x - 2\}$, show that the inverse of f^{-1} is f.

8
systems of linear equations

The statement "3 times a number less 2 is equal to 7" can be expressed in algebraic language as the equation $3x - 2 = 7$, and we know by the methods of Chap. 4 that the solution set of this equation is $\{3\}$. In order to express the more general statement "the sum of two numbers is 9" as an equation, we must have two variables, x and y, and then the equation is $x + y = 9$. It is evident that we can obtain as many *pairs* of numbers as we please that will satisfy the equation $x + y = 9$, because if we assign any value to y and then subtract that value from 9, we obtain the corresponding value of x. For example, if y is 1, x is 8; if y is 5, x is 4. If, however, we also require that the difference of the two numbers be 1, we have the two equations $x + y = 9$ and $x - y = 1$ that must be satisfied *by the same pair of numbers*.

Solving Equations Simultaneously

In this chapter we shall explain methods that will show that there is only one pair of numbers, $x = 5$ and $y = 4$, that satisfies both $x + y = 9$ and $x - y = 1$. The process of finding this pair of numbers is called *solving the equations simultaneously*. The problem of solving two equations in two variables simultaneously occurs frequently in all fields in which algebra is used. It is the purpose of this chapter to present the methods most often used when the two equations are linear.

8.1 DEFINITIONS AND GRAPHICAL SOLUTION OF TWO LINEAR EQUATIONS IN TWO VARIABLES

Linear Equation in Two Variables

An equation of the type $ax + by = c$, where a, b, and c are arbitrary constants and a and b are not both 0, is a *linear equation in two variables*.

Solution

An ordered pair of numbers, the first a value for x and the second a value for y, that satisfies an equation in two variables is a *solution* of the equation. For example, each of the ordered pairs (1, 9), (3, 3), and (4, 0) is a solution of $3x + y = 12$.

Simultaneous Solution

Simultaneous Solution Set

An ordered pair of numbers that satisfies each of two equations in two variables is a *simultaneous solution* of the two equations, and the set of all such pairs is the *simultaneous solution set*. For example, $(5, -3)$ is a simultaneous solution of the equations $3x + y = 12$ and $x - 3y = 14$.

Equivalent Equations

Two equations in two variables are *equivalent* if every solution of each is a solution of the other.

In the above definitions we have used the letters x and y as the variables. This is the usual custom, but frequently other pairs of letters such as u and v or z and w are used.

In order to solve $ax + by = c$ for y, we first write

$by = -ax + c$, and then get

$$y = \frac{-ax}{b} + \frac{c}{b} \qquad b \neq 0 \tag{1}$$

which is equivalent to the original equation

$$ax + by = c \tag{2}$$

In Chap. 7 we discussed graphs and saw that graphing linear equations is very easy. All that is necessary is to:

1 Find at least two solutions to the equation.
2 Plot these solutions as points on a plane.
3 Draw a line through these points.

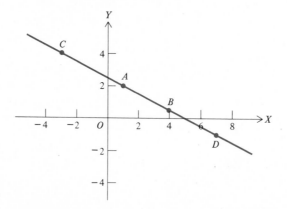

FIGURE 8.1

Example 1 For the equation $x + 2y = 5$, if $x = 1$, then $1 + 2y = 5$, so that $2y = 4$ and $y = 2$. Thus, $(1, 2)$ is a solution. If $x = 4$, then $4 + 2y = 5$, so that $2y = 5 - 4 = 1$ and $y = \frac{1}{2}$; thus, $(4, \frac{1}{2})$ is also a solution. Other solutions are $(-3, 4)$ and $(7, -1)$. For convenience we call these points $A(1, 2)$, $B(4, \frac{1}{2})$, $C(-3, 4)$, and $D(7, -1)$. The graph then consists of the line through these four points. See Fig. 8.1.

If the graph of a second equation

$$Ax + By = C \tag{3}$$

intersects the graph of (2), the coordinates of the point of intersection are the elements of the simultaneous solution pair of (2) and (3), since that point is on the graph of each equation. Consequently, we can use the graphical method to obtain the simultaneous solution pair of two linear equations in two variables. The possibility that the two equations may not have a simultaneous solution is dealt with in the next section.

Since the graph of a linear equation in two variables is a straight line, it is completely determined by two points, but it is advisable to obtain a third point as a check. We can obtain the coordinates of a point on the graph either by assigning a value to x and solving for y, or by assigning a value to y and solving for x. If $c \neq 0$, two points are easily obtained by first setting $x = 0$ and then setting $y = 0$. A third point is obtained by setting x or y equal to some value not zero.

Example 2   Solve the equations

$$3x - 4y = 7 \tag{4}$$
$$x + 6y = 6 \tag{5}$$

simultaneously by the graphical method.

Solution   We assign 0 to x and 0 and 2 to y in each of the equations, solve for the other unknown, and thus get the following tables of corresponding values:

For (4)

x	0	$\frac{7}{3}$	5
y	$-\frac{7}{4}$	0	2

For (5)

x	0	6	-6
y	1	0	2

We next plot the points determined by the pairs of corresponding values in each table, and draw a straight line through each set of points to get the lines in Fig. 8.2. By observation we see that these lines intersect at a point whose coordinates are $(3, \frac{1}{2})$. Therefore, we say that the graphical simultaneous solution of Eqs. (4) and (5) is $(3, \frac{1}{2})$. We check this solution by substituting these values in the left members of Eqs. (4) and (5) and get

From (4)   **From (5)**

$3(3) - 4(.5)$ $3 + 6(.5)$

$9 - 2 = 7$ $3 + 3 = 6$

Sometimes graphing gives the exact solution as in this example, but often it gives only an approximation. This approximation may be used as a starting point in more refined methods.

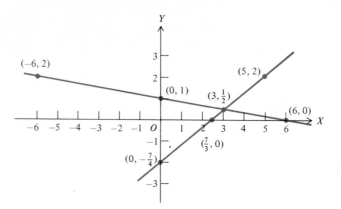

FIGURE 8.2

8.2 INDEPENDENT, INCONSISTENT, AND DEPENDENT EQUATIONS

If the graphs of two linear equations are parallel, the equations have no simultaneous solution. Furthermore, if the graphs are coincident (the same), every solution of one is a solution of the other. If the graphs are neither parallel nor coincident, they intersect in one and only one point. This illustrates the following definition:

Two linear equations in two variables are *independent* if their simultaneous solution set contains only one ordered pair of numbers, *inconsistent* if their simultaneous solution set is the empty set $\varnothing$, and *dependent* if every solution pair of one is a solution pair of the other.

We shall next derive a simple criterion that will enable us to decide whether two linear equations in two variables are independent, inconsistent, or dependent. We shall consider the equations

$$ax + by = c \tag{1}$$

$$Ax + By = C \tag{2}$$

The graphs of (1) and (2) intersect the X axis at $R(c/a, 0)$ and $S(C/A, 0)$, respectively, and the Y axis at the points $T(0, c/b)$ and $U(0, C/B)$, respectively, as illustrated in Fig. 8.3. We can see that the graphs of (1) and (2) are parallel if the segments RT and SU are parallel, and these two segments are parallel if and only if the triangles ORT and OSU are similar. The triangles are similar if and only if $OR/OS = OT/OU$.

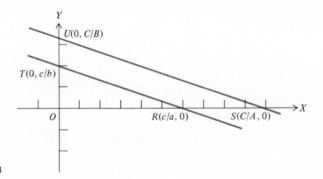

FIGURE 8.3

Now $OR/OS = c/a \div C/A = Ac/aC$, and $OT/OU = c/b \div C/B = Bc/bC$. Consequently, $OR/OS = OT/OU$ if and only if $Ac/aC = Bc/bC$. If we multiply each member of the last equation by C/c, we obtain $A/a = B/b$. Therefore, the graphs of (1) and (2) are parallel if and only if $A/a = B/b$. If in addition to $A/a = B/b$ we have $A/a = B/b = C/c = k$, then $A = ak$ and $C = ck$. Hence, $C/A = ck/ak = c/a$, and the points R and S coincide. Therefore, since the graphs are parallel and have one point in common, they coincide. Hence, we have this theorem:

Two linear equations $ax + by = c$ and $Ax + By = C$ are

Independent,
Inconsistent,
and Dependent
Systems

independent if and only if $A/a \neq B/b$ (8.1)

inconsistent if and only if $A/a = B/b \neq C/c$ (8.2)

dependent if and only if $A/a = B/b = C/c$ (8.3)

The following three examples illustrate the application of the above theorem.

Example 1   The equations

$$2x - 3y = 4$$
$$5x + 2y = 8$$

are independent by (8.1) since $2/5 \neq -3/2$. They intersect in one point.

Example 2   The equations

$$3x - 9y = 1$$
$$2x - 6y = 2$$

are inconsistent, since $3/2 = (-9)/(-6) \neq 1/2$. The lines are parallel.

Example 3 The equations

$$2x - 4y = 12$$
$$3x - 6y = 18$$

are dependent, since $2/3 = (-4)/(-6) = 12/18$. The two graphs are the same line.

EXERCISE 8.1 Graphical Solution of Systems of Equations

In Probs. 1 to 12, classify each pair of equations as independent, inconsistent, or dependent.

1 $2x + 5y = 3$
   $3x + 6y = 7$

2 $4x - y = 1$
   $2x - 3y = 2$

3 $-x + 3y = 6$
   $-2x + 6y = 12$

4 $2x - 3y = 14$
   $4x - 6y = 7$

5 $4x + 7y = -2$
   $-8x - 14y = 4$

6 $-12x - 15y = 9$
   $8x + 10y = -6$

7 $3x - 12y = 18$
   $4x - 16y = 42$

8 $8x + 3y = 16$
   $16x + 3y = 32$

9 $4x + 10y = 8$
   $6x - 15y = 12$

10 $4x + 4y = 8$
    $5x + 5y = 10$

11 $3x + y = 6$
    $6x + 2y = 10$

12 $5x - 15y = -15$
    $-7x + 21y = -14$

The equations in Probs. 13 to 16 are dependent. Verify this by graphing them and observing that the two lines in each problem are the same.

13 $2x - 2y = 4$
    $-3x + 3y = -6$

14 $8x + 16y = 20$
    $-6x - 12y = -15$

15 $-8x + 20y = 24$
    $10x - 25y = -30$

16 $16x + 12y = 28$
    $4x + 3y = 7$

The equations in Probs. 17 to 20 are inconsistent. Verify this by graphing them and observing that the two lines in each problem are parallel, but not coincident.

17 $8x + 4y = 6$
    $20x + 10y = 13$

18 $3x - 2y = 7$
    $-9x + 6y = 21$

19 $-12x + 15y = 6$
    $-16x + 20y = 10$

20 $49x + 21y = 77$
    $21x + 9y = 44$

The equations in Probs. 21 to 32 are independent. Solve the equations in each problem by graphing, and estimate each solution to the nearest integer in each coordinate.

21 $2x + 3y = 23$
    $4x - 3y = 1$

22 $4x - 3y = 11$
    $x + 6y = -4$

23 $3x + 6y = 0$
    $4x - y = -18$

24 $2x - 5y = -1$
    $-4x + 11y = -1$

25 $5x + 6y = 5$
    $-10x + 6y = -7$

26 $4x - 6y = 1$
    $-8x + 3y = 37$

27 $3x - 5y = 9$
    $3x + 5y = 17$

28 $-3x + 4y = 16$
    $6x + 4y = 31$

29 $9x - 2y = -7$
    $18x + 2y = 31$

30 $5x - 4y = 10$
    $10x + 12y = -85$

31 $3x + 3y = -2$
    $6x - 3y = -37$

32 $5x + 6y = -13$
    $10x - 6y = 7$

Problems 33 to 40 are like Probs. 21 to 32, except you are to estimate each solution to the nearest half integer in each coordinate.

33 $2x + 4y = 17$
$4x - 2y = -1$

34 $6x - 11y = 4$
$2x + y = 6$

35 $5x + 2y = 10$
$x - 4y = -9$

36 $2x + 3y = 5$
$4x - y = -4$

37 $5x + 10y = 7$
$10x - 5y = 4$

38 $8x + 5y = 15$
$16x - 5y = 8$

39 $6x + 5y = -1$
$12x - 5y = 16$

40 $7x + 2y = -13$
$14x + 13y = -8$

8.3 ELIMINATION BY ADDITION OR SUBTRACTION

To solve two independent linear equations in two variables algebraically, we combine the two equations in such a way as to obtain one equation in one variable whose root is one of the numbers in the simultaneous solution pair of the two given equations. This process is called *eliminating a variable*. After one of the numbers in the solution pair is found, the other is obtained by substituting the first number in one of the given equations and solving the resulting equation for the remaining variable.

Eliminating a Variable

One of the methods for eliminating a variable is a process called *elimination by addition or subtraction*. We shall explain the method by means of two examples.

Example 1 Find the simultaneous solution set of the equations

$$2x + 3y = 8 \tag{1}$$

$$x - 3y = -5 \tag{2}$$

by using the method of elimination by addition or subtraction.

Solution In order to solve (1) and (2) simultaneously by the method of elimination by addition or subtraction, we reason in this way: If the right and left members of each of (1) and (2) are equal for some pair of values of x and y, the equation obtained by adding the corresponding members will be valid for the same pair of values. This is just using the fact that if equals are added to equals, the results are equal. If we add the corresponding members of (1) and (2), we get $3x = 3$, and then if we divide each member of the last equation by 3, we get $x = 1$. Therefore, $x = 1$ is one number in the solution. We now substitute 1 for x in Eq. (1) and solve for y as follows:

$2 + 3y = 8$ **substituting 1 for x in Eq. (1)**

$3y = 6$ **adding −2 to each member**

$y = 2$ **dividing each member by 3**

Consequently, the simultaneous solution set of Eqs. (1) and (2) is $\{(1, 2)\}$. We check this solution by substituting 1 for x and 2 for y in the left member of Eq. (2) and obtain $1 - 6 = -5$. Therefore, since the right member of Eq. (2) is also -5, then $x = 1$, $y = 2$ satisfies the equation.

Example 2   Solve the equations

$$3x + 4y = -6 \tag{3}$$
$$5x + 6y = -8 \tag{4}$$

by the method of elimination by addition or subtraction.

Solution   In Eqs. (3) and (4) neither variable is eliminated if we add or subtract the corresponding members. We notice, however, that if we multiply the members of Eq. (3) by 3 and the members of Eq. (4) by 2, we get Eqs. (5) and (6), in which the coefficients of y are equal.

$$9x + 12y = -18 \qquad \text{multiplying Eq. (3) by 3} \tag{5}$$
$$\underline{10x + 12y = -16} \qquad \text{multiplying Eq. (4) by 2} \tag{6}$$
$$-x \qquad\quad = -2 \qquad \text{subtracting Eq. (6) from Eq. (5)}$$
$$\qquad x = 2 \qquad \text{dividing each member by } -1$$

By substituting 2 for x in Eq. (3) and solving for y, we get

$$6 + 4y = -6 \qquad \text{substituting 2 for x in Eq. (3)}$$
$$4y = -12 \qquad \text{adding } -6 \text{ to each member}$$
$$y = -3 \qquad \text{dividing each member by 4}$$

Therefore, the simultaneous solution set of Eqs. (3) and (4) is $\{(2, -3)\}$. To check the solution, we substitute 2 for x and -3 for y in the left member of Eq. (4) and obtain

$$5(2) + 6(-3) = 10 - 18 = -8$$

Hence, since the right member of Eq. (4) is also -8, $x = 2$, $y = -3$ satisfies the equation.

The reader should note that an alternative procedure is to multiply Eq. (3) by 5 and Eq. (4) by 3 and then solve the resulting equation for y after subtracting.

8.4   ELIMINATION BY SUBSTITUTION

If one of the equations in a pair of linear equations in two variables is easily solvable for one of the variables in terms of the other, a very efficient method for eliminating one of the variables is the *method of substitution*. We shall first list the steps in the process and then illustrate the procedure.

STEPS IN ELIMINATION BY SUBSTITUTION

1 Solve one of the equations for one variable in terms of the other.
2 Substitute the solution obtained in step 1 for that variable in the second equation and thus obtain an equation in one variable.
3 Solve the equation obtained in step 2.
4 Substitute the value found in step 3 in the solution obtained in step 1, and solve for the other variable.
5 Write the solution set in the form $\{(x, y)\}$, replacing x and y by the appropriate values found in steps 3 and 4.
6 Check the solution by substituting the values for x and for y from step 5 in the given equation not used in step 1.

Example Solve the equations

$$3x + 5y = 5 \tag{1}$$
$$x + 4y = 11 \tag{2}$$

simultaneously by the method of elimination by substitution.

Solution To solve Eqs. (1) and (2) simultaneously by the substitution method, we first notice that (2) is readily solvable for x in terms of y, and the solution is

$$x = 11 - 4y \qquad \text{step 1, adding } -4y \text{ to each member of (2)} \tag{3}$$

We now proceed as directed in steps 2 and 3 and obtain

$$
\begin{aligned}
3(11 - 4y) + 5y &= 5 && \text{step 2, substituting } 11 - 4y \text{ for } x \text{ in Eq. (1)}\\
33 - 12y + 5y &= 5 && \text{by distributive axiom}\\
-12y + 5y &= 5 - 33 && \text{adding } -33 \text{ to each member}\\
-7y &= -28 && \text{combining terms}\\
y &= 4 && \text{step 3, solve for } y \text{ by dividing each member by } -7
\end{aligned}
$$

We now substitute 4 for y in Eq. (3) as directed in step 4, and get

$$x = 11 - 16$$
$$x = -5$$

Therefore, the solution set (step 5) is $\{(-5, 4)\}$. As in step 6, we check the solution by substituting the values for x and y in the left member of Eq. (1). Thus, we get $-15 + 20 = 5$, and since the right member of Eq. (1) is also 5, $x = -5$, $y = 4$ satisfies the equation.

EXERCISE 8.2   Algebraic Solution of Systems of Equations

Solve the equations in Probs. 1 to 20 simultaneously by addition and subtraction.

1 $6x - y = 21$
      $2x + y = 11$

2 $3x + y = 11$
      $5x - y = 13$

3 $x - 3y = -1$
      $x + 4y = 6$

4 $x + 2y = 2$
      $x - 5y = -5$

5 $x + 4y = -2$
      $3x - 4y = 10$

6 $2x - y = -8$
      $-2x - 5y = -4$

7 $3x + 7y = 19$
      $-x + 7y = 31$

8 $4x - 9y = 25$
      $4x + y = 15$

9 $3x + 7y = 26$
      $-2x + 5y = 2$

10 $-4x + 9y = -59$
       $2x + 5y = 1$

11 $4x - 3y = -3$
       $3x + 2y = 19$

12 $6x + 11y = -9$
       $-8x - 3y = -23$

13 $7x + 5y = 1$
       $-4x - 3y = 1$

14 $-6x + 5y = 1$
       $-11x + 9y = 1$

15 $12x + 5y = 1$
       $5x + 2y = 1$

16 $6x + 7y = 1$
       $13x + 15y = 1$

17 $4x + 11y = 87$
       $-9x - 8y = -112$

18 $2x + 9y = 11$
       $4x - 3y = 1$

19 $14x + 9y = 37$
       $11x - 8y = 14$

20 $31x + 18y = 44$
       $-26x + 15y = -67$

For Probs. 21 to 40, solve the equations in Probs. 1 to 20 simultaneously by substitution. In Probs. 41 to 44, solve for $1/x$ and $1/y$ algebraically, then find x and y.

41 $\dfrac{3}{x} + \dfrac{1}{y} = 9$

     $\dfrac{4}{x} - \dfrac{3}{y} = -1$

42 $\dfrac{4}{x} - \dfrac{3}{y} = 31$

     $\dfrac{6}{x} - \dfrac{5}{y} = 49$

43 $\dfrac{5}{x} - \dfrac{8}{y} = 12$

     $\dfrac{3}{x} + \dfrac{2}{y} = 14$

44 $\dfrac{6}{x} - \dfrac{5}{y} = -37$

     $\dfrac{7}{x} + \dfrac{2}{y} = -4$

8.5   THREE LINEAR EQUATIONS IN THREE VARIABLES

An equation that is of the form $ax + by + cz = d$, where a, b, c, and d are constants and a, b, and c are not all zero, is a linear equation in three variables. A set of three values, one for x, one for y, and one for z, in the order (x, y, z), that satisfies the equation is called a *solution* of the equation.

If a simultaneous solution set of a system of three linear equations in three variables exists, we obtain it by means of the following steps.

Solving a System of Three Linear Equations in Three Variables

1 Eliminate one variable from the given equations, and thereby get two equations in two variables.

2 Solve the two equations in step 1 simultaneously for the other two variables.

3 Substitute the values obtained in step 2 in one of the given equations, and solve for the third variable.

Step 1 may be done in two ways. Example 1 illustrates the method of addition and subtraction, while Example 2 uses substitution.

Example 1 Solve the following system of equations simultaneously:

$$2x + 3y + 4z = 4 \tag{1}$$
$$3x - 2y - 6z = 7 \tag{2}$$
$$5x + 7y + 8z = 9 \tag{3}$$

Solution In order to solve the given system simultaneously, we shall arbitrarily select z as the first variable to be eliminated. In Eqs. (1) and (2) the lcm of the coefficients of z is 12. Therefore, as step 1, we multiply Eq. (1) by 3 and Eq. (2) by 2 and add the resulting equations.

$$6x + 9y + 12z = 12 \qquad \text{Eq. (1) times 3} \tag{4}$$
$$\underline{6x - 4y - 12z = 14} \qquad \text{Eq. (2) times 2} \tag{5}$$
$$12x + 5y \qquad = 26 \qquad \text{adding corresponding members of (4) and (5)} \tag{6}$$

Since the coefficient of z in Eq. (3) is twice the coefficient of z in Eq. (1), we multiply the members of Eq. (1) by 2 and subtract from Eq. (3) as follows. This is the second part of step 1.

$$5x + 7y + 8z = 9 \qquad \text{Eq. (3) copied} \tag{3}$$
$$\underline{4x + 6y + 8z = 8} \qquad \text{Eq. (1) times 2} \tag{7}$$
$$x + y \qquad = 1 \qquad \begin{array}{l}\text{subtracting each member of Eq. (7) from the} \\ \text{corresponding member of Eq. (3)}\end{array} \tag{8}$$

As step 2, we next solve Eqs. (6) and (8) simultaneously for x and y.

$$12x + 5y = 26 \qquad \text{Eq. (6) copied} \tag{6}$$
$$\underline{5x + 5y = 5} \qquad \text{Eq. (8) times 5} \tag{9}$$
$$7x \qquad = 21 \qquad \begin{array}{l}\text{subtracting each member of Eq. (9) from the} \\ \text{corresponding member of Eq. (6)}\end{array}$$
$$x \qquad = 3$$

We can now obtain the value of y by substituting $x = 3$ in either Eq. (6) or (8). We use the latter and get

$$3 + y = 1 \qquad \text{or} \qquad y = -2$$

Finally, as step 3, we can get the value of z by substituting $x = 3$ and $y = -2$ in any one of the three given equations. Using Eq. (1), we get

$$2(3) + 3(-2) + 4z = 4 \qquad \text{substituting } x = 3 \text{ and } y = -2 \text{ in Eq. (1)}$$
$$6 - 6 + 4z = 4$$
$$4z = 4$$
$$z = 1$$

Therefore, the solution is $(3, -2, 1)$.

We check the solution by substituting these values in either Eq. (2) or Eq. (3). Using Eq. (3), we have

$$5(3) + 7(-2) + 8(1) = 15 - 14 + 8 = 9$$

and since the right member of Eq. (3) is also 9, $x = 3$, $y = -2$, $z = 1$ satisfies the equation.

Example 2 Solve the following system simultaneously:

$$3x - y + 2z = 4 \tag{10}$$
$$4x + 2y - 5z = 11 \tag{11}$$
$$-5x - 3y + 8z = -14 \tag{12}$$

Solution The first equation solved for y is

$$y = 3x + 2z - 4 \tag{10a}$$

Substituting this in (11) and (12) gives

$$4x + 2(3x + 2z - 4) - 5z = 11 \tag{11a}$$
$$-5x - 3(3x + 2z - 4) + 8z = -14 \tag{12a}$$

Collecting terms in (11a) and (12a) gives

$$10x - z = 19 \tag{13}$$
$$-14x + 2z = -26 \tag{14}$$

Equations (13) and (14) may be solved by either addition and subtraction or by substitution. We shall use substitution. Solving (13) for z gives

$$z = 10x - 19 \tag{13a}$$

Putting this in (14) leads to

$$-14x + 2(10x - 19) = -26 \tag{14a}$$
$$-14x + 20x - 38 = -26$$
$$6x = 12$$
$$x = 2$$

We now get $z = 10x - 19 = 10(2) - 19 = 20 - 19 = 1$ from (13a), and $y = 3x + 2z - 4 = 3(2) + 2(1) - 4 = 4$ from (10a). Hence the solution is (2, 4, 1), and it should be checked in each of the original equations, (10) through (12).

EXERCISE 8.3 Systems of Three Linear Equations

Solve the following systems of equations.

1
$$2x + y - 2z = 4$$
$$4x - 3y + 3z = 19$$
$$8x + 6y - 3z = 35$$

2
$$6x + 2y - 3z = 4$$
$$3x - 3y - 4z = -28$$
$$9x + 4y - z = 25$$

3
$$10x - 2y + 3z = -5$$
$$5x + y - 4z = -15$$
$$10x + 3y - 4z = -16$$

4
$$14x + 3y - 5z = 23$$
$$14x - 8y + z = 27$$
$$7x + 2y - 4z = 15$$

5
$$x + 2y - 3z = -4$$
$$2x + 4y + 3z = 19$$
$$5x - 6y + 5z = 24$$

6
$$2x + 5y + 6z = 9$$
$$3x - 2y - 3z = 11$$
$$5x + y + 9z = 28$$

7
$$7x - 7y + 2z = 31$$
$$3x + 7y - 5z = 35$$
$$x - 14y + 6z = -20$$

8
$$3x + 4y - 5z = 22$$
$$2x + y - 5z = 13$$
$$-3x + 6y + 10z = -7$$

9
$$x + y - z = 7$$
$$-x + y + z = 1$$
$$x - y - z = -1$$

10
$$x - y + z = 11$$
$$-x + y + z = 1$$
$$-x - y + z = 17$$

11
$$x + y + z = -1$$
$$x + y - z = 5$$
$$-x + y - z = -9$$

12
$$-x - y + z = -7$$
$$-x - y - z = -1$$
$$x - y - z = -11$$

13
$$2x - y + z = 7$$
$$x - 2y - z = 2$$
$$3x + 2y + z = 2$$

14
$$3x + y + 2z = 1$$
$$2x - y + 3z = -6$$
$$x + y + 2z = -3$$

15
$$x + y + 2z = 3$$
$$x + 2y + 4z = 3$$
$$x - 3y - 5z = 5$$

16
$$3x - 2y + z = -1$$
$$2x + 3y + 2z = 17$$
$$4x - 4y - z = -1$$

17
$$3x + 5y + 2z = -7$$
$$2x + 4y + 3z = -2$$
$$5x + 7y + 5z = 3$$

18
$$2x - 3y + 2z = 13$$
$$3x + 5y - 3z = 31$$
$$5x + 2y - 5z = 20$$

19
$$2x + y - 3z = 5$$
$$3x - 5y + 7z = -7$$
$$5x + 3y - 5z = 17$$

20
$$2x + 3y + 4z = 6$$
$$3x - 6y + 2z = -1$$
$$4x + 9y - 8z = 2$$

21
$$6x - 5y - 3z = 3$$
$$2x - 5y + 3z = 5$$
$$2x + 15y - 9z = -3$$

22
$$8x - 6y + 4z = 5$$
$$4x + 9y - 8z = 5$$
$$6x + 3y + 3z = 10$$

23
$$4x + 3y + 2z = 6$$
$$2x - 6y + z = -7$$
$$2x + 3y - z = 0$$

24
$$10x + 5y - 4z = 6$$
$$x + y + 4z = 2$$
$$6x + y - 8z = 1$$

25
$$3x + 4y - 2z = 4$$
$$2x + 7y - 8z = -13$$
$$5x + 3y + 5z = 28$$

26
$$4x + 3y + 6z = 7$$
$$5x + 5y - 4z = 1$$
$$3x - 2y + 7z = -5$$

27
$$3x + 4y - 2z = 4$$
$$3x - 4y + 2z = 0$$
$$9x + 12y + 2z = 16$$

28
$$5x + 3y - 6z = -1$$
$$5x - 3y + 6z = 7$$
$$15x + 9y - 6z = 7$$

29
$$x + 2y = 14$$
$$3y + z = 8$$
$$x - 2y + 3z = -27$$

30
$$3x - y = 9$$
$$2x + 4y - 3z = 5$$
$$2x + z = 13$$

31
$$2x + 3y = 13$$
$$4y - z = 6$$
$$5x - 2z = -2$$

32
$$3x + 5z = 22$$
$$4x - 7y = -19$$
$$4y + 3z = 26$$

33
$$x + y + z = 2a$$
$$x - y - z = 0$$
$$x + 2y + z = 2a + b$$

34
$$ax + by - z = 2ab$$
$$bx + ay + z = 2a^2$$
$$x + y - z = 2a - a^2 + b^2$$

35
$$2x - y - z = -b - c$$
$$x - y + z = a - b + c$$
$$x + 2y + z = 4a + 2b + c$$

36
$$ax - ay + cz = a^2 + c^2$$
$$bx - ay + az = a^2 + b^2$$
$$cx - by - cz = -b^2 - c^2$$

8.6 PROBLEMS LEADING TO SYSTEMS OF LINEAR EQUATIONS

Many stated problems contain more than one unknown quantity, and often the equation for solving such a problem can be more easily obtained if more than one unknown is introduced. However, before the problem can be completely solved, the number of equations formed must be equal to the number of unknowns used. The general procedure for obtaining the equations is the same as that in Sec. 4.6, and the student is advised to reread that section before studying the following examples or attempting the problems in Exercise 8.4.

Example 1   A real estate dealer received $1200 in rents on two dwellings in 1 year; one of them brought $10 per month more than the other. How much did the dealer receive per month for each if the more expensive house was vacant for 2 months?

Solution   If we let

x = the monthly rental on the more expensive house

and

y = the monthly rental on the other

then

$$x - y = 10 \tag{1}$$

since one rented for $10 more per month than the other. Furthermore, since the first of the above houses was rented for 10 months and the other was rented for 12 months, we know that $10x + 12y$ is the total amount received in rentals. Hence,

$$10x + 12y = 1200 \tag{2}$$

We now have the two equations (1) and (2) in the unknowns x and y, and we shall solve them simultaneously by eliminating y. The solution follows:

$$
\begin{aligned}
12x - 12y &= 120 \qquad &&\text{Eq. (1) times 12} \tag{3}\\
10x + 12y &= 1200 \\
\hline
22x &= 1320 \qquad &&\text{Eq. (3) plus Eq. (2)} \tag{4}
\end{aligned}
$$

Hence,

$$x = 60$$

Substituting 60 for x in Eq. (1), we get

$$60 - y = 10$$

Hence,

$$-y = 10 - 60 = -50$$

and

$$y = 50$$

Therefore, the monthly rentals were $60 and $50, respectively.

Example 2   A tobacco dealer mixed one grade of tobacco worth $1.40 per pound with another worth $1.80 per pound in order to obtain 50 pounds of a blend that sold for $1.56 per pound.   How much of each grade was used?

Solution   We shall let

$x =$ the number of pounds of the $1.40 grade used

and

$y =$ the number of pounds of the $1.80 grade used

Then

$$x + y = 50 \tag{1}$$

since there were 50 pounds in the mixture.   Furthermore, $1.40x$ is the value in dollars of the first grade, $1.80y$ is the value in dollars of the second, and $(1.56)50 = 78$ is the value in dollars of the mixture.   Therefore,

$$1.40x + 1.80y = 78 \tag{2}$$

Hence (1) and (2) are the two required equations, and we shall solve them by eliminating x.

$$
\begin{array}{ll}
1.40x + 1.40y = 70 & \text{Eq. (1) times 1.40} \tag{3} \\
1.40x + 1.80y = 78 & \tag{2} \\
\hline
 - 0.40y = -8 & \text{Eq. (3) minus Eq. (2)}
\end{array}
$$

Therefore,

$$y = \frac{-8}{-0.40} = 20$$

Substituting 20 for y in Eq. (1), we have

$$x + 20 = 50$$

$$x = 30$$

Hence, the dealer used 30 pounds of the $1.40 grade and 20 pounds of the $1.80 grade in the mixture.

Example 3   Two airfields A and B are 400 miles apart, and B is due east of A. A plane flew from A to B in 2 hours and then returned to A in $2\frac{1}{2}$ hours.   If the wind blew with a constant velocity from the west during the entire trip, find the speed of the plane in still air and the speed of the wind.

Solution   Let

x = the speed of the plane in still air

and

y = the speed of the wind

Then, since the wind was blowing from the west,

$x + y$ = the speed of the plane from A to B

and

$x - y$ = the speed of the plane on the return trip

Hence,

$\dfrac{400}{x + y}$ = the time required for the first half of the trip

and

$\dfrac{400}{x - y}$ = the time required to return

Therefore,

$$\frac{400}{x + y} = 2 \tag{1}$$

$$\frac{400}{x - y} = 2\tfrac{1}{2} = \tfrac{5}{2} \tag{2}$$

Now we multiply both members of Eq. (1) by $x + y$, and both members of Eq. (2) by $2(x - y)$ and get

$$400 = 2x + 2y \tag{3}$$

and

$$800 = 5x - 5y \tag{4}$$

We shall solve Eqs. (3) and (4) simultaneously by first eliminating y.

$$2000 = 10x + 10y \qquad \text{Eq. (3) times 5} \tag{5}$$
$$\underline{1600 = 10x - 10y} \qquad \text{Eq. (4) times 2} \tag{6}$$
$$3600 = 20x \qquad \text{Eq. (5) plus Eq. (6)}$$

Hence,

$x = 180$

Substituting 180 for x in Eq. (3), we have

$$400 = 2(180) + 2y$$
$$400 = 360 + 2y$$
$$2y = 40$$
$$y = 20$$

Hence, the speed of the plane in still air was 180 miles per hour, and the speed of the wind was 20 miles per hour.

Example 4   A cash drawer contains $50 in nickels, dimes, and quarters. There are 802 coins in all, and 10 times as many nickels as dimes. How many coins of each denomination are in the drawer?

Solution   Let

q = the number of quarters
d = the number of dimes
n = the number of nickels

We now form the following three linear equations in q, d, and n:

$25q + 10d + 5n = 5000$	since $50 = 5000 cents	(1)
$q + d + n = 802$	since there are 802 coins in all	(2)
$n = 10d$	since there are 10 times as many nickels as dimes	(3)

If we substitute $10d$ for n [given by Eq. (3)] in Eqs. (1) and (2), we obtain two linear equations in q and d.   From Eq. (1) we get

$$25q + 10d + 5(10d) = 5000$$

which reduces to

$$25q + 60d = 5000$$

Furthermore, from Eq. (2) we have

$$q + d + 10d = 802$$

or

$$q + 11d = 802 \qquad (5)$$

We may eliminate q from Eqs. (4) and (5) as follows:

$$
\begin{array}{lll}
25q + 60d = 5,000 & & (4) \\
25q + 275d = 20,050 & \text{Eq. (5) times 25} & (6) \\
\hline
-215d = -15,050 & \text{Eq. (4) minus Eq. (6)} & \\
d = 70 & &
\end{array}
$$

Now, substituting 70 for d in Eq. (3), we get

$n = 10(70) = 700$

Finally, substituting $d = 70$ in Eq. (5), we have

$q + 11(70) = 802$

Hence,

$q = 802 - 770 = 32$

Consequently, there are 32 quarters, 70 dimes, and 700 nickels in the cash drawer.

EXERCISE 8.4   Problems Leading to Systems of Linear Equations

1 The sum of two numbers is three times the smaller, and their difference exceeds one-half the smaller by 12. Find the numbers.

2 The sum of two numbers is twice their difference, and the larger exceeds twice the smaller by 6. Find the numbers.

3 The quotient of the sum and the difference of the same two numbers is 5, and three times the larger number exceeds twice the smaller by 60. Find the numbers.

4 The sum of two numbers is four times the smaller. If the smaller number is increased by 15 and the larger is decreased by 13, the results are equal. Find the numbers.

5 A man invested $10,900 for stock in two companies at $55 and $36 per share, respectively. The more expensive stock yielded an annual dividend of $2.20 per share, and the other an annual dividend of $1.20. If the total income from the two was $400 per year, find the number of shares of each stock that was bought.

6 During a certain year some apartments in a building rented for $125 per month, and the remainder for $160 per month. The total monthly rental was $4900. The next year the monthly rent on the cheaper apartments was increased by $5, and on the others by $6. If the monthly income was thereby increased by $190, how many apartments of each type are in the building?

7 The dwellings in a new housing tract were priced at $30,000 and $35,000, respectively, and the value of the tract was $3,200,000. At the end of 6 months, one-half of the more expensive houses and two-thirds of the others had been sold. If the amount received from the sales was $1,900,000, how many dwellings of each type are in the tract?

8 The monthly rental for a beach cottage is higher during the 3 summer months than during the remainder of the year. If it is occupied during the entire year, the rental amounts to $2700. In a certain year, however, because of fire damage, the cottage was occupied for 2 summer months and 5 off-season months, and the rental amounted to $1600. Find the monthly rental for each portion of the year.

9 Two girls paddled a canoe 6 miles downstream in 1 hour. On the return trip they were joined by a companion and, with the three paddling, the rate of the canoe in still

water was increased by 1 mile per hour. If the return trip required 2 hours, find the rate of the current and the rate the first two girls could paddle in still water.

10 Airfield B is 960 miles north of field A. A pilot left A to fly to B, and 30 minutes later a pilot left B to fly to A. The two met 1 hour and 50 minutes after the departure of the first plane, and the first pilot reached B 1 hour and 10 minutes later. If the airspeeds of the two planes were the same, and the wind was blowing from the south at a constant velocity during the flight, find the airspeed of the two planes and the velocity of the wind.

11 Airfield B is 990 miles due north of field A. A pilot flew from A to B and back, and there was a north wind with a constant velocity during the entire flight. The northward trip required $5\frac{1}{2}$ hours, and the return trip required $4\frac{1}{2}$ hours. Find the airspeed of the plane and the velocity of the wind.

12 Airfild B is due west of field A, and field C is due south of B. A pilot left A at 8 A.M. and flew to B, delayed 3 hours, and then flew to C, arriving at 3 P.M. The wind blew from the west at 20 miles per hour during the westward flight, but changed to the north at 30 miles per hour during the delay. If the airspeed of the plane was 270 miles per hour and the total distance flown was 1080 miles, find the distances from A to B and from B to C.

13 A student traveled from the campus to an airport on a bus that averaged 45 miles per hour, and then flew home on a plane at 400 miles per hours. The time required for the trip, including a 24-minute wait at the airport, was 3 hours. The bus fare was 5 cents per mile, the plane fare was 8 cents per mile, and the cost of the trip was $65.35. Find the distance traveled by each method.

14 A rancher rode a horse at 6 miles per hour to his ranch headquarters. He then drove his car at 50 miles per hour to an airport, where he boarded a plane and flew at 300 miles per hour to a rancher's convention. The total distance traveled was 340 miles, and the traveling time was 4 hours. If he was on the plane twice as long as in the car, find the time required for each part of the trip.

15 One morning a sales representative left her home and called on customers in towns A and B. In the afternoon she returned to her home from B on a modern highway. Her average speed in the morning was 30 miles per hour, and in the afternoon it was 65 miles per hour. If she traveled a total of 7 hours and was on the road 3 hours longer in the morning than in the afternoon, find the distance she traveled on each part of the trip.

16 A tour bus and a car left a resort hotel at the same time, but traveled in opposite directions around a scenic loop. When they met, the bus had traveled 32 miles, and the car 48 miles. Find the average speed of each if the car reached the hotel 1 hour and 20 minutes ahead of the bus.

17 A man fenced a rectangular plot with one of the shorter sides along a highway. At the same time, he divided it into two parts with a fence parallel to the highway. The cost of the fence along the highway was 60 cents per foot, and elsewhere it was 50 cents per foot. The total cost of the fencing was $620, and the fence along the highway cost $380 less than the remainder. Find the dimensions of the plot.

18 A building that is 80 feet wide with the longer side fronting a street is divided into three parts by partitions perpendicular to the front wall. The sum of the areas of the

two smaller parts is equal to the area of the larger, and the perimeter of the building is 400 feet. If the areas of the two smaller parts are equal, find the dimensions of the two smaller parts and of the larger part.

19 The owners of a shopping center in which 70 percent of the area was used for parking bought an adjacent tract, set aside 85 percent for parking, and used the remainder for buildings. The enlarged area contained 30,000 square yards with 75 percent used for parking. Find the original area and the area of the land purchased.

20 A rectangular tract of land with the southern boundary running east and west and the eastern boundary 600 yards long is bordered on the north by another rectangular tract whose northern boundary is 500 yards long, and the eastern boundaries of the two tracts form a straight line. The combined areas of the two tracts is 680,000 square yards. They are enclosed with 3600 yards of fencing, with no fence along the common boundary. Find the unknown dimension of each tract.

21 An automobile dealer has in stock 45 cars, made up of sedans, sports cars, and station wagons. There are twice as many sports cars as station wagons. The sedans are priced at $4000, the sports cars at $3500, and the station wagons at $4200. If the retail value of the cars is $172,000, how many cars of each type are in stock?

22 A cotton buyer declined an offer of $175 per bale for her consignment of baled cotton. Two months later, when the price had increased by $5 per bale, she sold the consignment for $140 more than she would have received on the first offer. If, in the meantime, 2 bales had been destroyed by fire, find the number of bales in the original consignment and the amount she would have received if she had accepted the first offer.

23 Mr. Smith had 52 shares of stock A and 32 shares of stock B. On a certain day his holdings were worth $2810. The next day, the price of stock B was 1 percent lower, the price of stock A was 2 percent higher, and his holdings were worth $31.96 more. Find the price of each stock on the first day.

24 A family on vacation traveled an average of 220 miles a day at a cost of 5 cents per mile. Their meals averaged $14 per day, and their motel costs averaged $15 per night. The total cost of the vacation was $505, and the motel costs were $37 more than the mileage costs. Find the number of miles traveled and the number of nights spent in a motel.

25 On a certain day when his irrigation reservoir was full, a farmer opened the inlet pipe and the outlet pipe, and the reservoir was empty after 24 hours. On another day, when the reservoir was empty, he opened the inlet pipe and allowed it to run 3 hours. Then he opened the outlet, and the reservoir was drained in 9 hours. How long does it take the inlet to fill the reservoir if the outlet is closed, and how long will it take the outlet to drain the full reservoir if the inlet is closed?

26 Two sisters mowed a lawn together in $2\frac{2}{3}$ hours. The next week the older girl worked alone for 3 hours, and the younger girl finished the job in $1\frac{1}{4}$ hours. How long will it take each girl to mow the lawn alone?

27 A tobacoo dealer mixed two grades of tobacco worth $3 and $3.50 per pound, respectively, and obtained 20 pounds of a mixture worth $3.20 per pound. How many pounds of each grade were used?

28 A traveler left his home with his 5 gallon radiator filled with a solution that was 20

percent antifreeze. When he stopped for gasoline, he discovered that the radiator was leaking, and the attendant added water until it was full. He then went to the nearest repair shop, where the solution was tested. After repairs, $\frac{1}{2}$ gallon of a solution that was 56 percent antifreeze was added to bring the radiator solution up to the original strength. Find the number of gallons lost before the first stop and the percent of antifreeze revealed by the test.

29 On Monday, Sue, Tom, and Bill, working together, polished Sue's car in $1\frac{1}{2}$ hours. On Tuesday, Sue helped Bill polish his car in 2 hours, and on Wednesday, Bill and Tom polished Tom's car in $2\frac{1}{4}$ hours. If all cars were the same make and model with the same surface conditions, how long would it take each of them to polish one car alone?

30 There are 52 desks in three offices, and the number in the second office is one-half the number in the first. The first office has 48 square feet of floor space per desk, the second has 46 square feet per desk, and the third has 45 square feet per desk. If there are 2424 square feet in three offices, how many desks are in each?

31 A salesman was allowed 10 cents per mile for the use of his car, $10 per day for meals, and $15 per night for a hotel room. On a certain trip his expense bill was $170; he averaged 120 miles per day, and his hotel bill was $10 more than he spent for meals. Find the number of days and the number of miles traveled, and the number of nights he stayed in a hotel.

32 A woman had a total of $3600 in three savings banks. The interest rates were $5\frac{1}{2}$, 5, and $4\frac{3}{4}$ percent, respectively. Her total yearly income from the three accounts was $186. She withdrew her money from the third bank and deposited half of it in each of the others and increased her yearly income by $4. Find the amount of each of her original deposits.

8.7 DETERMINANTS OF THE SECOND ORDER AND CRAMER'S RULE

We have already seen how to solve systems of linear equations graphically, by addition and subtraction, and by substitution. In the remainder of the chapter we shall present Cramer's rule, a fourth method for solving systems of linear equations. In this section we shall solve two equations in two unknowns, and in the next section we shall solve three equations in three unknowns.

A notation called a *determinant* was discovered independently by the Japanese mathematician Kiowa in 1683 and by Leibnitz in 1693. In 1750, Cramer devised the method for using determinants to obtain the solution sets of systems of linear equations.

The square array

$$\begin{vmatrix} a & b \\ c & d \end{vmatrix}$$

Determinant of the Second Order is a *determinant of the second order*. It stands for the binomial $ad - bc$, and this binomial is called the *value* or the *expansion* of the determinant. The letters a, b, c, and d are the *elements* of the determinant. The terms in the binomial $ad - bc$ are the products of the elements connected by the arrows in the following diagram, and each product is preceded by the sign at the point of the arrow.

Expansion of a
Determinant
$$\begin{vmatrix} a & b \\ c & d \end{vmatrix} = ad - bc \tag{8.4}$$

If in Eq. (8.4) we replace a, b, c, and d by 3, 2, 5, and 7, respectively, we have

$$\begin{vmatrix} 3 & 2 \\ 5 & 7 \end{vmatrix} = 21 - 10 = 11$$

$-10 \qquad + 21$

Similarly,

$$\begin{vmatrix} -3 & 6 \\ -4 & 7 \end{vmatrix} = (-3)(7) - (6)(-4) = -21 + 24 = 3$$

We now consider the system of linear equations

$$ax + by = e \tag{1}$$

$$cx + dy = f \tag{2}$$

We obtain the solution using addition and subtraction as follows:

$adx + bdy = ed$	multiplying Eq. (1) by d	(3)
$bcx + bdy = bf$	multiplying Eq. (2) by b	(4)
$adx - bcx = ed - bf$	subtracting	
$x(ad - bc) = ed - bf$	by the distributive axiom	

$$x = \frac{ed - bf}{ad - bc} \qquad ad - bc \neq 0 \tag{5}$$

Note that by (8.1) if $ad - bc = 0$, the equations are not independent.

Similarly, solving for y we get

$$y = \frac{af - ec}{ad - bc} \tag{6}$$

We next show that each of the values of x and y in Eqs. (5) and (6) is the quotient of two determinants. For this purpose, we write the determinant

$$D = \begin{vmatrix} a & b \\ c & d \end{vmatrix} = ad - bc$$

whose elements are the coefficients of x and y in Eqs. (1) and (2). The expansion of D is the denominator in each of Eqs. (5) and (6).

Now we replace the column of coefficients of x in D by the column of constant terms e and f and get

$$D_x = \begin{vmatrix} e & b \\ f & d \end{vmatrix} = ed - bf$$

Finally, we replace the column of coefficients of y in D by e and f and obtain

$$D_y = \begin{vmatrix} a & e \\ c & f \end{vmatrix} = af - ec$$

Now referring to Eqs. (5) and (6), we see that D_x is the numerator of the fraction in (5) and D_y is the numerator in (6). Consequently,

$$x = \frac{D_x}{D} \quad \text{and} \quad y = \frac{D_y}{D}$$

Cramer's Rule   The procedure just given for obtaining the simultaneous solution set of a system of linear equations is known as *Cramer's rule*, and it consists of the following steps:

1   Arrange the terms in the equations so that the x terms are first, the y terms are second, and the constant terms are at the right of the equality sign.
2   Write the determinant D, whose elements are the coefficients of x and y, in the order in which they occur in the equations.
3   Replace the column of coefficients of x in D by the column of constant terms, and obtain the determinant D_x.
4   Replace the column of coefficients of y in D by the column of constant terms, and get the determinant D_y.
5   Then the elements in the solution set are

$$x = \frac{D_x}{D} \quad \text{and} \quad y = \frac{D_y}{D}$$

Example   Use Cramer's rule to get the simultaneous solution set of

$$3x - 5y = -7$$
$$2x + 3y = 8$$

Solution   The terms are arranged as in step 1, so

$$D = \begin{vmatrix} 3 & -5 \\ 2 & 3 \end{vmatrix} = 9 + 10 = 19 \qquad \text{step 2}$$

$$D_x = \begin{vmatrix} -7 & -5 \\ 8 & 3 \end{vmatrix} = (-7)(3) - (-5)(8) = -21 + 40 = 19 \qquad \text{step 3}$$

$$D_y = \begin{vmatrix} 3 & -7 \\ 2 & 8 \end{vmatrix} = 3(8) - (-7)(2) = 24 + 14 = 38 \qquad \text{step 4}$$

$$x = \frac{D_x}{D} = \frac{19}{19} = 1 \qquad y = \frac{D_y}{D} = \frac{38}{19} = 2 \qquad \text{step 5}$$

Hence the solution set is $\{(1, 2)\}$.

EXERCISE 8.5   Determinants of Order 2 and Cramer's Rule

Find the value of each determinant in Probs. 1 to 20.

1 $\begin{vmatrix} 2 & 5 \\ 3 & 8 \end{vmatrix}$ **2** $\begin{vmatrix} 3 & 7 \\ 1 & -4 \end{vmatrix}$ **3** $\begin{vmatrix} 4 & 8 \\ -2 & 5 \end{vmatrix}$ **4** $\begin{vmatrix} 1 & 6 \\ -3 & -4 \end{vmatrix}$

5 $\begin{vmatrix} 2 & -5 \\ 4 & 7 \end{vmatrix}$ **6** $\begin{vmatrix} 3 & -8 \\ 8 & -5 \end{vmatrix}$ **7** $\begin{vmatrix} 4 & -2 \\ -3 & 1 \end{vmatrix}$ **8** $\begin{vmatrix} 6 & -5 \\ -2 & -2 \end{vmatrix}$

9 $\begin{vmatrix} -1 & 4 \\ 2 & 5 \end{vmatrix}$ **10** $\begin{vmatrix} -2 & 4 \\ 2 & -5 \end{vmatrix}$ **11** $\begin{vmatrix} -4 & 5 \\ -2 & 5 \end{vmatrix}$ **12** $\begin{vmatrix} -3 & 7 \\ -2 & -5 \end{vmatrix}$

13 $\begin{vmatrix} -1 & -4 \\ 2 & 3 \end{vmatrix}$ **14** $\begin{vmatrix} -2 & -3 \\ -3 & 4 \end{vmatrix}$ **15** $\begin{vmatrix} -3 & -7 \\ 3 & -6 \end{vmatrix}$ **16** $\begin{vmatrix} -4 & -1 \\ -1 & -2 \end{vmatrix}$

17 $\begin{vmatrix} 4 & 10 \\ 5 & 12 \end{vmatrix}$ **18** $\begin{vmatrix} 2 & 6 \\ 6 & 19 \end{vmatrix}$ **19** $\begin{vmatrix} 3 & 4 \\ 4 & 6 \end{vmatrix}$ **20** $\begin{vmatrix} 9 & 5 \\ 5 & 3 \end{vmatrix}$

Verify each of the equations in Probs. 21 to 28 by expanding the determinants and collecting like terms.

21 $\begin{vmatrix} a & 3b \\ c & 3d \end{vmatrix} = 3 \begin{vmatrix} a & b \\ c & d \end{vmatrix}$ **22** $\begin{vmatrix} a & b \\ 5c & 5d \end{vmatrix} = 5 \begin{vmatrix} a & b \\ c & d \end{vmatrix}$

23 $\begin{vmatrix} 3a & 3b \\ 3c & 3d \end{vmatrix} = 9 \begin{vmatrix} a & b \\ c & d \end{vmatrix}$ **24** $\begin{vmatrix} a & b \\ 2a & 2b \end{vmatrix} = 0$

25 $\begin{vmatrix} a & b+3a \\ c & d+3c \end{vmatrix} = \begin{vmatrix} a & b \\ c & d \end{vmatrix}$ **26** $\begin{vmatrix} a+1 & b+2 \\ c & d \end{vmatrix} = \begin{vmatrix} a & b \\ c & d \end{vmatrix} + \begin{vmatrix} 1 & 2 \\ c & d \end{vmatrix}$

27 $\begin{vmatrix} a+2 & b \\ c-3 & d \end{vmatrix} = \begin{vmatrix} a & b \\ c & d \end{vmatrix} + \begin{vmatrix} 2 & b \\ -3 & d \end{vmatrix}$

28 $\begin{vmatrix} a+1 & b+2 \\ c+3 & d+4 \end{vmatrix} = \begin{vmatrix} a & b \\ c & d \end{vmatrix} + \begin{vmatrix} 1 & b \\ 3 & d \end{vmatrix} + \begin{vmatrix} a & 2 \\ c & 4 \end{vmatrix} + \begin{vmatrix} 1 & 2 \\ 3 & 4 \end{vmatrix}$

By use of Cramer's Rule, simultaneously solve the equations in each of Probs. 29 to 48.

29 $3x + y = 13$
$\quad\;\; x - 2y = 2$

30 $\quad x - 3y = 5$
$\quad 3x + 2y = 4$

31 $\quad 4x - y = -14$
$\quad -x + 2y = 7$

32 $-3x - 5y = 2$
$-2x + 7y = -40$

33 $6x + 7y = 17$
$3x - 5y = 17$

34 $3x + 5y = 3$
$x + 3y = 5$

35 $8x + 11y = 17$
$-3x + 7y = 27$

36 $6x + y = 51$
$4x + 3y = 55$

37 $6x + y = -1$
$4x - y = -9$

38 $3x + 5y = 14$
$7x - 4y = -30$

39 $3x + 4y = 8$
$-5x - 2y = 10$

40 $8x + 7y = 11$
$5x + 3y = 11$

41 $4x + 5y = 5$
$6x - 5y = 0$

42 $3x + 4y = 5$
$9x - 16y = 8$

43 $8x + 9y = 8$
$-4x + 9y = 5$

44 $4x + 5y = 7$
$-4x + 10y = 5$

45 $12x + 8y = 7$
$8x - 4y = 7$

46 $4x - 7y = -4$
$8x + 14y = -4$

47 $3x + 5y = 8$
$9x - 10y = 9$

48 $3x + 8y = -5$
$-15x + 64y = 64$

8.8 DETERMINANTS OF THE THIRD ORDER AND CRAMER'S RULE

The square array

$$D = \begin{vmatrix} a_1 & b_1 & c_1 \\ a_2 & b_2 & c_2 \\ a_3 & b_3 & c_3 \end{vmatrix}$$

is a determinant of order 3. The expansion is defined to be the polynomial whose terms are of the type $a_i b_j c_k$ as follows:

Expansion of a Determinant

$$D = a_1 b_2 c_3 + a_2 b_3 c_1 + a_3 b_1 c_2 - a_3 b_2 c_1 - a_2 b_1 c_3 - a_1 b_3 c_2 \qquad (1)$$

The expansion in this form is difficult to remember and cumbersome to apply. Fortunately, we can express it in terms of second-order determinants by a method that requires little memorization. The method involves the *minor* of an element.

Minor of an Element

The *minor* of a specified element of a determinant D is the determinant whose elements are all elements of D that are not in the same row or column as the specified element.

We shall designate the minors of a_i, b_i, and c_i by $m(a_i)$, $m(b_i)$, and $m(c_i)$, respectively, where $i = 1$, 2, or 3.

According to the above definition,

$$m(c_2) = \begin{vmatrix} a_1 & b_1 & c_1 \\ a_2 & b_2 & c_2 \\ a_3 & b_3 & c_3 \end{vmatrix} = \begin{vmatrix} a_1 & b_1 \\ a_3 & b_3 \end{vmatrix} = a_1 b_3 - a_3 b_1$$

Similarly, since a_3 is in the third row and first column,

$$m(a_3) = \begin{vmatrix} b_1 & c_1 \\ b_2 & c_2 \end{vmatrix} = b_1 c_2 - b_2 c_1$$

We now show how the expansion of D in (1) can be expressed with the help of minors. We consider the second row of D, group the terms of (1) and factor each group as follows:

$$D = -(b_1c_3 - b_3c_1)a_2 + (a_1c_3 - a_3c_1)b_2 - (a_1b_3 - a_3b_1)c_2$$

Now we note that

$$b_1c_3 - b_3c_1 = \begin{vmatrix} b_1 & c_1 \\ b_3 & c_3 \end{vmatrix} = m(a_2)$$

Likewise,

$$a_1c_3 - a_3c_1 = m(b_2) \qquad \text{and} \qquad a_1b_3 - a_3b_1 = m(c_2)$$

Consequently,

$$D = -m(a_2) \cdot a_2 + m(b_2) \cdot b_2 - m(c_2) \cdot c_2$$

Now we consider the third column of D and group and factor the terms of (1) as follows:

$$D = (a_2b_3 - a_3b_2)c_1 - (a_1b_3 - a_3b_1)c_2 + (a_1b_2 - a_2b_1)c_3$$
$$= m(c_1) \cdot c_1 - m(c_2) \cdot c_2 + m(c_3) \cdot c_3$$

By repeated application of this procedure, we can prove that the expansion of D can be expressed in terms of the minors of the elements of any row or column in any of the following ways:

First row: $\qquad m(a_1)[a_1] - m(b_1)[b_1] + m(c_1)[c_1]$ $\qquad\qquad$ (2)

Second row: $\quad -m(a_2)[a_2] + m(b_2)[b_2] - m(c_2)[c_2]$ $\qquad\qquad$ (3)

Third row: $\qquad m(a_3)[a_3] - m(b_3)[b_3] + m(c_3)[c_3]$ $\qquad\qquad$ (4)

First column: $\quad m(a_1)[a_1] - m(a_2)[a_2] + m(a_3)[a_3]$ $\qquad\qquad$ (5)

Second column: $-m(b_1)[b_1] + m(b_2)[b_2] - m(b_3)[b_3]$ $\qquad\qquad$ (6)

Third column: $\quad m(c_1)[c_1] - m(c_2)[c_2] + m(c_3)[c_3]$ $\qquad\qquad$ (7)

It is convenient to use the following diagram to determine the sign that must be placed before the minor of the corresponding element:

$$\begin{matrix} + & - & + \\ - & + & - \\ + & - & + \end{matrix} \qquad \begin{vmatrix} a_1 & b_1 & c_1 \\ a_2 & b_2 & c_2 \\ a_3 & b_3 & c_3 \end{vmatrix} \qquad\qquad (8)$$

Example 1 Expand the determinant

$$D_1 = \begin{vmatrix} 3 & 2 & 4 \\ 1 & 5 & 2 \\ 4 & 7 & 6 \end{vmatrix}$$

in terms of the elements of the first row.

Solution $\quad$ In order to use (2), we need to know a_1, b_1, c_1, and $m(a_1)$, $m(b_1)$, and $m(c_1)$. In D_1 we have $a_1 = 3$, $b_1 = 2$, and $c_1 = 4$, and

$$m(a_1) = m(3) = \begin{vmatrix} 5 & 2 \\ 7 & 6 \end{vmatrix} \qquad m(b_1) = m(2) = \begin{vmatrix} 1 & 2 \\ 4 & 6 \end{vmatrix}$$

$$m(c_1) = m(4) = \begin{vmatrix} 1 & 5 \\ 4 & 7 \end{vmatrix}$$

Hence, by (2),

$$D_1 = \begin{vmatrix} 5 & 2 \\ 7 & 6 \end{vmatrix}(3) - \begin{vmatrix} 1 & 2 \\ 4 & 6 \end{vmatrix}(2) + \begin{vmatrix} 1 & 5 \\ 4 & 7 \end{vmatrix}(4)$$
$$= (30 - 14)(3) - (6 - 8)(2) + (7 - 20)(4)$$
$$= (16)(3) - (-2)(2) + (-13)(4) = 48 + 4 - 52 = 0$$

Example 2   Expand the determinant

$$D_2 = \begin{vmatrix} -2 & 4 & 3 \\ 1 & -5 & -6 \\ 3 & 1 & -2 \end{vmatrix}$$

in terms of the elements of the third column, and check the result by expanding in terms of the second row.

Solution   Since

$$c_1 = 3 \qquad c_2 = -6 \qquad c_3 = -2$$

and the signs in the third column of the diagram in (8) are $+$, $-$, $+$, we have

$$D_2 = \begin{vmatrix} 1 & -5 \\ 3 & 1 \end{vmatrix}(3) - \begin{vmatrix} -2 & 4 \\ 3 & 1 \end{vmatrix}(-6) + \begin{vmatrix} -2 & 4 \\ 1 & -5 \end{vmatrix}(-2)$$
$$= (1 + 15)(3) - (-2 - 12)(-6) + (10 - 4)(-2)$$
$$= 48 - 84 - 12 = -48$$

The signs in the second row of the diagram are $-$, $+$, $-$, and the elements are 1, -5, and -6. Hence,

$$D_2 = -\begin{vmatrix} 4 & 3 \\ 1 & -2 \end{vmatrix}(1) + \begin{vmatrix} -2 & 3 \\ 3 & -2 \end{vmatrix}(-5) - \begin{vmatrix} -2 & 4 \\ 3 & 1 \end{vmatrix}(-6)$$
$$= -(-8 - 3)(1) + (4 - 9)(-5) - (-2 - 12)(-6)$$
$$= 11 + 25 - 84 = -48$$

If one or more of the elements of a determinant are zero, it is advisable to expand the determinant in terms of the elements of the row or column that contains the greatest number of zeros.

Example 3   Expand the determinant

$$D_3 = \begin{vmatrix} 3 & 2 & 4 \\ 0 & 2 & 0 \\ 1 & 3 & 2 \end{vmatrix}$$

Solution   Here, the second row contains two zeros. Hence, we shall use this row to get the expansion. According to (3) and (8), we have

$$D_3 = - \begin{vmatrix} 2 & 4 \\ 3 & 2 \end{vmatrix} (0) + \begin{vmatrix} 3 & 4 \\ 1 & 2 \end{vmatrix} (2) - \begin{vmatrix} 3 & 2 \\ 1 & 3 \end{vmatrix} (0)$$

$$= 0 + (6 - 4)(2) - 0 = 4$$

Consider the system of equations

$$a_1 x + b_1 y + c_1 z = d_1$$
$$a_2 x + b_2 y + c_2 z = d_2$$
$$a_3 x + b_3 y + c_3 z = d_3$$

In Sec. 8.7 we solved a system, Eqs. (3) and (4), of two equations in two unknowns by addition and subtraction. If we do a similar thing here, we get, assuming $D \neq 0$,

Cramer's
Rule
$$x = \frac{D_x}{D} \qquad y = \frac{D_y}{D} \qquad z = \frac{D_z}{D} \tag{8.5}$$

where D, D_x, D_y, and D_z are the third-order determinants

$$D = \begin{vmatrix} a_1 & b_1 & c_1 \\ a_2 & b_2 & c_2 \\ a_3 & b_3 & c_3 \end{vmatrix} \qquad D_x = \begin{vmatrix} d_1 & b_1 & c_1 \\ d_2 & b_2 & c_2 \\ d_3 & b_3 & c_3 \end{vmatrix}$$

$$D_y = \begin{vmatrix} a_1 & d_1 & c_1 \\ a_2 & d_2 & c_2 \\ a_3 & d_3 & c_3 \end{vmatrix} \qquad D_z = \begin{vmatrix} a_1 & b_1 & d_1 \\ a_2 & b_2 & d_2 \\ a_3 & b_3 & d_3 \end{vmatrix}$$

Notice that D_x is the same as D except that to get D_x from D, the first column of D, which is $\begin{matrix} a_1 \\ a_2 \\ a_3 \end{matrix}$, is replaced by the column of constants $\begin{matrix} d_1 \\ d_2 \\ d_3 \end{matrix}$. The determinants D_y and D_z are obtained from D similarly (by replacing the second and third columns of D with the column of constants).

Example 4   Solve the equations

$$2x - 3y + 4z = 19 \tag{9}$$
$$x + 2y - 2z = -6 \tag{10}$$
$$3x + y + z = 8 \tag{11}$$

Solution   In order to use Cramer's rule, we need the expansions of D, D_x, D_y, and D_z. Expanding by the first row gives

$$D = \begin{vmatrix} 2 & -3 & 4 \\ 1 & 2 & -2 \\ 3 & 1 & 1 \end{vmatrix} = \begin{vmatrix} 2 & -2 \\ 1 & 1 \end{vmatrix}(2) - \begin{vmatrix} 1 & -2 \\ 3 & 1 \end{vmatrix}(-3) + \begin{vmatrix} 1 & 2 \\ 3 & 1 \end{vmatrix}(4)$$

$$= (2+2)(2) - (1+6)(-3) + (1-6)(4) = 8 + 21 - 20 = 9$$

Using the second row in D_x gives

$$D_x = \begin{vmatrix} 19 & -3 & 4 \\ -6 & 2 & -2 \\ 8 & 1 & 1 \end{vmatrix}$$

$$= -\begin{vmatrix} -3 & 4 \\ 1 & 1 \end{vmatrix}(-6) + \begin{vmatrix} 19 & 4 \\ 8 & 1 \end{vmatrix}(2) - \begin{vmatrix} 19 & -3 \\ 8 & 1 \end{vmatrix}(-2)$$

$$= -(-3-4)(-6) + (19-32)(2) - (19+24)(-2)$$

$$= -42 - 26 + 86 = 18$$

Using the second column in D_y gives

$$D_y = \begin{vmatrix} 2 & 19 & 4 \\ 1 & -6 & -2 \\ 3 & 8 & 1 \end{vmatrix}$$

$$= -\begin{vmatrix} 1 & -2 \\ 3 & 1 \end{vmatrix}(19) + \begin{vmatrix} 2 & 4 \\ 3 & 1 \end{vmatrix}(-6) - \begin{vmatrix} 2 & 4 \\ 1 & -2 \end{vmatrix}(8)$$

$$= -(1+6)(19) + (2-12)(-6) - (-4-4)(8)$$

$$= -133 + 60 + 64 = -9$$

Using the third column in D_z gives

$$D_z = \begin{vmatrix} 2 & -3 & 19 \\ 1 & 2 & -6 \\ 3 & 1 & 8 \end{vmatrix}$$

$$= \begin{vmatrix} 1 & 2 \\ 3 & 1 \end{vmatrix}(19) - \begin{vmatrix} 2 & -3 \\ 3 & 1 \end{vmatrix}(-6) + \begin{vmatrix} 2 & -3 \\ 1 & 2 \end{vmatrix}(8)$$

$$= (1-6)(19) - (2+9)(-6) + (4+3)(8)$$

$$= -95 + 66 + 56 = 27$$

By Cramer's rule, in (8.5), we have

$$x = \frac{18}{9} = 2 \qquad y = \frac{-9}{9} = -1 \qquad z = \frac{27}{9} = 3$$

Verification

$2(2) - 3(-1) + 4(3) = 4 + 3 + 12 = 19$ **from Eq. (9)**

$2 + 2(-1) - 2(3) = 2 - 2 - 6 = -6$ **from Eq. (10)**

$3(2) + (-1) + 3 = 6 - 1 + 3 = 8$ **from Eq. (11)**

Example 5   Use Cramer's rule to get the solution set of

$$2x + 5y + 3z = 3$$
$$x + y + z = 0$$
$$3x - y + 2z = -5$$

Solution The determinant of the coefficients is

$$D = \begin{vmatrix} 2 & 5 & 3 \\ 1 & 1 & 1 \\ 3 & -1 & 2 \end{vmatrix} = \begin{vmatrix} 1 & 1 \\ -1 & 2 \end{vmatrix}(2) - \begin{vmatrix} 1 & 1 \\ 3 & 2 \end{vmatrix}(5) + \begin{vmatrix} 1 & 1 \\ 3 & -1 \end{vmatrix}(3)$$

$$= (2 + 1)(2) - (2 - 3)(5) + (-1 - 3)(3)$$
$$= 6 + 5 - 12 = -1$$

The expansion of D was obtained in terms of the elements of the first row.

By replacing the appropriate column of D by the column of constant terms, we get D_x, D_y, and D_z. Since the column of constant terms contains one zero, we expand each of these determinants in terms of this column, and hence each expansion will contain only two terms. Each of these determinants and its expansion follows:

$$D_x = \begin{vmatrix} 3 & 5 & 3 \\ 0 & 1 & 1 \\ -5 & -1 & 2 \end{vmatrix} = \begin{vmatrix} 1 & 1 \\ -1 & 2 \end{vmatrix}(3) + \begin{vmatrix} 5 & 3 \\ 1 & 1 \end{vmatrix}(-5)$$

$$= (2 + 1)(3) + (5 - 3)(-5) = 9 - 10 = -1$$

$$D_y = \begin{vmatrix} 2 & 3 & 3 \\ 1 & 0 & 1 \\ 3 & -5 & 2 \end{vmatrix} = -\begin{vmatrix} 1 & 1 \\ 3 & 2 \end{vmatrix}(3) - \begin{vmatrix} 2 & 3 \\ 1 & 1 \end{vmatrix}(-5)$$

$$= -(2 - 3)(3) - (2 - 3)(-5) = 3 - 5 = -2$$

$$D_z = \begin{vmatrix} 2 & 5 & 3 \\ 1 & 1 & 0 \\ 3 & -1 & -5 \end{vmatrix} = \begin{vmatrix} 1 & 1 \\ 3 & -1 \end{vmatrix}(3) + \begin{vmatrix} 2 & 5 \\ 1 & 1 \end{vmatrix}(-5)$$

$$= (-1 - 3)(3) + (2 - 5)(-5) = -12 + 15 = 3$$

Hence, the values of x, y, and z are

$$x = \frac{D_x}{D} = \frac{-1}{-1} = 1 \qquad y = \frac{D_y}{D} = \frac{-2}{-1} = 2 \qquad z = \frac{D_z}{D} = \frac{3}{-1} = -3$$

Therefore, the solution is $(1, 2, -3)$.

EXERCISE 8.6 Third-Order Determinants and Cramer's Rule

Find the values of the determinants in Probs. 1 to 20.

1 $\begin{vmatrix} 2 & 4 & 1 \\ -3 & -3 & -1 \\ 2 & 1 & 2 \end{vmatrix}$ **2** $\begin{vmatrix} 5 & 4 & 2 \\ -1 & 2 & -1 \\ 3 & -2 & 4 \end{vmatrix}$ **3** $\begin{vmatrix} -4 & -3 & -1 \\ 3 & 2 & 3 \\ 4 & 1 & 4 \end{vmatrix}$

4
$$\begin{vmatrix} 2 & 3 & 5 \\ 4 & 3 & 4 \\ 1 & 1 & 3 \end{vmatrix}$$

5
$$\begin{vmatrix} 3 & 1 & -2 \\ 4 & 8 & 7 \\ -2 & 6 & 11 \end{vmatrix}$$

6
$$\begin{vmatrix} -5 & -1 & 2 \\ -4 & 2 & 0 \\ -2 & -1 & 3 \end{vmatrix}$$

7
$$\begin{vmatrix} 3 & 2 & 3 \\ 2 & 3 & 2 \\ 1 & -1 & 1 \end{vmatrix}$$

8
$$\begin{vmatrix} 4 & 3 & 8 \\ 3 & 2 & 4 \\ 1 & 3 & 5 \end{vmatrix}$$

9
$$\begin{vmatrix} 2 & 2 & 1 \\ 3 & 0 & 1 \\ 5 & -1 & 3 \end{vmatrix}$$

10
$$\begin{vmatrix} -7 & -4 & -2 \\ 0 & 2 & -1 \\ 2 & 0 & 3 \end{vmatrix}$$

11
$$\begin{vmatrix} 1 & 3 & 1 \\ 1 & 2 & 3 \\ 4 & 4 & 5 \end{vmatrix}$$

12
$$\begin{vmatrix} 1 & 1 & 3 \\ -3 & -2 & -1 \\ 1 & 3 & 4 \end{vmatrix}$$

13
$$\begin{vmatrix} 4 & 3 & -4 \\ 2 & 0 & 1 \\ 1 & 1 & -3 \end{vmatrix}$$

14
$$\begin{vmatrix} 8 & 7 & 4 \\ -1 & 3 & -2 \\ 10 & 1 & 8 \end{vmatrix}$$

15
$$\begin{vmatrix} 7 & 0 & 2 \\ 4 & 2 & 6 \\ 2 & -1 & 0 \end{vmatrix}$$

16
$$\begin{vmatrix} 8 & 6 & 5 \\ 4 & -2 & -3 \\ 12 & 14 & 13 \end{vmatrix}$$

17
$$\begin{vmatrix} 4 & 0 & 1 \\ 3 & 4 & -2 \\ -8 & 0 & 3 \end{vmatrix}$$

18
$$\begin{vmatrix} -2 & 5 & 3 \\ 6 & 5 & 0 \\ 5 & 4 & 0 \end{vmatrix}$$

19
$$\begin{vmatrix} 0 & 0 & 5 \\ 3 & -1 & -2 \\ -4 & -5 & 2 \end{vmatrix}$$

20
$$\begin{vmatrix} 6 & 18 & 1 \\ 11 & 3 & 2 \\ 0 & 2 & 0 \end{vmatrix}$$

Verify the equations in each of Probs. 21 to 28 by expanding the determinants and collecting like terms.

21
$$\begin{vmatrix} 3 & 8 & 1 \\ 2 & 4 & 6 \\ a & b & c \end{vmatrix} = \begin{vmatrix} 3 & 8 & 1 \\ 2 & 4 & 6 \\ 2+a & 4+b & 6+c \end{vmatrix}$$

22
$$\begin{vmatrix} a & b & c \\ 3 & 2 & 1 \\ 6 & 7 & 9 \end{vmatrix} = \begin{vmatrix} a & 3 & 6 \\ b & 2 & 7 \\ c & 1 & 9 \end{vmatrix}$$

23
$$\begin{vmatrix} a & b & c \\ 1 & 2 & 3 \\ 4 & 5 & 6 \end{vmatrix} = \begin{vmatrix} a & b & c \\ 1 & 2 & 3 \\ 4+2a & 5+2b & 6+2c \end{vmatrix}$$

24
$$\begin{vmatrix} a & b & c \\ 1 & -4 & 8 \\ 2a & 2b & 2c \end{vmatrix} = 0$$

25
$$\begin{vmatrix} 3 & 1 & 3 \\ 3a & 3b & 3c \\ 4 & 1 & 2 \end{vmatrix} = 3 \begin{vmatrix} 3 & 1 & 3 \\ a & b & c \\ 4 & 1 & 2 \end{vmatrix}$$

26
$$\begin{vmatrix} a & b & c \\ 4 & 2 & 4 \\ 3 & 1 & 7 \end{vmatrix} = - \begin{vmatrix} -a & -b & -c \\ 4 & 2 & 4 \\ 3 & 1 & 7 \end{vmatrix}$$

27
$$\begin{vmatrix} a & b & c \\ d & e & f \\ 1 & 2 & 3 \end{vmatrix} = \begin{vmatrix} d & e & f \\ a & b & c \\ -1 & -2 & -3 \end{vmatrix}$$

28
$$\begin{vmatrix} a & b & c \\ d & e & f \\ a+d & b+e & c+f \end{vmatrix} = 0$$

By use of Cramer's rule, simultaneously solve the equations in each of Probs. 29 to 48.

29
$$x + 3y - z = 6$$
$$2x - y + 3z = 3$$
$$3x + 2y + 4z = 11$$

30
$$3x - y + 2z = 5$$
$$-x + 4y + 2z = 11$$
$$2x + 3y + 2z + 12$$

31
$$4x + y - 2z = 1$$
$$3x + y + 3z = 10$$
$$6x + 2y + z = 10$$

32
$$3x + 2y - z = 7$$
$$x - 3y + 5z = 4$$
$$4x + 3y + 4z = 15$$

33
$$2x + 7y - 3z = 19$$
$$3x - 4y + 6z = 6$$
$$7x - y + 3z = 17$$

34
$$3x - y + z = 7$$
$$4x + 2y - 7z = -4$$
$$10x - 3z = 14$$

35
$$3x + 8y + 4z = 12$$
$$x + 5y + 2z = 6$$
$$7x + 21y + 11z = 33$$

36
$$8x + 2y + 3z = 11$$
$$x + y + z = 4$$
$$6x + y + z = 4$$

37
$$2x + y + 3z = 5$$
$$x + 2y + z = 5$$
$$5x + 4y + 8z = 16$$

38
$$3x - y - 2z = 2$$
$$7x - 4y + 3z = 9$$
$$4x - 3y + 4z = 6$$

39
$$4x - 3y + 5z = 17$$
$$6x + 2y - 3z = 15$$
$$14x - 4y + 9z = 51$$

40
$$2x - y + 3z = 4$$
$$4x - 3y + 6z = 7$$
$$10x - 6y + 7z = 11$$

41
$$3x + 4y + 3z = 3$$
$$3x - 8y + 6z = 1$$
$$6x - 4y + 3z = 2$$

42
$$4x + 2y + 3z = 3$$
$$8x - 6y + 9z = 2$$
$$4x - 8y - 3z = -4$$

43
$$3x + 2y + 3z = 4$$
$$-6x + 2y - 9z = -6$$
$$-3x + 4y - 3z = -1$$

44
$$5x + 3y + 5z = 4$$
$$5x - 6y + 10z = 2$$
$$5x + 5z = 3$$

45
$$8x + 5y + 2z = 8$$
$$8x - 15y + 4z = -7$$
$$-10y + 6z = -5$$

46
$$3x + 3y + 2z = 10$$
$$9x - 6y + 8z = 8$$
$$9x - 3y + 10z = 16$$

47
$$3x + 7y + 7z = 4$$
$$-6x + 7y - 14z = -5$$
$$3x + 14y - 7z = 3$$

48
$$4x + 5y + 6z = 3$$
$$-4x - 5y + 12z = 0$$
$$6z = 1$$

It can be shown that the area of a triangle with vertices at (a, b), (c, d), and (e, f) is the absolute value of the determinant

$$\frac{1}{2} \begin{vmatrix} a & b & 1 \\ c & d & 1 \\ e & f & 1 \end{vmatrix}$$

Use this formula in each of Probs. 49 to 52 to find the area of the triangle whose vertices are given.

49 $(2, 3), (-4, -1), (6, -5)$

50 $(2, 4), (-3, -2), (12, 16)$

51 $(2, 4), (-1, 6), (5, -3)$

52 $(-1, 5), (3, 0), (7, 2)$

8.9 SUMMARY

Systems of two linear equations in two unknowns, and three linear equations in three unknowns, are discussed. The methods of solution presented include addition and subtraction, substitution, and Cramer's rule. In addition, graphing is discussed for two equations in two unknowns. Determinants of order 2 and 3 are included as a prelude to Cramer's rule. Problems from various disciplines which lead to a system of linear equations are also presented.

EXERCISE 8.7 Review

1 Show that $4x - 14y = 6$ and $6x = 21y + 9$ are dependent.

2 Show that $9x + 12y = -18$ and $-12x - 16y = 30$ are inconsistent.

Solve Probs. 3 to 6 graphically.

3 $2x + 5y = 3$
  $x - 2y = 6$

4 $3x - 5y = 1$
  $-x - y = 5$

5 $4x + 3y = 27$
  $2x - y = 1$

6 $2x + 5y = 15$
  $3x + 7y = 21$

Solve Probs. 7 to 10 by addition and subtraction.

7 $2x + 5y = 41$
  $2x - 7y = -43$

8 $3x + 5y = 23$
  $4x - 5y = 19$

9 $5x + 3y = 1$
  $8x - 5y = -67$

10 $4x - 5y = 4$
  $6x - 7y = 10$

Solve Probs. 11 to 14 by substitution.

11 $x + 8y = 13$
  $3x - 2y = 13$

12 $5x + y = 18$
  $3x - 2y = -10$

13 $3x - 4y = -4$
  $5x - 4y = 12$

14 $4x - 5y = 3$
  $3x - 4y = 0$

Solve Probs. 15 to 20 by Cramer's rule.

15 $6x + y = 26$
  $3x - y = 1$

16 $3x - 10y = -4$
  $x + 2y = 20$

17 $7x + 2y = -1$
  $5x + 3y = 15$

18 $4x + 5y = 0$
  $3x - 4y = 31$

19 $3x + 2y - z = 3$
  $4x + y - 2z = -2$
  $7x + 3y + z = 17$

20 $2x + y - 5z = 13$
  $3x - 2y - 3z = 1$
  $7x + 10z = 4$

Solve Probs. 21 to 24 by any method.

21 $6x + 4y - 3z = -1$
  $4x + 5y - 3z = -1$
  $2x - y + z = 5$

22 $3x + 4y + z = 3$
  $6x - 4y - 2z = -1$
  $-3x + 8y + 6z = 7$

23 $3x + 4y - 2z = 5$
  $-3x + 8y + 3z = -3$
  $12y + 3z = 0$

24 $\dfrac{3}{x-1} - \dfrac{1}{2y+1} = 1$

  $\dfrac{-4}{x-1} + \dfrac{3}{2y+1} = 7$

25 Solve the system

$$3(x - y) + (-2x + y) = 2$$
$$5(x - y) + 2(-2x + y) = 1$$

in two ways: (*a*) Multiply and collect terms, and then use substitution. (*b*) Solve the equations for $x - y$ and $-2x + y$, and then find x and y.

Verify the equations in Probs. 26 to 30.

26 $\begin{vmatrix} 3 & 5 \\ 2 & 4 \end{vmatrix} = \begin{vmatrix} 3 & 11 \\ 2 & 8 \end{vmatrix}$

27 $\begin{vmatrix} a & b \\ 5 & 6 \end{vmatrix} = \begin{vmatrix} a & 5 + 2a \\ b & 6 + 2b \end{vmatrix}$

28 $\begin{vmatrix} 2 & a & 2 + 3a \\ 4 & b & 4 + 3b \\ 5 & c & 5 + 3c \end{vmatrix} = 0$

29 $\begin{vmatrix} 6 & 4 & -2 \\ -5 & -2 & -1 \\ 1 & 3 & -5 \end{vmatrix} = 0$

30
$$\begin{vmatrix} 6 & 1 & 2 \\ 8 & 2 & 3 \\ 5 & 3 & 1 \end{vmatrix} = \begin{vmatrix} 2 & 1 & 2 \\ 3 & 2 & 3 \\ 2 & 3 & 1 \end{vmatrix} + \begin{vmatrix} 4 & 1 & 2 \\ 5 & 2 & 3 \\ 3 & 3 & 1 \end{vmatrix}$$

31 A landlord received $2400 in rent on two apartments in 1 year, and one of them rented for $20 per month more than the other. How much did he receive per month for each if the more expensive apartment was vacant for 2 months?

32 A druggist mixed one compound worth $1.40 per gram with another worth $1.80 per gram in order to obtain 50 grams of a mixture that sold for $1.56 per gram. How much of each compound did she use?

33 For a charity party, 802 tickets, priced at $5, $10, and $25, were sold. If 10 times as many $5 tickets as $10 tickets were sold, and the total sales were $5000, how many tickets of each type were sold?

34 Find the area of the triangle with vertices $(7, 2)$, $(4, -1)$, and $(3, 6)$.

9
systems of quadratic equations in two variables

The general quadratic equation in two variables is an equation of the type

$$Ax^2 + Bxy + Cy^2 + Dx + Ey + F = 0$$

where at least one of A, B, and C is not zero.

Conic Section   It is proved in analytic geometry that the graph of a quadratic equation in two variables is a member of the important class of curves called *conic sections*, which includes the circle, the ellipse, the hyperbola, and the parabola (see Fig. 9.1). In the next

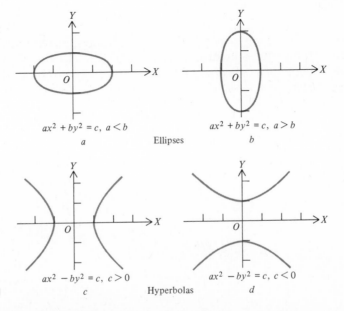

$ax^2 + by^2 = c,\ a < b$
a

Ellipses

$ax^2 + by^2 = c,\ a > b$
b

$ax^2 - by^2 = c,\ c > 0$
c

Hyperbolas

$ax^2 - by^2 = c,\ c < 0$
d

FIGURE 9.1

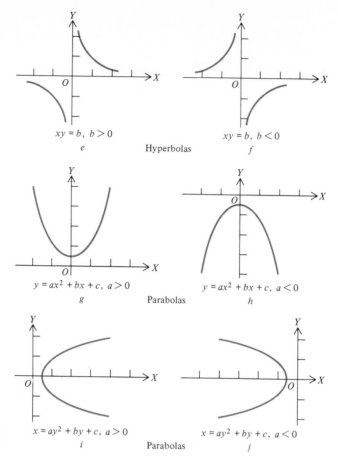

FIGURE 9.1
(continued)

two sections we shall discuss the graphs of special cases of a quadratic equation in two variables. In the remainder of the chapter we discuss methods for solving systems of two equations, at least one of which is a quadratic equation in two variables.

9.1 GRAPHS OF QUADRATIC EQUATIONS IN TWO VARIABLES

We shall discuss the graphs of the following special cases of the general quadratic:

Equation		Graph
$ax^2 + by^2 = c$ where $\begin{aligned} a &> 0 \\ b &> 0 \\ c &> 0 \end{aligned}$		An ellipse if $a \neq b$ and a circle if $a = b$. It may be long right and left or up and down. See Fig. 9.1a and b.

Equation			Graph
$ax^2 - by^2 = c$	where	$a > 0$ $b > 0$ $c \neq 0$	A hyperbola. It opens right and left if $c > 0$, and up and down if $c < 0$. See Fig. 9.1c and d.
$xy = b$	where	$b \neq 0$	A hyperbola. It is in quadrants I and III if $b > 0$, and in quadrants II and IV if $b < 0$. See. Fig. 9.1e and f.
$y = ax^2 + bx + c$	where	$a \neq 0$	A parabola. It opens up if $a > 0$, and down if $a < 0$. See Fig. 9.1g and h.
$x = ay^2 + by + c$	where	$a \neq 0$	A parabola. It opens to the right if $a > 0$, and to the left if $a < 0$. See Fig. 9.1i and j.

As in Chap. 7, the steps followed in constructing the graph are as follows:

1 Solve† the equation for y in terms of x.
2 Assign several values to x, compute each corresponding value of y, and arrange the associated pairs of values in tabular form.
3 Plot the points determined by the pairs of values, and then draw a smooth curve through them.

We shall illustrate the method with several examples.

Example 1 Construct the graph of

$$x^2 + y^2 = 25 \tag{1}$$

Solution 1 If we perform the operations suggested by the steps listed above, we have

$$y = \pm\sqrt{25 - x^2}$$

2 Assign to x the integers from -5 to 5, inclusive, since that is the domain, and compute each corresponding value of y. For example, if $x = -5$, then

$$y = \pm\sqrt{25 - (-5)^2} = \pm\sqrt{25 - 25} = 0$$

Similarly, if $x = 2$, then

$$y = \pm\sqrt{25 - (2)^2} = \pm\sqrt{25 - 4} = \pm\sqrt{21} = \pm4.6$$

---

†If the equation is easier to solve for x than for y, we solve it for x. Then, in reading the succeeding steps, we interchange x and y.

When a similar computation is performed for each of the other values assigned to x and the results are arranged in tabular form, we have the following:

x	-5	-4	-3	-2	-1	0	1	2	3	4	5
y	0	± 3	± 4	± 4.6	± 4.9	± 5	± 4.9	± 4.6	± 4	± 3	0

3 Note that in this table we have two values of y for each x except $x = -5$ and $x = 5$. The pair of values $x = 3$, $y = \pm 4$ determines the two points $(3, 4)$ and $(3, -4)$.

With this understanding, if we plot the points determined by the table and join them by a smooth curve, we have the graph in Fig. 9.2.

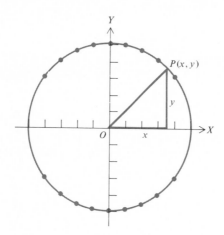

FIGURE 9.2

We can readily see that the curve is a circle, since the coordinates (x, y) of any point P on it satisfy Eq. (1); that is, the sum of their squares is 25. Furthermore, by looking at the figure, we see that the square of the distance OP of P from the center is $x^2 + y^2$. Hence, any point whose coordinates satisfy Eq. (1) is at a distance of 5 from the origin.

In general, by similar reasoning, we conclude that the graph of $x^2 + y^2 = r^2$ is a circle of radius r, and the graph of $ax^2 + ay^2 = c$ is a circle of radius $\sqrt{c/a}$ if a and c have the same sign.

Example 2   Construct the graph of $4x^2 + 9y^2 = 36$.

Solution   **1**   Solving for y, we have

$$y = \pm\sqrt{\frac{36 - 4x^2}{9}} = \pm\tfrac{2}{3}\sqrt{9 - x^2}$$

2 We note here that if $x^2 > 9$, the radicand is negative and y is imaginary. Hence, the graph exists only for values of x from -3 to 3, inclusive. Therefore, we assign to x the integers, 0, ± 1, ± 2, ± 3, compute each corresponding value of y, arrange the results in a table, and get

x	-3	-2	-1	0	1	2	3
y	0	± 1.5	± 1.9	± 2	± 1.9	± 1.5	0

3 When we construct the graph determined by this table, we get the curve in Fig. 9.3.

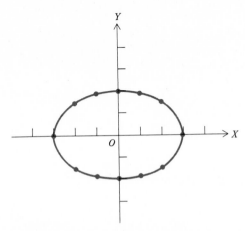

FIGURE 9.3

By referring to Fig. 9.1, we surmise that this curve is an ellipse. The proof that the equation

$$ax^2 + by^2 = c$$

with a, b, and c positive always defines an ellipse is beyond the scope of this book. However, the statement is true, and it is helpful to remember this fact when dealing with such an equation.

Example 3 Construct the graph of the equation $3x^2 - 4y^2 = 12$.

Solution **1** Proceeding as before, we have

$$y = \pm\sqrt{\frac{3x^2 - 12}{4}} = \pm\tfrac{1}{2}\sqrt{3(x^2 - 4)}$$

2 In this case, we notice that if $x^2 < 4$, the radicand is negative and y is imaginary. Hence, the graph does not exist between $x = -2$ and $x = 2$. If, however, x is either 2 or -2, y is zero. Thus, the curve must extend to the right from $(2, 0)$

and to the left from $(-2, 0)$.   Hence, we assign the values $\pm2, \pm3, \pm4, \pm5, \pm7, \pm9$, to x, proceed as in the previous example, and get the following table:

x	-9	-7	-5	-4	-3	-2	2	3	4	5	7	9
y	±7.6	±5.8	±4	±3	±1.9	0	0	±1.9	±3	±4	±5.8	±7.6

3 When the above points are plotted and the graph is drawn, we obtain the curve in Fig. 9.4.

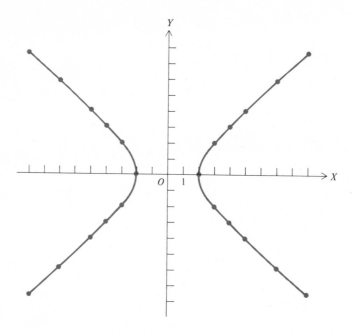

FIGURE 9.4

Again, by referring to Fig. 9.1, we conclude that this curve is a hyperbola.

This example illustrates the fact that if $a > 0$ and $b > 0$, an equation of the type

$$ax^2 - by^2 = c$$

defines a hyperbola.   If c is positive, the curve is in the same general position as that in Fig. 9.4.   However, if c is negative, the two branches of the curve cross the Y axis instead of the X axis and open upward and downward.

EQUATIONS OF THE TYPE $y = ax^2 + bx + c$ **or** $x = ay^2 + by + c$

Example 4   Construct the graph of $x^2 - 4x - 4y - 4 = 0$.

Solution   By solving the equation

$$x^2 - 4x - 4y - 4 = 0$$

for y, we obtain

$$y = \tfrac{1}{4}x^2 - x - 1$$

and this is of the first type mentioned above. We avoid fractions here if we substitute only even values for x. If we use the values $-4, -2, 0, 2, 4, 6, 8$ for x and proceed as before, we get the following table of corresponding values of x and y:

x	-4	-2	0	2	4	6	8
y	7	2	-1	-2	-1	2	7

Plotting the above points and drawing the graph, we get the curve in Fig. 9.5.

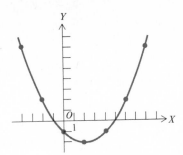

FIGURE 9.5

Example 5 Construct the graph of $2y^2 + 1 = x + 4y$.

Solution Since this equation contains only one term in x, the algebra is easier if we solve for x in terms of y and get

$$x = 2y^2 - 4y + 1$$

Now we assign values to y and compute each corresponding value of x. The following table was obtained by using the values $-2, -1, 0, 1, 2, 3, 4$ for y:

x	17	7	1	-1	1	7	17
y	-2	-1	0	1	2	3	4

Now we plot the graph and obtain the curve in Fig. 9.6.

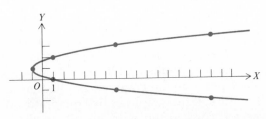

FIGURE 9.6

The curves in Figs. 9.5 and 9.6 are parabolas. It is proved in analytic geometry that an equation of the type $y = ax^2 + bx + c$

defines a parabola opening upward if a is positive, and a parabola opening downward if a is negative. Furthermore, an equation of the type

$$x = ay^2 + by + c$$

defines a parabola opening to the right if a is positive, and a parabola opening to the left if a is negative.

EXERCISE 9.1   Graphs of Equations in Two Variables

Sketch the graph of the function or relation defined by the equation in each of Probs. 1 to 52.

1 $2x + y = 6$	2 $x + 3y = 5$	3 $x - y = 2$
4 $3x - 2y = 12$	5 $y^2 = 2x$	6 $y^2 = -4x$
7 $x^2 = -6y$	8 $x^2 = 8y$	9 $x^2 - 2x = 2y + 1$
10 $x^2 - 4x - 4y = -8$	11 $y^2 + 6y - 4x = -1$	12 $y^2 + 4y + 6x = -10$
13 $x^2 + y^2 = 9$	14 $x^2 + y^2 = 25$	15 $x^2 + y^2 - 2x - 4y + 1 = 0$
16 $x^2 + y^2 + 6x - 2y = 6$	17 $x^2 + 4y^2 = 4$	18 $9x^2 + y^2 = 9$
19 $4x^2 + y^2 = 4$	20 $x^2 + 16y^2 = 16$	21 $4x^2 + 9y^2 = 36$
22 $9x^2 + 4y^2 = 36$	23 $9x^2 + 16y^2 = 144$	24 $16x^2 + 9y^2 = 144$
25 $x^2 - 4y^2 = 4$	26 $x^2 - 9y^2 = 9$	27 $16x^2 - y^2 = 16$
28 $4x^2 - 9y^2 = 36$	29 $xy = 1$	30 $xy = 5$
31 $xy = -4$	32 $xy = -6$	33 $2x + xy = 10$
34 $3x + xy = 12$	35 $4x^2 + xy = 5$	36 $2y^2 - xy = 4$
37 $y = x^2$	38 $y = 2x^2$	39 $y = -3x^2$
40 $y = -6x^2$	41 $y = x^2 + 1$	42 $y = x^2 + 5$
43 $y = 2x^2 - 3$	44 $y = 2x^2 - 7$	45 $y = x^2 + x + 1$
46 $y = x^2 + 3x + 1$	47 $y = x^2 - 3x + 1$	48 $y = x^2 - 6x + 1$
49 $y = (x - 1)^2$	50 $y = (x - 1)^2 + 4$	51 $y = (x + 3)^2$
52 $y = (x + 3)^2 - 5$		

9.2   GRAPHICAL SOLUTION OF TWO QUADRATIC EQUATIONS IN TWO VARIABLES

A simultaneous solution set of a system of two quadratic equations in two variables is the set of all pairs of corresponding values of x and y that satisfies both equations. Now the coordinates of each point on the graph of an equation satisfy the equation. Therefore, if the graphs of two quadratic equations in two unknowns are constructed with respect to the same axes, the coordinates of each point of intersection form a simultaneous solution pair of the equations, since these points are on both

graphs.   Consequently, to obtain the simultaneous solution set
of a system of two quadratic equations in two variables graphically, we construct the graphs of the equations with respect to the
same axes and then estimate the coordinates of their points of
intersection.   It is usually easy and worthwhile to find the x and
y intercepts, that is, the points where $y = 0$ and $x = 0$.

Example   Obtain the simultaneous solution set of the system of equations

$$y = x^2 - 4 \tag{1}$$
$$9x^2 + 25y^2 = 225 \tag{2}$$

by the graphical method.

Solution   We construct the graph of Eq. (1) by use of the following table:

x	0	±1	±2	±3	±4
y	−4	−3	0	5	12

We thus obtain the parabola in Fig. 9.7.
    The graph of Eq. (2) is an ellipse, and we obtain this graph by
use of the following table:

x	−5	−4	−2	−1	0	1	2	4	5
y	0	±1.8	±2.7	±2.9	±3	±2.9	±2.7	±1.8	0

The ellipse is also shown in Fig. 9.7.

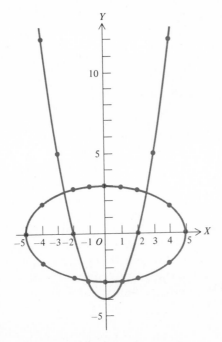

FIGURE 9.7

The two graphs intersect at the points whose coordinates to one decimal place appear to be $(2.5, 2.5)$, $(-2.5, 2.5)$, $(1, -3)$, and $(-1, -3)$. Therefore, the simultaneous solution set of (1) and (2) is $\{(2.5, 2.5), (-2.5, 2.5), (1, -3), (-1, -3)\}$ where the elements are accurate to one decimal place.

EXERCISE 9.2   Graphical Solution

Find the exact solution in Probs. 1 to 8, the solution to the nearest half unit in Probs. 9 to 16, and the number of solutions in Probs. 17 to 28.

1 $x + y = 3$
   $y^2 = 4x$

2 $y^2 = 5x - 35$
   $x^2 + y^2 = 169$

3 $4x^2 - 5y^2 = -16$
   $y^2 = 4x$

4 $y^2 = 8x$
   $x^2 = y$

5 $y^2 = x$
   $4x^2 - 37y^2 = -9$

6 $2x^2 + y^2 = 11$
   $9x^2 - 4y^2 = -27$

7 $4x^2 + y^2 = 100$
   $12x^2 - 5y^2 = 12$

8 $x^2 + 4y^2 = 4$
   $x^2 + 9y^2 = 9$

9 $2x - y = 1$
   $y^2 = 2x$

10 $2x + 3y = 11$
    $x^2 + y^2 = 14$

11 $4x - 3y = 0$
    $x^2 + 4y^2 = 41$

12 $2x + y = 6$
    $9x^2 - y^2 = 1$

13 $x - 3y = 1$
    $x^2 - 4y^2 = 6$

14 $x + 5y = 5$
    $y^2 - x^2 = 1$

15 $y = 5x$
    $y^2 - 9x^2 = 9$

16 $x - 3y = -1$
    $xy = 3$

17 $y = 5 - x$
    $xy = 5$

18 $3x + 4y = 3$
    $xy = -1$

19 $y = -3x$
    $xy = -8$

20 $y^2 = 2x$
    $9x^2 + y^2 = 9$

21 $y^2 = 6x$
    $x^2 + 4y^2 = 4$

22 $x^2 + y^2 = 1$
    $9x^2 - 4y^2 = 36$

23 $2y = x^2 - 12$
    $y^2 = x$

24 $x^2 = y + 3$
    $y^2 = 3x$

25 $3x^2 + 2y^2 = 35$
    $y^2 = 2(x + 5)$

26 $4x^2 - 5y^2 = -16$
    $y^2 = 5x$

27 $3x^2 - 13y^2 = -108$
    $y^2 = 3x$

28 $x^2 + y^2 = 9$
    $x^2 = 2y + 1$

9.3   ALGEBRAIC SOLUTIONS

As we stated in Sec. 9.2, a simultaneous solution of a system of two quadratic equations in two variables is an ordered pair of numbers (x, y) that satisfies each equation. We shall call the set of pairs of numbers that satisfy both equations in the system the *Solution Set*   *solution set* of the system. As the example in Sec. 9.2 illustrates, there are usually four pairs in the solution set of a system if both equations are quadratic. If, however, one equation is linear and the other is quadratic, the graphs of the equations are a straight line and one of the conic sections. Since these two graphs can intersect in, at most, two points, there are at most two pairs of numbers in the solution set.

The general method for solving a system of equations in two variables, with at least one of the equations quadratic, is the same as that used in Chap. 5. The first step in the method is to combine the equations in such a way as to obtain one equation in one variable, each of whose roots is one member of a pair in the solution set. This process is called *eliminating a variable*. After one number in each pair of the solution set is obtained, the other is determined by substitution. Frequently, the elimination of a variable yields an equation of the fourth degree. In such cases the completion of the process of solving requires more advanced methods than those presented in this book. We shall confine our discussion and exercises, however, to problems that can be solved by the available methods. In the remainder of this chapter we shall discuss the algebraic methods generally used.

Eliminating a Variable

9.4   ELIMINATION BY SUBSTITUTION

If in a system of two equations in two variables, one equation can be solved for one variable in terms of the other, this variable can be eliminated by substitution. We shall assume that the variables are x and y, that one equation can be solved for y in terms of x, and that the solution is in the form $y = f(x)$. We then replace y in the other equation by $f(x)$ and obtain an equation involving only x. We shall designate this equation by $F(x) = 0$. Next, we solve $F(x) = 0$ for x, and each root obtained will be the first number in each number pair of the solution set. We complete the process by substituting each root of $F(x) = 0$ in $y = f(x)$ to obtain the corresponding values of y. Finally, we arrange the corresponding values of x and y in pairs in this way, (x_1, y_1), (x_2, y_2), (x_3, y_3), (x_4, y_4), and thus have the solution set.

We shall illustrate the method with three examples. The first shows the process of solving a system containing *a linear equation in two variables and a quadratic equation in two variables*. The second and third illustrate the method for solving a system of *two quadratic equations in two variables where one equation is easily solvable for one variable in terms of the other*.

Example 1   Solve the following system of equations:

$$x^2 + 2y^2 = 54 \tag{1}$$
$$2x - y = -9 \tag{2}$$

Solution   We first solve Eq. (2) for y and get

$$y = 2x + 9 \tag{3}$$

Next, we replace y in Eq. (1) by $2x + 9$ and get

$$x^2 + 2(2x + 9)^2 = 54 \tag{4}$$

which we solve as follows:

$x^2 + 2(4x^2 + 36x + 81) = 54$	squaring $2x + 9$
$x^2 + 8x^2 + 72x + 162 = 54$	by distributive axiom
$x^2 + 8x^2 + 72x + 162 - 54 = 0$	adding -54 to each member
$9x^2 + 72x + 108 = 0$	combining similar terms
$x^2 + 8x + 12 = 0$	dividing by 9
$(x + 6)(x + 2) = 0$	factoring
$x = -6$	setting $x + 6 = 0$ and solving
$x = -2$	setting $x + 2 = 0$ and solving

We now replace x in Eq. (3) by -6 and get

$$y = 2(-6) + 9 = -12 + 9 = -3$$

Similarly, by replacing x in Eq. (3) by -2, we obtain

$$y = 2(-2) + 9 = -4 + 9 = 5$$

Therefore, the solution set of the given system is $\{(-6, -3), (-2, 5)\}$.

The final step is to verify that each of the above pairs of numbers satisfies the given equations. This verification is as follows.

Verification

Replacement for (x, y)	Equation number	Left member	Right member
$(-6, -3)$	(1)	$(-6)^2 + 2(-3)^2$ $= 36 + 18 = 54$	54
	(2)	$2(-6) - (-3)$ $= -12 + 3 = -9$	-9
$(-2, 5)$	(1)	$(-2)^2 + 2(5^2)$ $= 4 + 50 = 54$	54
	(2)	$2(-2) - 5$ $= -4 - 5 = -9$	-9

The graphs of Eqs. (1) and (2), together with the geometrical significance of the solution set, are shown in Fig. 9.8.

Example 2 Solve the following system of equations:

$$4x^2 - 2xy - y^2 = -5 \tag{1}$$

$$y + 1 = -x^2 - x \tag{2}$$

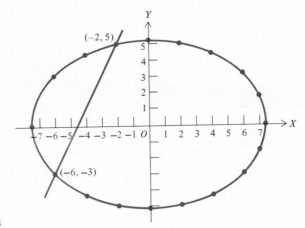

FIGURE 9.8

Solution We first solve Eq. (2) for y and get

$$y = -x^2 - x - 1 \tag{3}$$

Next, we substitute the right member of Eq. (3) for y in Eq. (1) and get

$$4x^2 - 2x(-x^2 - x - 1) - (-x^2 - x - 1)^2 = -5$$

By performing the indicated operations in the last equation and then adding 5 to each member of the resulting equation, we get

$$-x^4 + 3x^2 + 4 = 0 \tag{4}$$

This is an equation in quadratic form, and we solve it as follows:

$x^4 - 3x^2 - 4 = 0$	dividing each member of Eq. (4) by -1 $\qquad$ (5)
$(x^2 - 4)(x^2 + 1) = 0$	factoring left member of Eq. (5)
$x^2 - 4 = 0$	setting first factor equal to zero
$x^2 = 4$	adding 4 to each member
$x = \pm 2$	solving for x
$x^2 + 1 = 0$	setting second factor equal to zero
$x^2 = -1$	adding -1 to each member
$x = \pm\sqrt{-1} = \pm i$	solving for x

Finally, we substitute each of the above values of x in Eq. (3) and obtain the corresponding value of y. This procedure yields the following results.

$y = -2^2 - 2 - 1 = -7$	replacing x by 2 in Eq. (3)
$y = -(-2)^2 - (-2) - 1 = -3$	replacing x by -2 in Eq. (3)
$y = -i^2 - i - 1$	replacing x by i in Eq. (3)

$$= -(-1) - i - 1 = -i \qquad \text{since } i^2 = -1$$

$$y = -(-i)^2 - (-i) - 1 \qquad \text{replacing } x \text{ by } -i \text{ in Eq. (3)}$$

$$= -(-1) + i - 1 = i \qquad \text{since } (-i)^2 = i^2 = -1$$

Consequently, the possible solution set is $\{(2, -7), (-2, -3), (i, -i), (-i, i)\}$. These possible solutions can be checked in the usual manner.

Since the coordinates of any point in a cartesian plane are real numbers, $(i, -i)$ and $(-i, i)$ do not represent points on the graph of either equation, although they satisfy each equation. The graphs of Eqs. (1) and (2) are shown in Fig. 9.9. In the figure we see that $(2, -7)$ and $(-2, -3)$ are the points of intersection of the graphs, but $(i, -i)$ and $(-i, i)$ are not points on either graph.

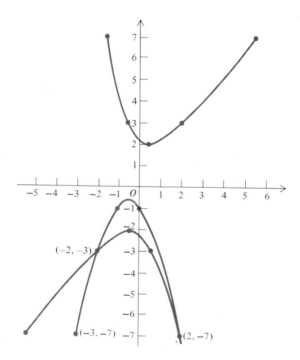

FIGURE 9.9

Example 3 Solve the following system of equations by substitution:

$$xy = 2 \tag{1}$$
$$15x^2 + 4y^2 = 64 \tag{2}$$

Solution   Equation (1) can be solved easily for y in terms of x. The solution is

$$y = \frac{2}{x} \tag{3}$$

We may now eliminate y by replacing it with $2/x$ in Eq. (2). We thus obtain

$$15x^2 + 4\left(\frac{2}{x}\right)^2 = 64 \tag{4}$$

which we solve for x as follows:

$15x^2 + 4\left(\dfrac{4}{x^2}\right) = 64$	**squaring 2/x**
$15x^4 + 16 = 64x^2$	**multiplying each member by x²**
$15x^4 - 64x^2 + 16 = 0$	**adding −64x² to each member and arranging terms**
$(x^2 - 4)(15x^2 - 4) = 0$	**factoring**
$x^2 - 4 = 0$	**setting first factor equal to zero**
$x^2 = 4$	
$x = \pm 2$	
$15x^2 - 4 = 0$	**setting second factor equal to zero**
$15x^2 = 4$	
$x^2 = \frac{4}{15}$	
$x = \pm \dfrac{2}{\sqrt{15}} = \pm \dfrac{2\sqrt{15}}{15}$	

We next substitute each of the above values for x in Eq. (3) and thereby obtain the corresponding value of y. This procedure yields:

With $x = 2$: $\qquad\qquad y = \dfrac{2}{2} = 1$

With $x = -2$: $\qquad\qquad y = \dfrac{2}{-2} = -1$

With $x = \dfrac{2\sqrt{15}}{15}$: $\qquad y = \sqrt{15}$

With $x = -\dfrac{2\sqrt{15}}{15}$: $\qquad y = -\sqrt{15}$

Consequently, the solution set is

$$\left\{(2,\ 1),\ (-2,\ -1),\ \left(\frac{2\sqrt{15}}{15},\ \sqrt{15}\right),\ \left(-\frac{2\sqrt{15}}{15},\ -\sqrt{15}\right)\right\}$$

The solution set may be checked by replacing x and y in Eqs. (1) and (2) by the appropriate member of each pair of numbers in the solution set. The graphs of Eqs. (1) and (2) are shown in Fig. 9.10.

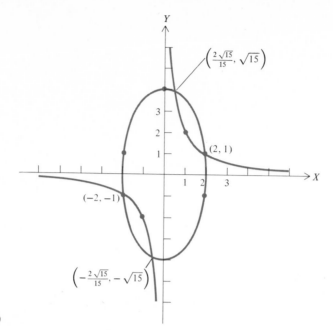

FIGURE 9.10

EXERCISE 9.3   Solution by Substitution

Solve the following pairs of equations by substitution.

1 $2x + y = 5$
$y^2 + 2y = 3x - 3$

2 $5x - y = 2$
$y^2 - 2y = x + 2$

3 $4x - y = 3$
$y^2 - 2y = x - 2$

4 $y^2 - 2x = -2$
$x + 2y = 7$

5 $5x^2 + 9y^2 = 6$
$2x - 3y = 1$

6 $x + 3y = -2$
$4x^2 + 3y^2 = 7$

7 $x - y = 4$
$5x^2 + y^2 = 24$

8 $16x^2 + 3y^2 = 91$
$8x + y = 13$

9 $ax^2 - by^2 = ab^2 - a^2b$
$ax - by = 0$

10 $x^2 + a^2y^2 = 10b^2$
$x + ay = 2b$

11 $m^2x^2 + y^2 = b^2$
$y = mx + b$

12 $2x^2 - 3y^2 = -a^2$
$2x + y = 3a$

13 $y = x - 1$
$xy = 2$

14 $y = 2x - 1$
$xy = 6$

15 $y = x + 4$
$xy = -3$

16 $y = -3x + 1$
$xy = -10$

17 $x^2 - 2xy + y^2 = 144$
$y = 3x^2 + x$

18 $x^2 - 2xy + y^2 = 1$
$x + 2 = y^2 + y$

19 $x^2 - y = 3$
$x^2 + 2x - 3y = 5$

20 $2x^2 - y = 3$
$x^2 + 3x - 2y = 6$

21 $x^2 - 4xy - 6y + 3x = 3$
$$x = 4y^2 + 2y$$

22 $x^2 - 8xy - 4y + x = -14$
$$x = y^2 + 4y$$

23 $x^2 - 6xy \qquad + 20 = 0$
$$x - 2y^2 - 3y + 4 = 0$$

24 $x^2 + 4xy + 2y^2 = 46$
$$y^2 - 2y - x = 1$$

25 $x^2 + 3y^2 = 7$
$$xy = 2$$

26 $2x^2 + 3y^2 = 21$
$$xy = 3$$

27 $2x^2 + y^2 = 9$
$$xy = 2$$

28 $2x^2 + 3y^2 = 29$
$$xy = 3$$

29 $x^2 + 2xy - y^2 = 1$
$$xy = 2$$

30 $4x^2 - 6xy + y^2 = 1$
$$2xy = 3$$

31 $x^2 + 2y^2 = 9$
$$xy = 2$$

32 $x^2 + xy - y^2 + 5 = 0$
$$xy = 3$$

9.5 ELIMINATION BY ADDITION OR SUBTRACTION

In this section we shall discuss two classes of pairs of equations in which the first step is to eliminate one of the variables by addition or subtraction. We justify the method by the following argument:

If $x = p$ and $y = q$ satisfy each of the equations

$$Ax^2 + Bxy + Cy^2 = F \qquad (9.1)$$
$$ax^2 + bxy + cy^2 = f \qquad (9.2)$$

then (p, q) also satisfies the equation

$$m(Ax^2 + Bxy + Cy^2) \pm n(ax^2 + bxy + cy^2) = mF \pm nf \qquad (9.3)$$

This statement follows from the fact that since (p, q) satisfies (9.1) and (9.2), we have

$$m(Ap^2 + Bpq + Cq^2) = mF \qquad \text{and} \qquad n(ap^2 + bpq + cq^2) = nf$$

Consequently, if (x, y) is replaced by (p, q), the left member of (9.3) is equal to $mF \pm nf$, and therefore (p, q) satisfies the equation. The object, then, is to first choose the variable to be eliminated. Then, if possible, we so choose m and n that (9.3) will not contain that variable.

We solve two equations of the type $Ax^2 + Cy^2 = F$ simultaneously by first eliminating one of the variables by addition or subtraction and then solving the resulting equation for the remaining variable. The value of the other variable is then found by substitution.

Example 1 Solve the following system of equations:

$$2x^2 + 3y^2 = 21 \qquad (1)$$
$$3x^2 - 4y^2 = 23 \qquad (2)$$

Solution

$8x^2 + 12y^2 = 84$	multiplying Eq. (1) by 4 (3)
$9x^2 - 12y^2 = 69$	multiplying Eq. (2) by 3 (4)
$17x^2 \qquad = 153$	adding corresponding members of Eqs. (3) and (4)
$x^2 = 9$	dividing by 17
$x = \pm 3$	

If either 3 or -3 is substituted for x in Eq. (1), we have

$$18 + 3y^2 = 21$$
$$3y^2 = 3$$
$$y^2 = 1$$
$$y = \pm 1$$

Therefore, if x is equal to either 3 or -3, then y is equal to ± 1. Hence, the solution set is

$$\{(3, 1), (3, -1), (-3, 1), (-3, -1)\}$$

Figure 9.11 shows the graph of the two given equations and the coordinates of the points of intersection.

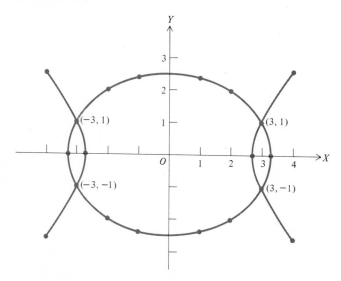

FIGURE 9.11

The first step in simultaneously solving two equations of the type $Ax^2 + Cy^2 + Dx = F$ is to eliminate y^2 by addition or subtraction. Then the process of solving can be completed by the method illustrated in Example 2.

Example 2   Solve the following system of equations:

$$3x^2 - 2y^2 - 6x = -23 \tag{1}$$
$$x^2 + y^2 - 4x = 13 \tag{2}$$

Solution   Since each of the given equations contains y^2 and no other term involving y, we eliminate y^2 and complete the solution as follows:

$3x^2 - 2y^2 - 6x = -23$ Eq. (1) copied   (1)

$\underline{2x^2 + 2y^2 - 8x = 26}$ multiplying Eq. (2) by 2   (3)

$5x^2 - 14x = 3$ adding corresponding members of Eqs. (1) and (3)

$5x^2 - 14x - 3 = 0$ adding −3 to each member

$x = \dfrac{14 \pm \sqrt{196 + 60}}{10}$ by quadratic formula

$ = \dfrac{14 \pm \sqrt{256}}{10}$

$ = \dfrac{14 \pm 16}{10}$

$ = 3, -\tfrac{1}{5}$

$9 + y^2 - 4(3) = 13$ replacing x by 3 in Eq. (2)

$9 + y^2 - 12 = 13$

$y^2 = 13 + 3$ adding 3 to each member

$y^2 = 16$

$y = \pm 4$

Hence, if $x = 3$, then $y = \pm 4$.

$(-\tfrac{1}{5})^2 + y^2 - 4(-\tfrac{1}{5}) = 13$ replacing x by $-\tfrac{1}{5}$ in Eq. (2)

$\tfrac{1}{25} + y^2 + \tfrac{4}{5} = 13$ since $(-\tfrac{1}{5})^2 = \tfrac{1}{25}$ and $-4(-\tfrac{1}{5}) = \tfrac{4}{5}$

$1 + 25y^2 + 20 = 325$ multiplying each member by 25

$25y^2 = 325 - 21$ adding −21 to each member

$25y^2 = 304$

$y^2 = \tfrac{304}{25}$.

$y = \pm \sqrt{\dfrac{304}{25}} = \pm \dfrac{4\sqrt{19}}{5}$

Consequently, if $x = -\tfrac{1}{5}$, then

$y = \pm \dfrac{4\sqrt{19}}{5}$

Therefore, the solution set is

$$\left\{ (3, 4), (3, -4), \left(-\frac{1}{5}, \frac{4\sqrt{19}}{5} \right), \left(-\frac{1}{5}, -\frac{4\sqrt{19}}{5} \right) \right\}$$

The graphs of Eqs. (1) and (2) are shown in Fig. 9.12.

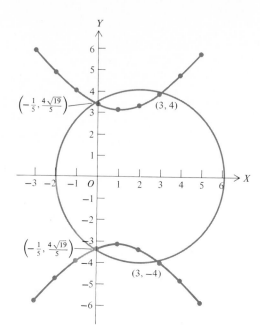

FIGURE 9.12

EXERCISE 9.4   Solution by Addition and Subtraction

Solve each system of equations given below by addition and subtraction.

1 $x^2 + y^2 = 1$
      $2x^2 + 3y^2 = 2$

2 $x^2 + y^2 = 2$
      $3x^2 + 4y^2 = 7$

3 $x^2 + y^2 = 5$
      $x^2 + 4y^2 = 8$

4 $x^2 + y^2 = 13$
      $4x^2 + y^2 = 40$

5 $x^2 + y^2 = 5$
      $2x^2 - 3y^2 = 5$

6 $x^2 + y^2 = 5$
      $3x^2 - 2y^2 = -5$

7 $x^2 + y^2 = 13$
      $5x^2 - 3y^2 = -7$

8 $x^2 + y^2 = 17$
      $9x^2 - y^2 = -7$

9 $x^2 + 4y^2 = 5$
      $9x^2 - y^2 = 8$

10 $x^2 + 9y^2 = 9$
       $4x^2 - 25y^2 = 36$

11 $2x^2 + 3y^2 = 35$
       $4x^2 + y^2 = 25$

12 $2x^2 + 3y^2 = 53$
       $3x^2 + 5y^2 = 80$

13 $x^2 + y^2 + 2x = 9$
       $4x^2 + y^2 - 3x = 11$

14 $x^2 + y^2 - 3y = 4$
       $9x^2 + y^2 - 15y = 0$

15 $x^2 + y^2 + 2x = 9$
       $x^2 + 4y^2 + 3x = 14$

16 $x^2 + y^2 - 4x = -2$
       $9x^2 + y^2 + 18x = 136$

17 $x^2 + y^2 - 4x = -7$
       $4x^2 - y^2 - 6x = 2$

18 $9x^2 - 8y^2 - 21x + 14 = 0$
       $x^2 + 3y^2 - 14x + 21 = 0$

19 $x^2 + y^2 - 6x = -4$
       $x^2 - y^2 + 5x = 5$

20 $x^2 + y^2 - 8x = -8$
       $x^2 - 4y^2 + 6x = 0$

21 $4x^2 + y^2 - 9y = -4$
       $4x^2 - 9y^2 - 2y = -219$

22 $9x^2 + 4y^2 - 17y = 21$
       $4x^2 - y^2 - 2y = 1$

23 $x^2 - 2y^2 + 5x = -4$
       $3x^2 + 4y^2 - 9x = 30$

24 $5x^2 + 5y^2 + 3x = 4$
       $2x^2 - 2y^2 + x = -1$

25 $x^2 - 2xy - x = 12$
       $x^2 - xy - 3x = 3$

26 $3y^2 - 5xy + y = 4$
       $2y^2 - 4xy + y = 2$

27 $x^2 + xy + 4x = 2$
       $3x^2 - 2xy + x = 10$

28 $2x^2 + 3xy + x = -2$
       $3x^2 - 2xy + 4x = 1$

29 $3y^2 - 5xy + y = 4$
       $2y^2 - 4xy + y = 2$

30 $y^2 + 3xy + 3y = -8$
       $3y^2 + xy + y = 8$

31 $x^2 - 3xy + 3x = -8$
       $3x^2 - xy + x = 8$

32 $2y^2 + 5xy - y = 0$
       $3y^2 - 2xy + 6y = 4$

9.6 SUMMARY

The first section is devoted to sketching the graphs of several special cases of the general quadratic $Ax^2 + Bxy + Cy^2 + Dx + Ey + F = 0$. We then show how to find an approximation to the simultaneous solution of a pair of quadratics by sketching the graphs and then estimating the coordinates of the points of intersection. In the next sections, we show how to solve quadratics of specified types by algebraic methods.

EXERCISE 9.5 Review

Solve the pair of equations in each of Probs. 1 to 6 graphically.

1 $y = 3x^2 - 7$
$y = -x^2 + 9$

2 $9x^2 + 5y^2 = 189$
$5y = x^2 + 6x + 23$

3 $3x^2 + y^2 = 12$
$3x^2 - 2y^2 = -15$

4 $2x^2 + y^2 = 33$
$x^2 + y^2 = 29$

5 $2x^2 + y^2 = 17$
$-3x^2 + 2y^2 = 6$

6 $x^2 + y^2 = 25$
$x^2 = 8y - 8$

Solve the pair of equations in each of Probs. 7 to 12 by substitution.

7 $2x^2 - y^2 = -25$
$y - 3x = 5$

8 $9x^2 + 5y^2 = 1$
$3x - 2y = 1$

9 $x^2 - 6xy = -20$
$x - 2y^2 - 3y = -4$

10 $x^2 - 2xy + y^2 = 9$
$x^2 + x - y = 4$

11 $x^2 + 4xy + 2y^2 = 46$
$y^2 - 2y - x = 1$

12 $2x^2 + 3xy - y^2 = -16$
$xy = -3$

Solve the pair of equations in each of Probs. 13 to 18 by addition and subtraction.

13 $2x^2 + 3y^2 = 35$
$4x^2 + y^2 = 25$

14 $4x^2 - 5y^2 = 15$
$2x^2 + 7y^2 = 17$

15 $x^2 - 2y^2 + 5x = -4$
$3x^2 + 4y^2 - 9x = 30$

16 $3x^2 + y^2 - 8x = 0$
$5x^2 + 3y^2 - 4x = 24$

17 $y^2 + 3xy + 3y = -8$
$3y^2 + xy + y = 8$

18 $3y^2 - 5xy + y = 4$
$2y^2 - 4xy + y = 2$

10
ratio, proportion, and variation

In education circles one frequently hears the term *IQ*, or *intelligence quotient*. In a machine shop one may hear references to the *gear ratio*. Highway engineers are interested in the *grades* of highways, and carpenters continually deal with the *pitch* of a roof. Each of the italicized words is the quotient of two numbers and is the measure of some particular thing. For example, IQ is a measure of mental ability, and both the grade of a highway and the pitch of a roof are measures of a slope. In this chapter we discuss ratios, and we shall be especially interested in pairs of numbers that vary in such a way that their ratios or products never change.

10.1 RATIO

The *ratio* of the number a to the number b, for $b \neq 0$, is defined as the quotient a/b, which can also be written as

$$a \div b \qquad \frac{a}{b} \qquad a{:}b$$

It represents the part of a that corresponds to 1 unit of b. Thus, 60 cents/12 eggs indicates that one egg costs 60/12 cents = 5 cents, and \$5700/10 months discloses the rate of \$570 per month.

If a and b are measurements of the same kind, they are ordinarily expressed in terms of the same unit before a ratio is formed. Thus, if we want the ratio of 3 feet to 9 inches, we must express 3 feet in terms of inches or 9 inches in terms of feet before forming their ratio. If we do the former, we have 36 inches/9 inches = 4.

10.2 PROPORTION

In applying ratios to problems, we often encounter situations in which two ratios are equal. The statement that two ratios are

Proportion equal is called a *proportion.* Thus, if a/b and c/d are equal, we write

$$\frac{a}{b} = \frac{c}{d} \qquad\qquad (10.1)$$

or

$$a{:}b = c{:}d \qquad\qquad (10.1')$$

and read the proportion as "*a* divided by *b* is equal to *c* divided by *d*," or "*a* is to *b* as *c* is to *d*." A proportion is in reality a fractional equation or an equation that involves fractions. Regardless of which of the forms is used, we say that *a* and *d* are

Extremes the **extremes**, and *b* and *c* are the **means**.
Means
   We shall now present several properties of proportions that facilitate work with proportions and their applications.
   If we multiply each member of (10.1) by bd, we get

$$\frac{abd}{b} = \frac{cbd}{d}$$

and, by dividing the members of each fraction by their common factor, we see that

$$ad = bc \qquad \text{if} \qquad \frac{a}{b} = \frac{c}{d} \qquad\qquad (10.2)$$

This can be expressed in words as follows:

   In any proportion, the product of the extremes is equal to the product of the means.

Example 1 Find x if $x/15 = 2/5$.

Solution If we apply (10.2) to the given equation, we have

$$5x = (15)(2) = 30$$
$$x = 6 \qquad \text{dividing each member by 5}$$

   We shall derive two other proportions from (10.1). If we add 1 to each member of (10.1), we get

$$\frac{a}{b} + 1 = \frac{c}{d} + 1$$

Then, since $a/b + 1 = (a + b)/b$ and $c/d + 1 = (c + d)/d$, we have

$$\frac{a + b}{b} = \frac{c + d}{d} \qquad \text{if} \qquad \frac{a}{b} = \frac{c}{d} \qquad\qquad (1)$$

Similarly, by subtracting 1 from each member of (10.1) and simplifying, we get

$$\frac{a-b}{b} = \frac{c-d}{d} \qquad \text{if} \qquad \frac{a}{b} = \frac{c}{d} \tag{2}$$

If we now equate the quotients of the corresponding members of (1) and (2), we see that

$$\frac{a+b}{a-b} = \frac{c+d}{c-d} \qquad \text{if} \qquad \frac{a}{b} = \frac{c}{d} \tag{3}$$

We can now state the following property:

If $\dfrac{a}{b} = \dfrac{c}{d}$, then $\dfrac{a+b}{b} = \dfrac{c+d}{d}, \dfrac{a-b}{b} = \dfrac{c-d}{d}$, and $\dfrac{a+b}{a-b} = \dfrac{c+d}{c-d}$

$$(10.3)$$

Example 2 Find a and b if

$$\frac{a}{b} = \frac{c}{d} \qquad a - b = 12 \qquad c = 3 \qquad d = 2$$

Solution If $a/b = c/d$, then by (2) we have

$$\frac{a-b}{b} = \frac{c-d}{d}$$

Consequently, substituting the given values for $a - b$, c, and d, we obtain

$$\frac{12}{b} = \frac{3-2}{2} = \frac{1}{2}$$

Consequently,

$$b = 24 \qquad \textbf{by (10.2)}$$

Finally, since $a - b = 12$ and $b = 24$, it follows that

$$a - 24 = 12$$
$$a = 36$$

Mean, Third, and Fourth Proportional   If $c = b$ in $a/b = c/d$, then b is called the *mean proportional* to or between a and d, and d is the *third proportional* to a and b. If $c \neq b$ in $a/b = c/d$, then d is the *fourth proportional* to a, b, and c.

Example 3 Find the mean proportional between 2 and 12.5.

Solution If we represent the desired mean proportional by x and make use of the definition, we have

$$\frac{2}{x} = \frac{x}{12.5}$$

$$x^2 = 25 \qquad \text{**by (10.2)**}$$
$$x = \pm 5 \qquad \text{**solving for x**}$$

Example 4 Find the third proportional to 2 and 6.

Solution If we represent the third proportional by x, we have

$$\frac{2}{6} = \frac{6}{x}$$

$$2x = 36 \qquad \text{**by (10.2)**}$$
$$x = 18 \qquad \text{**solving for x**}$$

Example 5 Find the fourth proportional to 3, 4, and 5.

Solution If we let x represent the fourth proportional, then

$$\frac{3}{4} = \frac{5}{x}$$

$$3x = 20 \qquad \text{**by (10.2)**}$$
$$x = \tfrac{20}{3} \qquad \text{**solving for x**}$$

EXERCISE 10.1 Ratio and Proportion

Express the indicated ratio as a fraction in each of Probs. 1 to 12, and simplify.

1 12 days to 3 days

2 10 cats to 2 cats

3 40 years to 25 years

4 78 feet to 12 feet

5 2 weeks to 4 days

6 12 yards to 9 feet

7 $1.75 to 7 nickels

8 6 quarters to 5 dimes

9 1.6 miles to 256 yards

10 10 quarts to 4 gallons

11 9 months to 3 years

12 2280 seconds to 1.3 hours

Find the value of the indicated ratio in each of Probs. 13 to 16, and interpret the result.

13 517 miles to 11 hours

14 $16,320 to 12 months

15 39 hot dogs to 13 boys

16 342 miles to 18 gallons

17 What is the specific heat of a substance if 8 calories increase the temperature of a 25-gram block by 1°C? The specific heat of a substance is the number of calories required to increase the temperature of 1 gram of it by 1°C.

18 Find the pitch of a roof if a 17-foot rafter is 8 feet higher on one end than on the other. The pitch of a roof is the distance the roof rises per unit of horizontal distance covered.

19 The tangent of an acute angle of a right triangle is the ratio of the lengths of the opposite and adjacent sides. Find the tangent of an acute angle if the opposite side is 5 inches long and the hypotenuse is 13 inches long.

20 The specific gravity of a body is the ratio of the weight of the body to the weight of an equal volume of water. Find the specific gravity of a substance if 1 cubic foot of it weighs 1281.25 pounds and 1 cubic foot of water weighs 62.5 pounds.

Find the value of x in each of Probs. 21 to 28.

21 $2:x = 4:3$

22 $3:x = 1:4$

23 $5:2 = 4:x$

24 $5:3 = 7:x$

25 $(5 - x):(x + 3) = 3:5$

26 $4:7 = (x + 1):(2x + 1)$

27 $(x + 1):4 = 6:(x - 1)$

28 $9:(x + 5) = (x - 1):3$

Find the fourth proportional to the numbers in each of Probs. 29 to 32.

29 $2, 5, 4$

30 $5, 2, 2.5$

31 $4, 7, 12$

32 $9, 12, 15$

Find the third proportional to the pair of numbers in each of Probs. 33 to 36.

33 $1, 2$

34 $3, 9$

35 $2, 3$

36 $8, 5$

Find the mean proportional to the pair of numbers in each of Probs. 37 to 40.

37 $1, 9$

38 $2, 8$

39 $12, 27$

40 $9, 16$

41 If $x:y = 5:1$, and $x - y = 12$, find x and y.

42 If $x:y = 2:1$, and $x - y = 7$, find x and y.

43 If $x:y = 5:3$, and $x + y = 16$, find x and y.

44 If $x:y = 12:7$, and $x + y = 19$, find x and y.

45 If the sound of a steam whistle was heard 3267 feet from the train after 3 seconds, how long before it was heard 5445 feet from the train?

46 How many grams of water can be vaporized by 6456 calories if 4304 calories can vaporize 8 grams?

47 How much cement should be used to make 216 cubic feet of concrete if 14 sacks are needed for 4 cubic yards?

48 If a house plan is drawn on a scale of ½ inch per foot, what are the dimensions of a rectangular house if the floor plan is shown on a 20-inch by 30-inch drawing?

10.3 VARIATION

Vary Directly If two variables are so related that one of them is always a constant times the other, then the one is said to *vary directly* as the other.

Constant of Variation This can be written symbolically as $y = kx$. The constant k is called the *constant of variation*.

Example 1 Put the fact that W varies directly as t in symbolic form, and find the constant of variation if $W = 20$ for $t = 4$.

Solution   From the above definition, we can write $W = kt$. Since $W = 20$ for $t = 4$, this can be put in the form $20 = k4$; hence, $k = 5$, and the equation of the variation is $W = 5t$.

Vary Inversely   If two variables are so related that one of them is a constant divided by the other, then the one is said to *vary inversely* as the other.

This can be put in symbolic form as $y = k/x$, where k is the constant of variation.

Example 2   State symbolically that L varies inversely as y; furthermore, find the constant of variation if $L = 7$ for $y = 4$.

Solution   By use of the definition of inverse variation, we can write $L = k/y$. Now, putting the corresponding values of L and y in this equation, we get $7 = k/4$. Therefore $k = 28$, and for the equation that expresses the variation we have $L = 28/y$.

Varies Jointly   If three variables are so related that one of them is a constant times the product of the other two, we say that the one *varies jointly* as the other two.

This statement can be written symbolically as $Z = kxy$.

Example 3   Express the statement that V varies jointly as L and W in symbolic form, and find k if $V = 420$ for $L = 10$ and $W = 7$.

Solution   By use of the definition of joint variation, we can write $V = kLW$. Putting the given values of V, L, and W in this equation gives $420 = k(10)(7) = 70k$. Therefore, $k = 6$, and the equation that states the variation becomes $V = 6LW$.

Example 4   The amount of gas used by a car traveling at a uniform rate varies jointly as the distance traveled and the square of the speed. If a car uses 5 gallons in going 100 miles at 40 miles per hour, how much will it use in going 80 miles at 55 miles per hour?

Solution   The first step is to write the equation of variation. In doing this, we shall represent the number of gallons of gas used by g, the number of miles traveled by s, and the speed by r; then

$$g = ksr^2 \qquad \text{**by definition of joint variation**}$$

Consequently,

$$5 = k(100)(40)^2 \qquad \text{since } g = 5 \text{ when } s = 100 \text{ and } r = 40$$

$$k = \frac{5}{100(40)^2} = \frac{1}{32,000} \qquad \text{solving for } k \text{ and simplifying}$$

Therefore,

$$g = \frac{1}{32,000} sr^2 \qquad \text{since } k = 1/32,000$$

Finally, we can find the desired value of g by substituting $s = 80$ and $r = 55$ in this equation, since they are a pair of corresponding values. Thus,

$$g = \frac{1}{32,000}(80)(55)^2 = 7.5625$$

Example 5 The safe load of a rectangular beam of given width varies directly as the square of the depth of the beam and inversely as the length between supports. If the safe load for a beam of a given width that is 10 feet long and 5 inches deep is 2400 pounds, determine the safe load for a beam of the same material and width if it is 8 inches deep and 16 feet between supports.

Solution We shall let

d = the number of inches in the depth of the beam

s = the number of feet between supports

L = the number of pounds in the safe load

Then,

$$L = \frac{kd^2}{s} \qquad \text{by definition of the variations}$$

If we now make use of the set of corresponding values of d, s, and L, we obtain

$$2400 = \frac{k5^2}{10}$$

$$k = 960 \qquad \text{solving for } k$$

Therefore, the equation that states the variation becomes

$$L = \frac{960d^2}{s}$$

and, substituting $d = 8$ and $s = 16$, we obtain

$$L = \frac{960(8^2)}{16}$$

$$= 3840$$

Therefore, the safe load is 3840 pounds.

EXERCISE 10.2   Variation

1   Express the following statements as equations: (a) p varies directly as q; (b) a varies inversely as b; (c) x varies jointly as y and z; (d) u varies directly as the square of w and inversely as v.

2   If y varies directly as x and is 18 when $x = 6$, find the value of y if $x = 2$.

3   If w varies directly as x and is 24 when $x = 4$, find the value of w if $x = 5$.

4   Given that y varies inversely as x and $y = 6$ when $x = 3$, find the value of y when $x = 9$.

5   If w varies inversely as y and is equal to 12 when $y = 3$, find the value of w if $y = 18$.

6   If y varies jointly as x and w and is 36 when $x = 3$, and $w = 2$, find the value of y if $x = 5$ and $w = 4$.

7   Given that x varies jointly as w and y and also that $x = 60$ when $w = 3$ and $y = 5$, find the value of x if $w = 4$ and $y = 7$.

8   Given that w varies directly as the product of x and y and inversely as the square of z, and that $w = 4$ when $x = 2$, $y = 6$, and $z = 3$, find the value of w when $x = 1$, $y = 4$, and $z = 2$.

We may change from one unit of measurement to another by making several successive changes between familiar units. For example, to change from miles to centimeters, we use

$$1 \text{ mile} = 1 \text{ mile} \, \frac{5280 \text{ feet}}{1 \text{ mile}} \, \frac{12 \text{ inches}}{1 \text{ foot}} \, \frac{2.54 \text{ centimeters}}{1 \text{ inch}} = 160{,}934.4 \text{ centimeters}$$

Change the measurement in each of Probs. 9 to 16 from the British or metric system to the other by making use of the fact that the number of centimeters in a length varies directly as the number of inches in it and is 2.54 for 1 inch. Also use 1 meter = 39.37 inches.

9 1 inch	10 6 inches	11 1 foot	12 1 yard
13 1500 meters	14 1 mile	15 100 meters	16 220 yards

Change the weight in each of Probs. 17 to 24 from the system in which it is given (metric or British) to the other by making use of the fact that the weight of an object in grams varies directly as the weight in pounds and is 453.6 grams for 1 pound and is 2.204 pounds for 1 kilogram.

17 1 pound	18 1 ounce	19 3 pounds	20 35.2734 ounces
21 31 kilograms	22 1 gram	23 787.3 grams	24 1984 grams

Use the fact that 1 quart = .9463 liter to change the number in each of Probs. 25 to 28 to another unit of measurement.

25 1 gallon	26 1 liter	27 15 quarts	28 21 liters

The Celsius temperature varies as the quantity "Fahrenheit minus 32" and is 10°C for 50°F [symbolically, $C = k(F - 32)$]. Change from F to C or C to F in Probs. 29 to 36.

29 77°F	30 22°F	31 −40°F	32 0°F
33 25°C	34 40°C	35 100°C	36 10°C

Use $f = 2C + 30$ as an approximation for $F = 1.8C + 32$, an exact relation, and find f and F for each value of C given in Probs. 37 to 40.

37 30°   **38** 20°   **39** 5°   **40** −5°

41 The area of a rhombus varies jointly as the lengths of the diagonals. If the area of a rhombus with diagonals 3 and 4 inches is 6 square inches, find the area of another rhombus whose diagonals are 2 and 5 inches.

42 When aluminum is added to an excess of hydrochloric acid, the amount of hydrogen produced varies directly as the amount of aluminum added. If 18 grams of aluminum produces 2 grams of hydrogen, how much hydrogen can be produced by adding 63 grams of aluminum to an excess of hydrochloric acid?

43 The simple interest earned in a given time varies jointly as the principal and the interest rate. If $600 earned $108 at 6 percent, find the interest earned in the same time by $810 at 5 percent.

44 The air resistance to a moving object is approximately proportional to the square of the speed of the object. Compare the air resistance of an automobile traveling 30 miles per hour with that of an automobile traveling 60 miles per hour.

45 According to Boyle's law, the volume of a confined mass of gas varies inversely as the pressure, provided that the temperature is constant. If a mass of gas has a volume of 1 liter under a pressure of 76 centimeters of mercury, what is its volume under a pressure of 38 centimeters of mercury?

46 The amount of paint needed to paint the sides of a cylindrically shaped fence post varies jointly as the radius and height of the post. Compare the amount of paint needed for a post 5 feet high of radius ½ foot with that needed for a post 6 feet high of radius ⅓ foot.

47 The volume of gas discharged in a given time from a horizontal pipe under constant pressure and specific gravity varies inversely as the square root of the length of the pipe. If 246 cubic feet of gas is discharged in 1 hour from a pipe 2500 feet long, what volume would be discharged in 1 hour from a pipe 3600 feet long?

48 The weight of a body above the surface of the earth varies inversely as the square of the distance from the body to the center of the earth. If a boy weighs 121 pounds on the surface, how much would he weigh 400 miles above the surface? Assume the radius of the earth to be 4000 miles.

49 The amount that a bar will bend under a given force varies inversely as the width of the bar when the thickness and length of the bar remain the same. If a bar 2.5 inches wide will bend 7° under a certain force, what is the width of a bar of the same material, length, and thickness that will bend 5° under the same force?

50 The horsepower that a rotating shaft can safely transmit varies jointly as the cube of the radius of the shaft and the number of revolutions through which the shaft turns per minute. Compare the safe load of a shaft of radius 2 inches which turns 600 revolutions per minute with that of another shaft which has a radius of 3 inches and revolves 800 times per minute.

51 The weight of a clothesline varies jointly as the length and the square of the diameter. If the weight of a 36-foot line ¼ inches in diameter is 5.4 pounds, find the weight of a 48-foot line ⅛ inches in diameter.

52 On the ocean, the square of the distance in miles to the horizon varies as the height in feet that the observer is above the surface of the water. If a 6-foot woman on a surfboard can see 3 miles, how far can she see if she is standing on the deck of a ship that is 48 feet above the water?

53 The power available in a jet of water varies jointly as the cube of the water's velocity and the cross-sectional area of the jet. Compare the power of a jet with that of another that is moving two times as fast through an opening one-half as large.

54 The centrifugal force at a point of a revolving body varies as the radius of the circle in which the point is revolving. If the centrifugal force is 450 pounds when the radius is 12 inches, at what radius is the force 375 pounds?

55 The force of a wind on a flat surface perpendicular to the direction of the wind varies as the area of the surface and the square of the wind velocity. When the wind is blowing 8 miles an hour, the force on a 4- by 6-foot signboard is 7.5 pounds. What is the force on a window 3 by 4 feet when the wind is blowing at 16 miles per hour?

56 The crushing load of a pillar varies as the fourth power of its diameter and inversely as the square of its height. If 1953 tons will crush a pillar 1 foot in diameter and 15 feet high, find the load that will crush a pillar 4 inches in diameter and 5 feet high.

57 One of Kepler's laws states that the square of the time in days required by a planet to make one revolution about the sun varies directly as the cube of the average distance of the planet from the sun. If Mars is 1½ times as far from the sun, on the average, as the earth, find the approximate number of days required for it to make a revolution about the sun.

58 The horsepower required to operate a fan varies as the cube of the speed. If 2 horsepower will drive a fan at 600 revolutions per minute, find the speed derived from ¼ horsepower.

59 The exposure time necessary to obtain a good photographic negative varies directly as the square of the f number of the camera lens. If an exposure of $\frac{1}{50}$ seconds produces a good negative at $f16$, what exposure time would be necessary at a lens setting of $f8$?

60 The amount of oil used by a ship traveling at a uniform speed varies jointly as the distance traveled and the square of the velocity. If a ship used 1500 barrels of oil traveling 480 miles at 25 knots, at what uniform rate did the ship travel if it covered 540 miles and used 1080 barrels of oil?

11

sequences and series

Set A collection of numbers is called a *set*. If each element of a set
Sequence has a definite place in the set, we have a *sequence*. The indi-
Series cated sum of a sequence is called a *series*. If we were interested
in sequences merely as a plaything, we could find many of their
intriguing properties. If we are interested in sequences because
of their use in mathematics and its applications, we can find
interesting properties and applications. This chapter will be
devoted to two types of sequences: *Arithmetic* sequences occur
in food production and in analyzing the notion of a body falling
from rest in a vacuum. *Geometric* sequences occur in popula-
tion growth, radioactive decay, and interest earned on money.

11.1 ARITHMETIC PROGRESSIONS

Arithmetic An *arithmetic progression* is a sequence of numbers so related
Progression that each element after the first can be obtained from the im-
Common mediately preceding one by adding a fixed number. This fixed
Difference number is called the *common difference*. Thus, the arithmetic
progression of seven terms with first term 3 and common difference 4 is the following: 3, 3 + 4 = 7, 7 + 4 = 11, 11 + 4 = 15,
15 + 4 = 19, 19 + 4 = 23, and 23 + 4 = 27. This progression can
be written as 3, 7, 11, 15, 19, 23, 27.

Most problems on arithmetic progressions deal with four or
more of the five quantities

- The first term in the progression, symbolized by a_1
- The nth or last term, represented by a_n
- The common difference d
- The number of terms n
- The sum of the terms s

We shall develop formulas that enable us to determine the
other two if three of these five quantities are known. The first of
these formulas gives the last (nth) term in terms of a_1, d, and n.
By use of the definition of an arithmetic progression of n terms

with first term a_1, and common difference d, we can construct the following table:

Number of the term	1	2	3	4	$\cdots$	n
The term	a_1	$a_1 + d$	$a_1 + 2d$	$a_1 + 3d$	$\cdots$	$a_1 + (n-1)d$

Consequently, the nth or last term of an arithmetic progression with first term a_1, and common difference d is

$$a_n = a_1 + (n-1)d \tag{11.1}$$

Example 1   Find the last term of an arithmetic progression with $a_1 = -2$, $d = 3$, and $n = 6$.

Solution   We could find the sixth term by use of the definition but shall use (11.1). Substituting the given values in it, we have

$$a_6 = -2 + (6-1)3 = -2 + 15 = 13$$

Example 2   If the first term of an arithmetic progression is -3 and the eighth term is 11, find d and write the eight terms of the progression.

Solution   In this problem, $a_1 = -3$, $n = 8$, and $a_8 = 11$. If these values are substituted in (11.1), we have

$$11 = -3 + (8-1)d \qquad \text{or} \qquad 11 = -3 + 7d$$

Hence,

$$-7d = -14 \qquad \text{and} \qquad d = 2$$

Therefore, since $a_1 = -3$, the first eight terms of the desired progression are $-3, -1, 1, 3, 5, 7, 9, 11$.

In order to obtain the formula for the sum s of the n terms of an arithmetic progression in which the first term is a_1 and the common difference is d, we note that the terms in the progression are $a_1, a_1 + d, a_1 + 2d$, and so on, until we reach the last term, which by formula (11.1) is $a_n = a_1 + (n-1)d$. Hence,

$$s = a_1 + (a_1 + d) + (a_1 + 2d) + \cdots + [a_1 + (n-1)d] \tag{1}$$

Since there are n terms in Eq. (1) and each term contains a_1, we may rearrange the terms and write s as

$$s = na_1 + [1 + 2 + \cdots + (n-1)]d \tag{2}$$

Now, if we reverse the order of the terms in the progression by writing a_n as the first term, then the second term is $a_n - d$, the third $a_n - 2d$, and so on to the nth term, which by (11.1) is $a_n - (n-1)(d)$. Hence we can write the sum as

$$s = a_n + (a_n - d) + (a_n - 2d) + \cdots + [a_n - (n-1)(d)]$$

Next, combining the a_n's and the d's, we get

$$s = na_n - [1 + 2 + \cdots + (n-1)]d \tag{3}$$

Finally, if we add the corresponding members of Eqs. (2) and (3), we have

$$2s = na_1 + na_n + 0 \quad \text{since coefficients of the term in brackets in Eqs. (2) and (3)}$$
$$\text{have opposite signs}$$

$$= n(a_1 + a_n)$$

Dividing by 2, we obtain the formula

$$s = \frac{n}{2}(a_1 + a_n) \tag{11.2}$$

This can be expressed in the form

$$s = \frac{n}{2}[2a_1 + (n-1)d] \tag{11.3}$$

by replacing a_n by $a_1 + (n-1)d$.

Example 3   Find the sum of all the even integers from 2 to 1000 inclusive.

Solution   Since the even integers 2, 4, 6, etc., taken in order, form an arithmetic progression with $d = 2$, we may use (11.2) with $a_1 = 2$, $n = 500$, and $a_n = 1000$ to obtain the desired sum. The substitution of these values in (11.2) yields

$$s = \tfrac{500}{2}(2 + 1000)$$
$$= 250(1002) = 250{,}500$$

Example 4   A man buys a used car for $1100 and agrees to pay $100 down and $100 per month plus interest at 6 percent on the outstanding indebtedness until the car is paid for. How much will the car cost him?

Solution   The rate of 6 percent per year is 0.5 percent per month. Hence, when he makes his first payment after the down payment, he will owe 1 month's interest on $1000, or $(0.005)(\$1000) = \$5.00$. Since he pays $100 on the principal monthly, his interest from month to month is reduced by 0.5 per cent of $100 or by $.50. The final payment will be $100 plus interest on $100 for 1 month, which is $100.50. Hence, his payments constitute an arithmetic progression with $a_1 = \$105$, $a_n = \$100.50$, and $n = 10$. Therefore, by (11.2), the sum of his payments is

$$s = \tfrac{10}{2} (\$105 + \$100.50)$$
$$= \tfrac{10}{2} (\$205.50) = \$1027.50$$

Thus, the total cost of the car will be $1127.50.

If any three of a_1, a_n, n, d, and s are known, the other two can be found by use of (11.1) and (11.2) or (11.3). If, after substituting the known quantities, there is only one unknown in either of (11.1), (11.2), or (11.3), we can use that equation to find that unknown. We can then find the other unknown by use of another of the three formulas.

Example 5 Find d and s if $a_1 = 4$, $n = 10$, and $a_n = 49$.

Solution Since there is only one unknown d remaining after the given values are substituted in (11.1), we can use it to solve for d. Thus, substituting gives

$$49 = 4 + (10 - 1)d = 4 + 9d$$

Therefore,

$$9d = 45 \quad \text{and} \quad d = 5$$

If we now substitute this value for d and the given values for a_1 and n in (11.3) or the given values for n, a_1, and a_n in (11.2), we have only s remaining as an unknown. Using (11.2), we get

$$s = \tfrac{10}{2} (4 + 49) = 265$$

Example 6 Find a_1 and n if $a_n = 23$, $d = 3$, and $s = 98$. Also find the terms.

Solution If we substitute the given values in either (11.1), (11.2), or (11.3), we obtain an equation in two variables; hence, we must solve the resulting pair of equations simultaneously after substituting in any two of (11.1), (11.2), and (11.3). Using the first two, we get

$$23 = a_1 + (n - 1)3 \qquad \text{from (11.1)} \tag{1}$$

$$98 = \frac{n}{2}(a_1 + 23) \qquad \text{from (11.2)} \tag{2}$$

We begin the simultaneous solution by solving (1) for a_1. Thus, we get

$$a_1 = 23 - (n - 1)3 = 26 - 3n \tag{3}$$

The next step is to substitute this for a_1 in (2). We thereby obtain

$$98 = \frac{n}{2}(26 - 3n + 23) = \frac{n}{2}(49 - 3n)$$

Hence,

$$196 = 49n - 3n^2 \quad \text{or} \quad 3n^2 - 49n + 196 = 0$$

Therefore, by the quadratic formula,

$$n = \frac{49 \pm \sqrt{49^2 - 4(3)(196)}}{2(3)} = \frac{49 \pm \sqrt{2401 - 2352}}{6}$$

$$= \frac{49 \pm 7}{6} = \frac{28}{3}, 7$$

We must discard $\frac{28}{3}$ since the number of terms cannot be a fraction. The value of a_1 can now be obtained from (3), and it is $26 - 3(7) = 5$. Consequently, the terms of the sequence are $5, 5 + 3 = 8, 11, 14, 17, 20, 23$.

EXERCISE 11.1 Arithmetic Progressions

Write the terms of the arithmetic progression that satisfies the conditions in each of Probs. 1 to 8.

1 $a_1 = 1, d = 2, n = 6$

2 $a_1 = 7, d = -2, n = 5$

3 $a_1 = 12, d = -3, n = 7$

4 $a_1 = -5, d = 4, n = 4$

5 $a_1 = 2$, second term 6, $n = 5$

6 $a_1 = 7$, third term 5, $n = 6$

7 Second term 6, fourth term 10, $n = 7$

8 Third term 7, seventh term 15, $n = 7$

In each of Probs. 9 to 32, find the value of whichever of a_1, a_n, n, d, and s that are not given.

9 $a_1 = 1, n = 7, d = 2$

10 $a_1 = 2, n = 8, d = 3$

11 $a_1 = 9, n = 6, d = -1$

12 $a_1 = 13, n = 7, d = -3$

13 $a_1 = 22, a_n = 2, n = 5$

14 $a_1 = 3, a_n = 17, n = 8$

15 $a_1 = 8, a_n = 26, n = 7$

16 $a_1 = 27, a_n = -9, n = 7$

17 $a_1 = -17, d = 4, a_n = 11$

18 $a_1 = 1, d = 10, a_n = 51$

19 $a_n = 1, n = 6, d = -10$

20 $a_n = -17, n = 8, d = -3$

21 $a_n = 27, a_1 = -9, s = 63$

22 $a_n = -8, a_1 = 10, s = 7$

23 $a_n = 3, n = 8, s = 80$

24 $a_n = 22, n = 5, s = 60$

25 $a_1 = -5, n = 7, s = 28$

26 $a_1 = 4, n = 6, s = 39$

27 $a_1 = 23, n = 8, s = 100$

28 $a_1 = 13, n = 7, s = 49$

29 $a_n = 13, d = 2, s = 49$

30 $a_n = 23, d = 3, s = 100$

31 $a_n = 4, d = -1, s = 39$

32 $a_n = -5, d = -3, s = 28$

33 Find the sum of the smallest 17 positive integral multiples of 3.

34 What is the sum of the integral multiples of 7 between 6 and 99?

35 Find the sum of all two-digit integral multiples of 9.

36 Find the sum of all integral multiples of 17 from 34 to 187 inclusive.

37 If a machine costs $8400, depreciates 29 percent during the first year, 24 percent

during the second year, 19 percent during the third year, and so on for its useful life of 6 years, find its scrap value.

38 An engineer made $3700 the first year after graduating in 1948. If she received an $800-per-year raise at the end of each year, what was her salary at the beginning of her twenty-first working year? How much income did she have during her first 20 years as an engineer?

39 An ill-prepared but capable and industrious student made 51 on the first of six algebra tests. What was his average if his grade was 9 better on each test than on the immediately preceding one?

40 The first of two runners travels 401 yards per minute, whereas the second runs at a uniform rate during each minute but runs 52 yards further each minute than in the immediately preceding minute. Which will win a mile race if the second goes 311.5 yards in the first minute?

41 The three digits of a number form an arithmetic progression. What is the number if it is increased by 594 by interchanging the units and hundreds digits?

42 Find x so that $x + 1$, $x + 7$, and $x + 13$ form an arithmetic progression.

43 Find x and y so that x, y, and $x + y + 1$ form an arithmetic progression with sum $13x$.

44 Determine x so that $3x + 2$, $x^2 - x$, and $2x^2 - 6x + 1$ form an arithmetic progression.

11.2 GEOMETRIC PROGRESSIONS

Geometric Progression

Common Ratio

A *geometric progression* is a sequence of numbers so related that each element after the first can be obtained from the immediately preceding one by multiplying by a fixed number. This fixed number is called the *common ratio*. Thus, the geometric progression of seven terms with first term 3 and common ratio 2 is 3, $3(2) = 6$, $6(2) = 12$, $12(2) = 24$, $24(2) = 48$, $48(2) = 96$, and $96(2) = 192$.

Most problems on geometric progressions deal with four or more of the five quantities

- The first term in the progression symbolized by a_1
- The nth or last term, represented by a_n
- The common ratio r
- The number of terms n
- The sum of the terms s

We shall develop formulas that enable us to determine the other two if three of these five quantities are known. The first of these formulas gives the last (nth) term in terms of a_1, r, and n. By use of the definition of a geometric progression of n terms with first term a_1 and common ratio r, we can construct the following table:

Number of the term	1	2	3	4	$\cdots$	n
The term	a_1	$a_1 r$	$a_1 r^2$	$a_1 r^3$	$\cdots$	$a_1 r^{n-1}$

Consequently, the nth or last term of a geometric progression with first term a_1 and common ratio r is

$$a_n = a_1 r^{n-1} \tag{11.4}$$

Example 1 Find the last term of a geometric progression with $a_1 = 3$, $r = 2$, and $n = 5$.

Solution We could find the fifth term by use of the definition but shall use (11.4). Substituting the given values in it, we have

$$a_5 = (3)(2^{5-1}) = (3)(2^4) = 48$$

In order to obtain the formula for the sum of a geometric progression with first term a_1 and common ratio r, we note that the terms in the progression are $a_1, a_1 r, a_1 r^2$, and so on until we reach the nth term $a_1 r^{n-1}$ as given by (11.4). Hence,

$$s = a_1 + a_1 r + a_1 r^2 + \cdots + a_1 r^{n-1} \tag{1}$$

and, multiplying by r,

$$rs = a_1 r + a_1 r^2 + a_1 r^3 + \cdots + a_1 r^{n-1} + a_1 r^n \tag{2}$$

Now, subtracting each member of (2) from the corresponding member of (1), we have

$$s - rs = a_1 - a_1 r^n$$
$$s(1 - r) = a_1 - a_1 r^n$$

Hence, the sum of a geometric progression of n terms with first term a_1 and common ratio r is

$$s = \frac{a_1 - a_1 r^n}{1 - r} \tag{11.5}$$

$$= \frac{a_1(1 - r^n)}{1 - r}$$

The form of this formula may be changed by writing $a_1 r^n$ as $r a_1 r^{n-1} = r a_n$ and thus obtaining, by (11.4)

$$s = \frac{a_1 - r a_n}{1 - r} \tag{11.6}$$

Example 2 Find the sum of the first six terms of the progression 2, -6, 18, ...

Solution In this progression $a_1 = 2$, $r = -3$, and $n = 6$. Hence, if we substitute these values in (11.5), we have

$$s = \frac{2 - 2(-3)^6}{1 - (-3)}$$

$$= \frac{2 - 2(729)}{1 + 3}$$

$$= \frac{2 - 1458}{4}$$

$$= \frac{-1456}{4}$$

$$= -364$$

Example 3   The first term of a geometric progression is 3; the fourth term is 24.   Find the tenth term and the sum of the first 10 terms.

Solution   In order to find either the tenth term or the sum, we must have the value of r.   We may obtain this value by considering the progression made up of the first four terms.   Then, we have $a_1 = 3$, $n = 4$, and $a_4 = 24$.   If we substitute these values in formula (11.4), we get

$$24 = 3r^{4-1} \quad \text{or} \quad 3r^3 = 24$$

Hence,

$$r^3 = 8 \quad \text{and} \quad r = 2$$

Now, again using (11.4) with $a = 3$, $r = 2$, and $n = 10$, we get

$$a_{10} = 3(2^{10-1}) = 3(2^9) = 3(512) = 1536$$

Hence, the tenth term is 1536.
    In order to obtain s, we use (11.5), with $a_1 = 3$, $r = 2$, and $n = 10$ and get

$$s = \frac{3 - 3(2)^{10}}{1 - 2} = \frac{3 - 3(1024)}{-1} = \frac{3 - 3072}{-1} = 3069$$

    If any three of s, n, a_1, r, and a_n are known, the other two can be found by use of (11.4) and (11.5) or (11.6).   If, after the known quantities are substituted, there is only one unknown in either (11.4), (11.5), or (11.6), we can use that equation to find that unknown.   We can then find the other unknown by use of another of the three formulas.

Example 4   Find r and n if $a_1 = 2$, $a_n = 162$, and $s = 242$.

Solution   If we substitute the given values in (11.6), r is the only unknown, and (11.6) becomes

$$242 = \frac{2 - r162}{1 - r}$$

$242 - 242r = 2 - 162r$ **multiplying by 1 – r**

$-80r = -240$ **combining similar terms**

$r = 3$

If we now substitute $a_1 = 2$ and $a_n = 162$ as given, along with $r = 3$, in (11.4), we have

$162 = (2)3^{n-1}$

$81 = 3^{n-1}$ **dividing by 2**

$3^4 = 3^{n-1}$ **since 81 = 3^4**

$4 = n - 1$

$n = 5$

Example 5   Find a_1 and a_n if $r = 2$, $s = 127$, and $n = 7$.

Solution   Since both unknowns enter in Eqs. (11.4) and (11.6) and only one in (11.5), we shall use (11.5) to determine a_1 and then use (11.4) to find a_n. If we substitute in (11.5), we get

$$127 = \frac{a_1 - a_1(2^7)}{1 - 2}$$

$127 = -a_1 + 128a_1$

$a_1 = 1$

Now substituting $r = 2$ and $n = 7$ as given, along with $a_1 = 1$, in (11.4) yields

$a_n = 1(2^6) = 64$

EXERCISE 11.2   Geometric Progressions

Write the terms of the geometric progression that satisfies the conditions in each of Probs. 1 to 8.

1 $a_1 = 1, r = 2, n = 6$ **2** $a_1 = 7, r = -2, n = 5$

3 $a_1 = 12, r = -3, n = 6$ **4** $a_1 = -5, r = 4, n = 4$

5 $a_1 = 2$, second term 6, $n = 5$ **6** $a_1 = 7$, third term 7, $n = 6$

7 Second term 6, fourth term 24, $n = 7$ **8** Third term 8, sixth term 64, $n = 8$

Find whichever of s, n, a_1, r, and a_n that are missing in each of Probs. 9 to 32.

9 $a_1 = 2, n = 5, r = 2$ **10** $a_1 = 2, n = 6, r = 3$

11 $a_1 = 1, n = 5, r = 4$ **12** $a_1 = 3, r = 2, n = 7$

13 $a_1 = 1, r = -2, a_n = 64$

14 $a_1 = 3, r = -2, a_n = -96$

15 $a_1 = \frac{1}{2}, r = -2, a_n = 32$

16 $a_1 = 1, r = -3, a_n = -243$

17 $a_1 = 243, n = 5, a_n = 3$

18 $a_1 = 4096, n = 8, a_n = -\frac{1}{4}$

19 $n = 9, r = \frac{1}{2}, a_n = 1$

20 $n = 7, r = \frac{1}{5}, a_n = \frac{1}{5}$

21 $a_1 = 2, a_n = 32, s = 62$

22 $a_1 = 2, a_n = 486, s = 728$

23 $a_1 = 1, r = 4, s = 341$

24 $a_1 = 3, r = 2, s = 381$

25 $a_n = 64, r = -2, s = 43$

26 $a_n = -96, r = -2, s = -63$

27 $a_1 = \frac{1}{2}, r = -2, s = 21.5$

28 $a_1 = 1, r = -3, s = -182$

29 $a_n = 3, n = 5, s = 363$

30 $a_n = -\frac{1}{4}, n = 8, s = 3276.75$

31 $a_n = 1, n = 9, s = 511$

32 $a_n = \frac{1}{5}, n = 7, s = 3906.2$

33 If it were possible to save 1 cent on the first day of the month, 2 cents on the second day, 4 cents on the third, and so on, on what day would the total savings amount to a million dollars? On what day would a million dollars be added to the total savings?

34 Eleven people are fishing from a pier. The first is worth $2000, the second $4000, the third $8000, and so on. How many millionaires are in the group?

35 The number of bacteria in a culture triples every 2 hours. If there were n at the beginning of one 24-hour period, how many were there at the beginning of the next?

36 A man willed one-third of his estate to one friend, one-third of the remainder to another, and so on until the fifth friend received $800. What was the value of the estate?

37 If 1, 4, and 19 are added to three consecutive terms of an arithmetic progression with $d = 3$, a geometric progression is obtained. Find the arithmetic progression and the common ratio of the geometric progression.

38 If, 3, 3, and 7 are subtracted from three consecutive terms of a geometric progression with common ratio 2, an arithmetic progression is obtained. Find the geometric progression and the common difference of the arithmetic progression.

39 For what values of k are k, $3k - 2$, and $5k - 2$ consecutive terms of a geometric progression?

40 If $1/(y - x)$, $1/2y$, and $1/(y - z)$ form an arithmetic progression, show that x, y, and z form a geometric progression.

11.3 INFINITE GEOMETRIC PROGRESSIONS

In this section we discuss the sum of a geometric progression in which the common ratio is between -1 and 1 and the number of terms increases indefinitely. As an example, we consider the sequence

$$1, \frac{1}{2}, \frac{1}{4}, \ldots, \frac{1}{2^{n-1}} \tag{1}$$

Here $a = 1$ and $r = \frac{1}{2}$. We shall let $s(n)$ stand for the sum of the first n terms in progression (1) and shall tabulate the corresponding values of n and $s(n)$ for $n = 1, 2, 3, \ldots, 7$.

n	1	2	3	4	5	6	7
s	1	$1\frac{1}{2}$	$1\frac{3}{4}$	$1\frac{7}{8}$	$1\frac{15}{16}$	$1\frac{31}{32}$	$1\frac{63}{64}$

The tabulated values show that as n increases from 1 to 7, $s(n)$ approaches nearer and nearer to 2. We now prove that this is true as n increases beyond 7 by replacing a_1 by 1 and r by $\frac{1}{2}$ in (11.5). This replacement yields

$$s(n) = \frac{1 - (\frac{1}{2})^n}{1 - \frac{1}{2}} = \frac{1}{\frac{1}{2}} - \frac{(\frac{1}{2})^n}{\frac{1}{2}}$$

$$= 2 - \frac{1}{2^{n-1}}$$

Now, by choosing n sufficiently large, we can make $1/2^{n-1}$ less than any number ϵ selected in advance. Hence, for the chosen value of n, $s(n)$ differs from 2 by a number less than ϵ. This situation is described in mathematical language by the statement, "the limit of $s(n)$ as n approaches infinity is 2," and the statement is abbreviated as

$$\lim_{n \to \infty} s(n) = 2$$

We now consider a progression with $|r| < 1$, and we shall prove that the limit of the sum s as n approaches infinity is given by the formula

$$s = \frac{a_1}{1 - r} \qquad |r| < 1 \tag{11.7}$$

Note that in progression (1), $a_1 = 1$ and $r = \frac{1}{2}$, and if we substitute these values in (11.7), we get $s = 2$.

The proof of (11.7) follows. We let $s(n)$ stand for the sum of the first n terms of a progression with $|r| < 1$. Then, by (11.5),

$$s(n) = \frac{a_1 - a_1 r^n}{1 - r}$$

and this can be expressed in the form

$$s(n) = \frac{a_1}{1 - r}(1 - r^n) \tag{2}$$

We next notice that the binomial $1 - r^n$ in (2) differs from 1 by r^n. Furthermore, if $|r| < 1$, then $|r| > |r^2| > |r^3| > \cdots > |r^n|$, and it can be proved† that by choosing n sufficiently large, $|r^n|$ can be made less than any preassigned positive number. Hence, for

---

†P. K. Rees, F. W. Sparks, and C. S. Rees, "Algebra and Trigonometry," 3d ed. p. 451, McGraw-Hill Book Company, New York, 1975.

this value and larger values of n, $1 - r^n$ differs from 1 by an amount that is less than the preassigned number. Hence,

$$\lim_{n \to \infty} (1 - r^n) = 1$$

Now we return to Formula (2) and complete the proof as follows:

$$\lim_{n \to \infty} s(n) = \lim_{n \to \infty} \frac{a_1}{1 - r}(1 - r^n)$$

$$= \frac{a_1}{1 - r} \lim_{n \to \infty} (1 - r^n)$$

$$= \frac{a_1}{1 - r} \qquad \text{since } \lim_{n \to \infty} (1 - r^n) = 1$$

The reader should reread the part of Sec. 1.3 on repeating decimals and rational numbers.

Any nonterminating decimal fraction with a sequence of digits repeated indefinitely is an infinite geometric progression with r equal to a negative integral power of 10. For example,

$$.3333 \cdots = .3 + .03 + .003 + .0003 + \cdots$$

Here

$$a_1 = .3 \qquad \text{and} \qquad r = .1$$

Example 1 Express $.999 \cdots$ as a rational number.

Solution This is the infinite geometric progression

$$.9 + .09 + .009 + \cdots$$

with $a_1 = .9$ and $r = .09/.9 = .1$; hence by (11.7)

$$s = \frac{.9}{1 - .1} = 1$$

Example 2 Find the rational number represented by $.351351351 \cdots$.

Solution This nonterminating, repeating decimal is equal to

$$.351 + .000351 + .000000351 + \cdots$$

It is now clear that $a_1 = .351$ and $r = .001$; hence,

$$s = \frac{.351}{1 - .001} = \frac{.351}{.999} = \frac{13}{37}$$

Example 3 A share in a mine was given to a college endowment fund. The dividend the first year was \$2500 and each year thereafter was 80 percent of the amount received the previous year. How much did the college receive?

Solution   In this problem $a_1 = \$2500$ and $r = .8$. Consequently,

$$s = \frac{\$2500}{1 - .8} = \$12,500$$

EXERCISE 11.3   Infinite Geometric Progressions

Write out the first four terms of the progressions in Probs. 1 to 4, and find the sum of the infinite geometric progressions described in Probs. 1 to 16.

1 $a_1 = 4, r = \frac{2}{3}$

2 $a_1 = 4, r = \frac{1}{3}$

3 $a_1 = 6, r = \frac{2}{5}$

4 $a_1 = 12, r = \frac{4}{7}$

5 $a_1 = 3, r = \frac{1}{4}$

6 $a_1 = 7, r = \frac{1}{8}$

7 $a_1 = 5, r = \frac{6}{11}$

8 $a_1 = 11, r = \frac{1}{12}$

9 $a_1 = 2, r = -\frac{1}{3}$

10 $a_1 = 22, r = -\frac{5}{6}$

11 $a_1 = 9, r = -\frac{2}{7}$

12 $a_1 = 5, r = -\frac{2}{3}$

13 $a_1 = 12$, second term -6

14 Second term 2, third term 1

15 Second term 9, fourth term 4

16 Third term 4, fifth term 1

Express each of the following repeating decimal fractions as a rational number.

17 $.333 \cdots$

18 $.777 \cdots$

19 $.555 \cdots$

20 $.888 \cdots$

21 $.757575 \cdots$

22 $.363636 \cdots$

23 $.272727 \cdots$

24 $.939393 \cdots$

25 $21.351351 \cdots$

26 $3.102102 \cdots$

27 $2.3078078 \cdots$

28 $7.8346346 \cdots$

Find the sum of all numbers of the form given in each of Probs. 29 to 36 provided n is a positive integer.

29 2^{-n}

30 3^{-n}

31 7^{-n}

32 6^{-n}

33 $\left(\frac{2}{3}\right)^n$

34 $\left(\frac{3}{7}\right)^n$

35 $\left(\frac{4}{5}\right)^n$

36 $\left(\frac{5}{9}\right)^n$

37 If a ball rebounds $\frac{4}{7}$ as far as it falls, how far will it travel if dropped from 9 feet?

38 If the midpoints of the sides of a square are joined in order, another square is formed. If we begin with a square of side 8 inches and form a sequence of squares in the manner mentioned, find the sum of their areas.

39 Find the sum of the perimeters of the squares in Prob. 38.

40 Find the sum of the diagonals of the squares in Prob. 38. Each square has two diagonals.

41 A child received $1200 on the day of her birth, and $\frac{3}{4}$ as much on each birthday as on the preceding one. About how much had she received from this source by the time the ladies in the neighborhood began considering her to be an old woman?

42 The seat of a swing describes an arc 10 feet long on its first swing, and each following arc is $\frac{7}{8}$ as long as the immediately preceding one. How far does the seat move before the swing comes to rest?

43 Assume that stored potatoes shrink $\frac{2}{3}$ as much each week as during the immediately preceding one. If a dealer stores 1000 pounds of potatoes when the price is n cents per pound, and if the weight decreases to 900 pounds the first week, for what range of n can he afford to hold them until the price rises to $n + 1$ cents per pound?

44 A subdivision is laid out to contain 96 lots. If the profit on the first lot sold is $800, and on each of the others the profit is 96 per cent of the amount on the immediately preceding one, about how much profit was made on the entire subdivision?

45 If $x > 0$, find the sum of $\dfrac{1}{x+1}, \dfrac{1}{(x+1)^2}, \dfrac{1}{(x+1)^3}, \cdots\cdots$

46 If $x > 3$, find the sum of $\dfrac{1}{x-2}, \dfrac{1}{(x-2)^2}, \dfrac{1}{(x-2)^3}, \cdots\cdots$

47 If $x > 1$, find the sum of $\dfrac{1}{2x-1}, \dfrac{1}{(2x-1)^2}, \dfrac{1}{(2x-1)^3}, \cdots\cdots$

48 If $x > \dfrac{-1}{3}$, find the sum of $\dfrac{1}{3x+2}, \dfrac{1}{(3x+2)^2}, \dfrac{1}{(3x+2)^3}, \cdots\cdots$

11.4 ARITHMETIC MEANS

The terms between the first and last terms of an arithmetic progression are called *arithmetic means*. If the progression contains only three terms, the middle term is called *the arithmetic mean* of the first and last term. We may obtain the arithmetic means between two numbers by first using (11.1) to find d; then the means can be computed by use of the definition of an arithmetic progression. If the progression consists of the three terms a_1, m, and a_3, then by (11.1),

$$a_3 = a_1 + (3 - 1)d = a_1 + 2d$$

Hence,

$$d = \frac{a_3 - a_1}{2} \quad \text{and} \quad m = a_1 + \frac{a_3 - a_1}{2} = \frac{a_1 + a_3}{2}$$

Therefore, the following is proved:

Arithmetic Mean The arithmetic mean of two numbers is equal to one-half their sum.

Example Insert five arithmetic means between 6 and -10.

Solution Since we are to find five means between 6 and -10, we shall have seven terms in all. Hence, $n = 7$, $a_1 = 6$, and $a_n = -10$. Thus, by (11.1), we have

$$-10 = 6 + (7 - 1)d$$

Hence,

$$6d = -16$$
$$d = -\tfrac{16}{6} = -\tfrac{8}{3}$$

and the progression consists of

$$6, \tfrac{10}{3}, \tfrac{2}{3}, -\tfrac{6}{3}, -\tfrac{14}{3}, -\tfrac{22}{3}, -\tfrac{30}{3}$$

11.5 GEOMETRIC MEANS

The terms between the first and last terms of a geometric progression are called the *geometric means*. If the progression contains only three terms, the middle term is called *the geometric mean* of the other two. In order to obtain the geometric means between a_1 and a_n, we use (11.4) to find the value of r; then the means can be computed by use of the definition of a geometric progression. If there are only three terms in the progression, then by (11.4),

$$a_3 = a_1 r^2$$

Hence,

$$r = \pm\sqrt{\frac{a_3}{a_1}}$$

Thus the second term, or geometric mean between a_1 and a_3, is

$$a_1\left(\pm\sqrt{\frac{a_3}{a_1}}\right) = \pm\sqrt{\frac{a_1^2 a_3}{a_1}} = \pm\sqrt{a_1 a_3}$$

Hence, the following is proved:

Geometric Means   The geometric means between two quantities are plus and minus the square root of their product.

Example 1   Find four geometric means between 3 and 96.

Solution   A geometric progression starting with 3, ending with 96, and with four intermediate terms contains six terms. Hence, $n = 6$, $a_1 = 3$, and $a_n = 96$. Therefore, by (11.4),

$$96 = 3r^{6-1}$$

Hence,

$$r^5 = \tfrac{96}{3} = 32 \quad \text{and} \quad r = \sqrt[5]{32} = 2$$

Consequently, the four geometric means between 3 and 96 are

6,   12,   24,   48

Example 2   Find the geometric means of $\frac{1}{2}$ and $\frac{1}{8}$.

Solution   By the statement just before Example 1, the geometric means of $\frac{1}{2}$ and $\frac{1}{8}$ are

$$\pm \sqrt{(\tfrac{1}{2})(\tfrac{1}{8})} = \pm \sqrt{\tfrac{1}{16}} = \pm \tfrac{1}{4}$$

11.6   HARMONIC PROGRESSIONS

Harmonic Progression   A *harmonic progression* is a sequence of numbers whose reciprocals form an arithmetic progression. Accordingly 1, $\frac{1}{3}$, $\frac{1}{5}$, $\frac{1}{7}$, $\frac{1}{9}$, and $\frac{1}{11}$ are the terms of a harmonic progression, since their reciprocals 1, 3, 5, 7, 9, and 11 form an arithmetic progression. The name "harmonic" progression is used since strings of a given material and size give off a pleasing sound if their lengths form a harmonic progression. A fuller discussion can be found in an encyclopedia or a book on harmonics.

Harmonic Means   The terms between any two terms of a harmonic progression are called the *harmonic means* between those two terms. They can be found by working with the associated arithmetic progression.

Example   Find three harmonic means between $\frac{1}{2}$ and $\frac{1}{14}$.

Solution   We shall use the reciprocals of the given terms as the first and fifth terms of an arithmetic progression, find the three arithmetic means between them, and obtain the desired harmonic means by taking the reciprocals of the arithmetic means. The terms of the associated arithmetic progression are 2 and 14, the reciprocals of $\frac{1}{2}$ and $\frac{1}{14}$. Since we need three arithmetic means between 2 and 14, we must use $a_1 = 2$, $n = 5$, and $a_n = 14$. Now, by use of (11.1), we have

$$14 = 2 + (5 - 1)d = 2 + 4d$$

Therefore, $d = 3$, and the arithmetic means are $2 + 3 = 5$, $5 + 3 = 8$, and $8 + 3 = 11$. Consequently, the desired harmonic means are the reciprocals $\frac{1}{5}$, $\frac{1}{8}$, and $\frac{1}{11}$ of 5, 8, and 11.

EXERCISE 11.4   Means and Harmonic Progressions

Classify the progression in each of Probs. 1 to 16, and give the next two terms of each.

1	1, 5, 9, 13	**2**	6, 2, −2, −6	**3**	2, 7, 12, 17	**4**	−2, 1, 4, 7
5	2, 6, 18, 54	**6**	3, 6, 12, 24	**7**	1, −2, 4, −8	**8**	2, −8, 32, −128
9	64, −32, 16, −8	**10**	243, −81, 27, −9	**11**	125, 25, 5, 1	**12**	343, 49, 7, 1
13	$1, \frac{1}{4}, \frac{1}{7}, \frac{1}{10}$	**14**	$\frac{1}{3}, \frac{1}{8}, \frac{1}{13}, \frac{1}{18}$	**15**	$2, \frac{6}{5}, \frac{6}{7}, \frac{2}{3}$	**16**	$\frac{10}{7}, \frac{5}{6}, \frac{10}{17}, \frac{5}{11}$

Insert the specified number of means between the terms given in each of Probs. 17 to 28.

17 Two arithmetic between 3 and 9

18 Three arithmetic between 1 and 13

19 Four arithmetic between 2 and 17

20 Five arithmetic between -4 and 14

21 One geometric between 2 and 8

22 Two geometric between 3 and 81

23 Three geometric between $\frac{1}{4}$ and 64

24 Four geometric between 3 and -486

25 Five geometric between 128 and 2

26 Three geometric between 162 and 2

27 Four geometric between 3125 and 1

28 Two geometric between 64 and $\frac{1}{8}$

29 Find the fourth term of a harmonic progression if the third and fifth terms are $\frac{1}{5}$ and $\frac{1}{9}$.

30 Find the third term of a harmonic progression if the first and fourth are $\frac{1}{2}$ and $\frac{1}{23}$.

31 Find the fifth term of a harmonic progression if the second and third are -1 and 1.

32 Find the seventh term of a harmonic progression if the first and fourth are $-\frac{1}{5}$ and $\frac{1}{7}$.

11.7 SUMMARY

We define arithmetic, geometric, infinite geometric, and harmonic progressions in this chapter. Then we develop formulas for the last term of each of the first two types, and the sum of each of the first three types. These formulas are

$$a_n = a_1 + (n-1)d \tag{11.1}$$

$$a_n = a_1 r^{n-1} \tag{11.4}$$

$$s = \frac{n}{2}(a_1 + a_n) \tag{11.2}$$

$$= \frac{n}{2}[2a_1 + (n-1)d] \tag{11.3}$$

$$s = \frac{a_1 - a_1 r^n}{1-r} \tag{11.5}$$

$$= \frac{a_1 - ra_n}{1-r} \tag{11.6}$$

$$s = \frac{a_1}{1-r} \tag{11.7}$$

Finally we define and determine arithmetic, geometric, and harmonic means.

EXERCISE 11.5 Review

1 What are the five terms of an arithmetic progression with $a_1 = 2$ and $d = \frac{1}{3}$?

2 What are the six terms of a geometric progression with $a_1 = \frac{9}{16}$ and $r = \frac{2}{3}$?

3 If the first two terms of a harmonic progression are 1 and $\frac{3}{5}$, what are the next four terms?

4 Show that if x, y, and z are consecutive terms of a geometric progression, then $1/x$, $1/y$, and $1/z$ are consecutive terms of another.

5 If -1 and 1 are the first two terms of an arithmetic progression, find the sixth term and the sum of the first six terms.

6 If -1 and 1 are the first two terms of a geometric progression, find the sixth term and the sum of the first six terms.

7 If -1 and 1 are the first two terms of a harmonic progression, find the first six terms.

8 If the second and fourth terms of an arithmetic progression are $\frac{3}{5}$ and $\frac{7}{5}$, and $n = 5$, find d, a_1, a_n, and s.

9 If the second and fifth terms of a geometric progression are 2 and $\frac{16}{27}$, and $n = 6$, find r, a_1, a_n, and s.

10 Find the value of k so that $2k + 1$, $3k - 2$, and $5k - 7$ form an arithmetic progression.

11 Find all values of k so that $k + 1$, $5k - 1$, and $11k + 5$ form a geometric progression.

12 Find all values of k so that $1/(2k + 1)$, $1/(3k - 1)$, and $1/(5k - 3)$ form a harmonic progression.

13 Find the sum of all integers from 3 to 47 inclusive.

14 Find the sum of the first seven multiples of 3, beginning with 9.

15 Find the sum of the first seven integral powers of 3, beginning with 9.

16 Find the sum of all fractions of the form $(\frac{2}{3})^n$ for n a positive integer.

17 Express $.4545\cdots$ as a rational number.

18 Show that if a, b, c is an arithmetic progression and x is any real number, then $a + x$, $b + x$, $c + x$ is also an arithmetic progression, as is ax, bx, cx.

19 Show a^2, b^2, c^2 is a geometric progression if a, b, c is.

20 Show that if a^2, b^2, c^2 is an arithmetic progression, then $a + b$, $a + c$, $b + c$ is a harmonic progression.

21 Show that $\frac{3}{5}$, $\frac{6}{5}$, $\frac{12}{5}$, $\frac{24}{5}$ is a geometric progression and that $\log \frac{3}{5}$, $\log \frac{6}{5}$, $\log \frac{12}{5}$, $\log \frac{24}{5}$ is an arithmetic progression (see Chap. 13).

22 Show that, for any two positive numbers a and b, their arithmetic mean is greater than their geometric mean, and that the difference is $(\sqrt{a} - \sqrt{b})^2/2$.

23 Find three geometric means between 2 and 32.

24 Find three arithmetic means between 2 and 32.

25 Find three harmonic means between $\frac{1}{2}$ and $\frac{1}{32}$.

Classify the progression in each of Probs. 26 to 28, and give the next two terms.

26 $\frac{3}{4}, \frac{3}{2}, \frac{9}{4}, 3$ **27** $\frac{3}{4}, \frac{3}{2}, 3, 6$ **28** $\frac{3}{4}, 3, -\frac{3}{2}, -\frac{3}{5}$

12

the binomial theorem

In earlier chapters we developed formulas for obtaining the square and the cube of a binomial. In this chapter we shall develop the binomial formula, which expresses the nth power of a binomial as a polynomial. This formula is known as the *binomial formula*, and it has many applications in more advanced mathematics and in all fields in which mathematics is applied. For example, it is used extensively in probability, calculus, and genetics. The polynomial yielded by the formula is called the *expansion of a power of a binomial*.

12.1 THE BINOMIAL FORMULA

By actual multiplication we may obtain the following expansions of the first, second, third, fourth, and fifth powers of $x + y$:

$(x + y)^1 = x + y$

$(x + y)^2 = x^2 + 2xy + y^2$

$(x + y)^3 = x^3 + 3x^2y + 3xy^2 + y^3$

$(x + y)^4 = x^4 + 4x^3y + 6x^2y^2 + 4xy^3 + y^4$

$(x + y)^5 = x^5 + 5x^4y + 10x^3y^2 + 10x^2y^3 + 5xy^4 + y^5$

By referring to these expansions, we may readily verify the fact that the following properties of $(x + y)^n$ exist when $n = 1, 2, 3, 4,$ and 5:

1 The first term in the expansion is x^n.
2 The second term is $nx^{n-1}y$.
3 The exponent of x decreases by 1 and the exponent of y increases by 1 as we proceed from term to term.
4 There are $n + 1$ terms in the expansion.
5 The $(n + 1)$st term, or the last term, is y^n.
6 The nth, or the next to the last, term of the expansion is nxy^{n-1}.

7   If we multiply the coefficient of any term by the exponent of x in that term and then divide the product by the number of the term in the expansion, we obtain the cofficient of the next term.

8   The sum of the exponents of x and y in any term is n.

If we assume that these properties hold for all integral values of n, we may write the first five terms in the expansion of $(x + y)^n$ as follows:

First term: x^n by property 1

Second term: $nx^{n-1}y$ by property 2

Third term: $\dfrac{n(n-1)}{2}x^{n-2}y^2$ by properties 7 and 3

Fourth term: $\dfrac{n(n-1)(n-2)}{(3)(2)}x^{n-3}y^3$ by properties 7 and 3

Fifth term: $\dfrac{n(n-1)(n-2)(n-3)}{(4)(3)(2)}x^{n-4}y^4$ by properties 7 and 3

We continue this process until we reach the nth term, which is

nth term: nxy^{n-1} by property 6

and finally we reach the last, or the

$(n + 1)$st term: y^n by property 5

The sum of these terms is the expansion of $(x + y)^n$. It is

$$(x + y)^n = x^n + nx^{n-1}y + \frac{n(n-1)x^{n-2}y^2}{2} + \frac{n(n-1)(n-2)x^{n-3}y^3}{(3)(2)}$$

$$+ \frac{n(n-1)(n-2)(n-3)x^{n-4}y^4}{(4)(3)(2)} + \cdots + nxy^{n-1} + y^n \tag{1}$$

Factorial   If, however, we introduce the factorial notation, we can write this equation in a more compact form. By n *factorial*, for n a positive integer, we mean the product of n and all postive integers less than n, and we write it $n!$. In keeping with this, $2! = 2 \times 1, 3! = 3 \times 2 \times 1$, and $4! = 4 \times 3 \times 2 \times 1$. Furthermore, $4 \times 3 \times 2 = 4 \times 3 \times 2 \times 1 = 4!, 3 \times 2 = 3 \times 2 \times 1 = 3!$ and $2 = 2 \times 1 = 2!$. We may now write (1) in the form

$$(x + y)^n = x^n + \frac{n}{1!}x^{n-1}y + \frac{n(n-1)x^{n-2}y^2}{2!}$$

$$+ \frac{n(n-1)(n-2)x^{n-3}y^3}{3!} + \cdots + nx\,y^{n-1} + y^n \tag{12.1}$$

Binomial
Formula   if we define 1! to be 1. Formula (12.1) is called the *binomial*

formula, and the statement that it is true is called the *binomial theorem.* The proof of the theorem will not be given.

Example 1 Find the expansion of $(a + b)^6$ by use of (12.1).

Solution We need only apply (12.1) with $x = a$, $y = b$, and $n = 6$ and then simplify. By (12.1), we have

$$(a + b)^6 = a^6 + \frac{6a^5b}{1!} + \frac{(6)(5)a^4b^2}{2!} + \frac{(6)(5)(4)a^3b^3}{3!}$$
$$+ \frac{(6)(5)(4)(3)a^2b^4}{4!} + \frac{(6)(5)(4)(3)(2)ab^5}{5!} + b^6$$
$$= a^6 + 6a^5b + 15a^4b^2 + 20a^3b^3 + 15a^2b^4 + 6ab^5 + b^6$$

which is obtained by simplifying the coefficients.

The coefficients can be computed in most cases by the use of property 7. If this is done, we must be very careful, since an error in any coefficient is automatically carried over to all that follow it.

Example 2 Expand $(a + 3b)^5$.

Solution To get the expansion we need only use (12.1) with $x = a$, $y = 3b$, and $n = 5$. Thus,

$$(a + 3b)^5 = a^5 + 5a^4(3b) + \frac{(5)(4)a^3(3b)^2}{2!} + \frac{(5)(4)(3)a^2(3b)^3}{3!}$$
$$+ 5a(3b)^4 + (3b)^5$$
$$= a^5 + 15a^4b + 90a^3b^2 + 270a^2b^3 + 405ab^4 + 243b^5$$

after simplifying the coefficients.

Example 3 Expand $(2a - 5b)^4$

Solution We need only use (12.1) with $x = 2a$, $y = -5b$, and $n = 4$. Doing this, we have

$$(2a - 5b)^4 = (2a)^4 + 4(2a)^3(-5b) + \frac{(4)(3)(2a)^2(-5b)^2}{2!}$$
$$+ 4(2a)(-5b)^3 + (-5b)^4$$
$$= 16a^4 - 160a^3b + 600a^2b^2 - 1000ab^3 + 625b^4$$

after simplifying the coefficients.

12.2 THE *r*th TERM OF THE BINOMIAL FORMULA

In the last section, we found how to obtain the expansion of a binomial, and property 7 showed how to obtain any term from the one immediately before it. We shall now turn our attention to obtaining any specified term without reference to any other term. To do this we shall examine the binomial formula (12.1) and make some observations. We first notice the coefficient n of the next to the last term may be written as $n(n - 1)(n - 2) \cdots (3)(2)/(n - 1)!$, and that the coefficient 1 of the last term may be put in the form $n(n - 1)(n - 2) \cdots (3)(2)(1)/n!$ It is then easy to see that after the first term x^n,

1   The exponent of the second term y of the binomial is always one less than the number of the term in the expansion.
2   The exponent of the first term x of the binomial is always n decreased by the exponent of the second term.
3   The denominator of each coefficient is the factorial of the exponent of the second term of the binomial.
4   The numerator of each coefficient is the product of one fewer consecutive integers than the number of the term, and it begins with n.

If we decide to call the general term the *r*th term and then make use of 1 to 4 above, we find that for the *r*th term of the expansion:

1   The exponent of the second term y of the binomial is $r - 1$.
2   The exponent of the first term x of the binomial is
$n - (r - 1) = n - r + 1$
3   The denominator of the coefficient is $(r - 1)!$
4   The numerator of the coefficient is
$n(n - 1)(n - 2) \cdots (n - r + 2)$

Consequently, *the rth term of the expansion of* $(x - y)^n$ *is*

$$\frac{n(n - 1)(n - 2) \cdots (n - r + 2)x^{n-r+1}y^{r-1}}{(r - 1)!} \tag{12.2}$$

The numerical coefficient has $r - 1$ factors in both the numerator and denominator. We define $0! = 1$.

Example   Find the fifth term in the expansion of $(a + 5b)^6$.

Solution   We can use (12.2) or the steps used in obtaining it, with $x = a$, $y = 5b$, $n = 6$, and $r = 5$. We have

$$\frac{(6)(5)(4)(3)}{4!}a^2(5b)^4 = 15a^2(625b^4)$$

$$= 9375a^2b^4$$

EXERCISE 12.1   The Binomial Formula

Expand the binomial in each of Probs. 1 to 40.

1 $(a + b)^5$ **2** $(b + c)^7$ **3** $(c + d)^4$ **4** $(d + a)^3$

5 $(b - d)^7$ **6** $(a - c)^4$ **7** $(a - b)^6$ **8** $(b - c)^5$

9 $(2a + x)^4$ **10** $(3b + c)^6$ **11** $(2a + b)^5$ **12** $(5c + d)^6$

13 $(a - 2b)^6$ **14** $(b - 3c)^5$ **15** $(c - 4a)^3$ **16** $(a - 5d)^4$

17 $(2a + 3b)^3$ **18** $(3a + 4b)^3$ **19** $(2a + 5b)^4$ **20** $(3b + 5c)^5$

21 $(x^2 - 3y)^4$ **22** $(a^3 - 2y)^6$ **23** $(b^2 - 4y)^5$ **24** $(b^4 - 3x)^3$

25 $(2a + 5y^2)^5$ **26** $(3a + 4y^3)^4$ **27** $(4b + 3y^4)^5$ **28** $(3x + 2y^2)^6$

29 $(2x^2 - 3y^3)^6$ **30** $(3a^3 - 4b^2)^5$ **31** $(4x^2 - 5y^3)^3$ **32** $(5x^2 - 2y^5)^4$

33 $\left(x + \dfrac{2}{x}\right)^5$ **34** $\left(x - \dfrac{3}{x}\right)^4$ **35** $\left(2x - \dfrac{1}{3x}\right)^4$ **36** $\left(3x + \dfrac{1}{3x}\right)^6$

37 $\left(a^2 - \dfrac{3}{a}\right)^4$ **38** $\left(b^3 - \dfrac{2}{b}\right)^5$ **39** $\left(x^2 - \dfrac{2}{x^3}\right)^3$ **40** $\left(2x^3 - \dfrac{1}{3x^2}\right)^4$

Approximate the number in each of Probs. 41 to 48 by adding the first four terms of a binomial expansion.

41 $1.03^4 = (1 + .03)^4$ **42** 1.05^5 **43** 1.07^6 **44** 1.06^4

45 $.98^5 = (1 - .02)^5$ **46** $.96^6$ **47** $.95^4$ **48** $.97^5$

Find the first four terms in the expansion of the binomial in each of Probs. 49 to 56.

49 $(a + 2b)^{10}$ **50** $(2a - b)^{12}$ **51** $(x - 3y)^9$ **52** $(3x + y)^7$

53 $(x^2 - x^{-1})^{14}$ **54** $(x^3 + x^{-2})^9$ **55** $(x^4 - x^{-4})^4$ **56** $(x^2 + x^{-3})^9$

Find the specified term of the expansion in each of Probs. 57 to 72.

57 Fourth term of $(x + 2y)^9$

58 Fifth term of $(2x - y)^{14}$

59 Sixth term of $(3x - 2a)^{11}$

60 Sixth term of $(4x + 3y)^{12}$

61 Eighth term of $(x^2 - 3y)^{10}$

62 Fourth term of $(2x^3 + y)^7$

63 Sixth term of $(2a + x^2)^7$

64 Fourth term of $(3a - y^3)^9$

65 Middle term of $(a + 3y)^8$

66 Middle term of $(2a - b)^6$

67 Middle term of $(2x - 5y)^4$

68 Middle term of $(3x + 2b)^{10}$

69 The term of $(2x - 5y)^5$ that involves x^3

70 The term of $(x + 3y)^7$ that involves y^5

71 The term of $(2x^2 - 3y)^6$ that involves x^4

72 The term of $(5x + 2y^3)^8$ that involves y^9

13
logarithms

Logarithms, invented in the seventeenth century, are useful and efficient in arithmetical computation, important in the application of mathematics to chemistry, physics, and engineering, and indispensable for some parts of advanced mathematics. In this chapter we shall develop some properties of logarithms and show how they are used in numerical computation. We also use logarithms to obtain the solution sets of certain types of equations. The theory of logarithms is based on the laws of exponents, and the reader is advised to review these laws now.

Perhaps the main use of logarithms originally was as an aid in computation. Today this use has largely been taken over by computers and pocket calculators, as we point out in this chapter at several points. However it is still important to study logarithms and their properties in order to be able to simplify complicated expressions, to solve equations like $5^x = 30$, to express natural laws (such as growth and decay, pH values in chemistry, severity of earthquakes by Richter numbers), and to understand the inverse relationship between logarithms and exponents.

13.1 APPROXIMATIONS

In this chapter we shall deal with computations that depend on approximations. Hence, it is desirable to know how to indicate the accuracy of an approximate number, and to know how many digits or decimal places to use in the result of a calculation that involves approximations.

Scientific Notation A number is in *scientific notation* if a decimal point is put after the first nonzero digit, and the resulting number is multiplied by whatever integral power of 10 is needed to make the number the proper size. Thus, 8.643 is in scientific notation as is every other number from 1 to 10, including 1 but not 10.

Example 1 Write 3498 and .01776 in scientific notation.

Solution   If we put a decimal point in 3498 just after the first nonzero digit, we have 3.498; this must be multiplied by 10^3 to obtain a number of the proper size. Hence, $3.498(10^3)$ is the scientific notation for 3498. Similarly, the scientific notation for .01776 is $1.776(10^{-2})$.

Rounding Off   If a number is given to more digits than we are to use, we need a method for reducing the number of digits, i.e., for *rounding off*. The usual procedure for rounding off a number of more than n digits to n digits is to write the number in scientific notation, delete all digits beyond the nth, and increase the nth by 1 if the $(n + 1)$st is 5, 6, 7, 8, or 9.

Example 2   Round off 38.47 and .25938 to three digits.

Solution   In scientific notation, $38.47 = 3.847(10)$; then, dropping the 7, noting that $7 \geq 5$, and increasing the third digit by 1 give $3.85(10)$. The scientific notation for .25938 is $2.5938(10^{-1})$. Now, dropping the 38, we get $2.59(10^{-1})$ as the three-digit approximation for .25938.

In finding the product of two approximate numbers that are both correct to n digits, we perform the multiplication and round off the result to n digits. If one factor is correct to n digits and the other is correct to more than $n + 1$ digits, we round off the second one to $n + 1$ digits, perform the multiplication, and round off the product to n digits.

Products
to n Digits

Example 3   Find the product of $F = 23.4$ and $S = 1.7869$.

Solution   Since F is correct to three digits and S to five, we round off S to $3 + 1 = 4$ digits and have $S = 1.787$. Hence, $FS = (23.4)(1.787) = 41.8158 = 4.18(10)$ after rounding off to three digits.

A similar procedure is followed in finding the quotient of two approximations.

Example 4   Find the quotient of the approximations $N = 38$ and $D = 58.714$.

Solution   Since N is given to two digits and D to five, we round D off to $2 + 1 = 3$ digits and have $D = 5.87(10)$. Therefore, $N/D = 38/5.87(10) = .647 = 6.5(10^{-1})$ after rounding off to two digits.

Addition of
Approximations   In adding approximate numbers, we are interested in the precision of the calculations; hence, we concentrate on the decimal portions of the numbers. If the decimal portions of the addends contain the same number of digits, we add and keep all of the decimal portion. If the one addend contains n digits in the decimal portion and others contain more than $n + 1$ digits, round off the latter to $n + 1$ digits, perform the addition, and round off the decimal portion of the sum to n digits.

Example 5 Find the sum of $F = 3.81$, $S = 17.6$, and $T = 2.736$.

Solution We must round off $T = 2.736$ to two digits in the decimal portion. We thereby get $T = 2.74$. Now $F + S + T = 3.81 + 17.6 + 2.74 = 24.15 = 24.2$ after rounding the decimal portion off to one digit. A similar procedure is followed in subtraction.

13.2 CALCULATORS AND FUNDAMENTAL OPERATIONS

Fundamental Operations

There are computers that are too large to fit into an ordinary room, and calculators that will fit into the palm of a hand. Many operations, including complex ones, can be performed on some of the latter. At the moment we are interested only in addition, multiplication, subtraction, and division. To add M and N with a calculator, we display M on the dial, punch the $+$ button, display N on the dial, and then punch the $=$ button. If this is done, the value of $M + N$ appears on the display dial. The other three operations are performed in a similar manner.

Example Find the product of $M = 2.31$ and $N = 48.3$ by use of a calculator if one is available.

Solution We perform the following operations in the indicated order: Display M, punch the $\times$ button, display N, punch the $=$ button, and the product $MN = 111.573$ appears on the display dial. This is the value if M and N are exact; but if they are approximations, we must round off the product to three digits and thus have $MN = 112$, since 2.31 and 48.3 are correct to 3 digits.

EXERCISE 13.1 Approximations

Express the number in each of Probs. 1 to 16 in scientific notation.

1	3284	2	731.5
3	685,713	4	1234.67
5	.5984	6	.004862
7	.0001405	8	.0707
9	8470; zero not significant	10	8470; zero significant
11	21,000; zeros significant	12	21,000; first zero significant
13	6480; zero significant	14	6480; zero not significant
15	2,786,000; zeros not significant	16	2,786,000; first two zeros significant

Round off the number in each of Probs. 17 to 32 to four significant digits and then to three significant digits.

17	48.4612	18	873.847	19	27.1573	20	.0187608
21	697,824	22	89,412	23	500,817.84	24	7,235,485

25 8968.805	**26** 407.6481	**27** 237,649.4	**28** 18,985.02
29 36.354	**30** 36.345	**31** 80.256	**32** 37.850

Perform the indicated operations in Probs. 33 to 56 under the assumption that the numbers are approximations. Then round off each result to the proper number of digits or decimal places, and express it in scientific notation if needed.

33 (8.1)(1.2)	**34** (.64)(9.7)	**35** (78)(235)
36 (493)(8.1)	**37** (4.7)(4754)	**38** (4.7)(4.745)
39 (23.9)(1.8763)	**40** (93.8)(.60305)	**41** $82 \div 27$
42 $7.9 \div 3.8$	**43** $576 \div 27$	**44** $89.3 \div 4.9$
45 $184.5 \div 3.7$	**46** $6.7 \div 27.31$	**47** $90.324 \div 25.9$
48 $4.83 \div 70.345$	**49** $4.3 + 5.74 - 1.87$	**50** $7.41 - 2.3 + 5.76$
51 $11.43 + 1.706 - 2.305$	**52** $23.9 - 13.76 + 4.61$	
53 $3.5 + 2.86 - 4.702$	**54** $7.896 - 2.73 + 4.9006$	
55 $28.1 - 5.963 + 3.147$	**56** $19.315 + 11.289 - 24.7$	

13.3 DEFINITION OF A LOGARITHM

In the introduction to this chapter we stated that the theory of logarithms is based on the laws of exponents. This results from the relationship between logarithms and exponents established by the following definition:

Logarithm

The *logarithm L* of a number N to the *base b* (where $b > 0$, $b \neq 1$) is the exponent that indicates the power to which the base b must be raised in order to produce N.

The abbreviation of the phrase "logarithm to the base b of N" is $log_b N$. In terms of this abbreviation, the preceding definition can be stated as follows:

$log_b N = L$ **if and only if** $b^L = N$, **for** $N > 0, b > 0, b \neq 1$ **(13.1)**

By use of this definition, we see that $b^{log_b N} = N$, since $log_b N = L$; hence,

$log_2 64 = 6$ since $2^6 = 64$

$log_4 64 = 3$ since $4^3 = 64$

$log_8 64 = 2$ since $8^2 = 64$

$log_{16} 64 = \frac{3}{2}$ since $16^{3/2} = (16^{1/2})^3 = 4^3 = 64$

$log_b 1 = 0$ since $b^0 = 1, b \neq 0$

By use of (13.1), we can also see that $\log_b b^L = L$; hence,

If $\log_3 N = 4$, then $N = 3^4 = 81$

If $\log_b 125 = 3$, then $b^3 = 125 = 5^3$ and $b = 5$

If $\log_{16} 4 = L$, then $16^L = 4 = 16^{1/2}$ and $L = \frac{1}{2}$

We see from these examples that a logarithm may be integral or fractional. In fact, it can be and often is irrational. Furthermore, in many cases, if two of the three letters in (13.1) are known, the third can be found by inspection.

Example 1   Find the value of N in $\log_7 N = 2$.

Solution   By use of (13.1), we have $7^2 = N$; hence, $N = 49$.

Example 2   Find the value of b if $\log_b 125 = 3$.

Solution   By use of (13.1), we have

$$b^3 = 125$$
$$b = \sqrt[3]{125} = 5 \qquad \textbf{solving for } b$$

Example 3   Find the value of a if $\log_{27} 3 = a$.

Solution   Again using (13.1), we have

$$27^a = 3$$

and, since $27^{1/3} = 3$, it follows that $27^a = 27^{1/3}$ and

$$a = \frac{1}{3}$$

13.4   PROPERTIES OF LOGARITHMS

We shall now use the laws of exponents and the definition of a logarithm to derive three important properties of logarithms. In the next article we shall show how to use these properties in numerical computation.

We first show how to find the logarithm of a product of two numbers in terms of the logarithms of the two numbers.

If we are given that

$$\log_b M = m \qquad \text{and} \qquad \log_b N = n \tag{1}$$

then, by (13.1), we have

$$M = b^m \qquad \text{and} \qquad N = b^n \tag{2}$$

Hence,

$$MN = (b^m)(b^n) = b^{m+n}$$

and

$$\log_b MN = m + n \qquad \text{**by (13.1)**}$$

Hence,

$$\log_b MN = \log_b M + \log_b N \qquad \text{**by (1)**} \qquad (13.2)$$

Consequently, we have proved the following theorem:

Logarithm of a Product	The logarithm of the product of two positive numbers is equal to the sum of the logarithms of the numbers.

Example 1 If

$$\log_{10} 3 = .4771 \qquad \text{and} \qquad \log_{10} 2 = .3010$$

then

$$\begin{aligned} \log_{10} 6 &= \log_{10} (3 \times 2) \\ &= \log_{10} 3 + \log_{10} 2 \\ &= .4771 + .3010 = .7781 \end{aligned}$$

This theorem can be extended to three or more numbers by the following process:

$$\begin{aligned} \log_b MNP &= \log_b (MN)(P) \\ &= \log_b MN + \log_b P \\ &= \log_b M + \log_b N + \log_b P \end{aligned}$$

Again, using relations (2), we have

$$\frac{M}{N} = \frac{b^m}{b^n} = b^{m-n}$$

Hence,

$$\log_b \frac{M}{N} = m - n \qquad \text{**by (13.1)**}$$

and

$$\log_b \frac{M}{N} = \log_b M - \log_b N \qquad \text{**by (1)**} \qquad (13.3)$$

Therefore, we have proved the following theorem:

Logarithm of a Quotient	The logarithm of the quotient of two positive numbers is equal to the logarithm of the dividend minus the logarithm of the divisor.

Example 2   If

$$\log_{10} 3 = .4771 \qquad \text{and} \qquad \log_{10} 2 = .3010$$

then

$$\log_{10} 1.5 = \log_{10} \tfrac{3}{2}$$
$$= \log_{10} 3 - \log_{10} 2$$
$$= .4771 - .3010 = .1761$$

Finally, if we raise both members of $M = b^m$ to the kth power, we have

$$M^k = (b^m)^k = b^{km}$$

Therefore,

$$\log_b M^k = km \qquad \text{by (13.1)}$$

and

$$\log_b M^k = k \log_b M \qquad \text{by (1)} \qquad\qquad (13.4)$$

Thus we have proved the following theorem:

Logarithm of a Power   The logarithm of a power of a positive number is equal to the product of the exponent of the power and the logarithm of the number.

Example 3   If

$$\log_{10} 3 = .4771$$

then

$$\log_{10} 3^2 = 2 \log_{10} 3 = 2(.4771) = .9542$$

Note   Since a root of a number can be expressed as a fractional power, the last theorem can be used to find the logarithm of a root of a number. Thus,

Logarithm of a Root $$\log_b \sqrt[r]{M} = \log_b (M)^{1/r} = \frac{1}{r} \log_b M \qquad\qquad (13.4a)$$

Example 4   If

$$\log_{10} 2 = .3010$$

then

$$\log_{10} \sqrt[3]{2} = \log_{10} 2^{1/3} = \tfrac{1}{3} \log_{10} 2 = \tfrac{1}{3}(.3010) = .1003$$

COMPUTATIONAL THEOREMS

For the convenience of the reader, we restate the preceding properties in symbolic form.

$$\log_b MN = \log_b M + \log_b N \qquad \begin{matrix} M > 0 \\ N > 0 \end{matrix} \qquad (13.2)$$

$$\log_b \frac{M}{N} = \log_b M - \log_b N \qquad \begin{matrix} M > 0 \\ N > 0 \end{matrix} \qquad (13.3)$$

$$\log_b M^k = k \log_b M \qquad M > 0 \qquad (13.4)$$

$$\log_b \sqrt[r]{N} = \frac{1}{r} \log_b N \qquad N > 0 \qquad (13.4a)$$

EXERCISE 13.2   Logarithms and Exponentials

Change the statement in each of Probs. 1 to 16 to logarithmic form by use of Eq. (13.1).

1 $2^3 = 8$ **2** $3^2 = 9$ **3** $5^4 = 625$ **4** $7^3 = 343$

5 $5^{-3} = \frac{1}{125}$ **6** $7^{-2} = \frac{1}{49}$ **7** $2^{-4} = \frac{1}{16}$ **8** $3^{-5} = \frac{1}{243}$

9 $8^{2/3} = 4$ **10** $4^{3/2} = 8$ **11** $16^{3/4} = 8$ **12** $125^{4/3} = 625$

13 $(\frac{1}{2})^{-3} = 8$ **14** $(\frac{1}{4})^{-2} = 16$ **15** $(\frac{1}{3})^{-5} = 243$ **16** $(\frac{1}{10})^{-4} = 10,000$

Change the statement in each of Probs. 17 to 32 to exponential form by use of (13.1).

17 $\log_3 81 = 4$ **18** $\log_2 8 = 3$ **19** $\log_5 25 = 2$

20 $\log_4 256 = 4$ **21** $\log_2 \frac{1}{16} = -4$ **22** $\log_5 \frac{1}{125} = -3$

23 $\log_7 \frac{1}{49} = -2$ **24** $\log_2 \frac{1}{32} = -5$ **25** $\log_8 4 = \frac{2}{3}$

26 $\log_{36} 216 = \frac{3}{2}$ **27** $\log_{32} 8 = \frac{3}{5}$ **28** $\log_{81} 27 = \frac{3}{4}$

29 $\log_{1/3} 9 = -2$ **30** $\log_{1/4} 64 = -3$ **31** $\log_{1/2} 16 = -4$

32 $\log_{1/3} 81 = -4$

Determine L, n, or b in each of Probs. 33 to 56 so that the statement is true.

33 $\log_2 8 = L$ **34** $\log_3 81 = L$ **35** $\log_5 25 = L$ **36** $\log_7 343 = L$

37 $\log_9 27 = L$ **38** $\log_{16} 8 = L$ **39** $\log_{25} 125 = L$ **40** $\log_{32} 8 = L$

41 $\log_2 N = 3$ **42** $\log_5 N = 2$ **43** $\log_3 N = 4$ **44** $\log_2 N = 5$

45 $\log_{27} N = \frac{2}{3}$ **46** $\log_{16} N = \frac{3}{4}$ **47** $\log_9 N = \frac{3}{2}$ **48** $\log_{64} N = \frac{4}{3}$

49 $\log_b 25 = 2$ **50** $\log_b 81 = 4$ **51** $\log_b 32 = 5$ **52** $\log_b 216 = 3$

53 $\log_b 4 = \frac{2}{3}$ **54** $\log_b 27 = \frac{3}{2}$ **55** $\log_b 32 = \frac{5}{6}$ **56** $\log_b 27 = \frac{3}{4}$

Use $\log_{10} 2 = .3010$, $\log_{10} 3 = .4771$, and $\log_{10} 5 = .6990$, along with Eqs. (13.2), (13.3), (13.4), and (13.4a) to find the logarithm to base 10 of the combination of numbers in each of Probs. 57 to 72.

57 $(2)(5)$ **58** $(3)(2)$ **59** $(5)(3)$ **60** 30

61 $\frac{3}{5}$ **62** $\frac{2}{3}$ **63** $(3)(5)/2$ **64** 2.5

65 2^3 **66** 3^2 **67** 5^4 **68** 16

69 $3^2 2^3/5$ **70** $5^4 3^3/2$ **71** $6/(2^2)(5^3)$ **72** $2^5/45$

13.5 COMMON, OR BRIGGS, LOGARITHMS

Common
or Briggs
Logarithms

We stated previously that logarithms can be used efficiently in numerical computation, and we shall use logarithms to the base 10 for this purpose. Logarithms to the base 10 are called *common*, or *Briggs*, logarithms.

The common logarithm of a number that is not an integral power or root of 10 cannot be computed by elementary methods, but tables have been prepared that enable us to obtain a decimal approximation to the common logarithm of any postive number. The use of the tables will be explained in the next three sections.

It is customary to omit the subscript indicating the base in the notation for a common logarithm. Consequently, in the statement $\log N = L$, it is understood that the base is 10.

If c is a real number, then 10^c is positive. Hence the common logarithm of zero or of a negative number does not exist as a real number.

13.6 CHARACTERISTIC AND MANTISSA

Reference
Position

If we express a positive number $N \neq 1$ in scientific notation, we say the decimal point is in reference position and have

$$N = N'(10^c) \tag{1}$$

where

$$1 < N' < 10$$

and c is an integer.

We consider the following three situations:

1 If $N \geq 10$, then in Eq. (1), $c \geq 1$. For example, $231 = 2.31(10^2)$.

2 If $N < 1$, then $c < 0$. For example, $.0231 = 2.31(10^{-2})$.

3 If $1 < N < 10$, then in Eq. (1), $N = N'$ and $c = 0$. For example, $2.31 = 2.31(10^0)$.

Now, if we equate the common logarithms of the members of Eq. (1), we have

$$\log N = \log N' + \log 10^c \qquad \textbf{by (13.2)}$$
$$= \log N' + c\log 10 \qquad \textbf{by (13.4)}$$
$$= \log N' + c \qquad \textbf{since log 10 = 1}$$

Thus, we have

$$\log N = c + \log N' \quad \textbf{by commutative axiom of addition} \tag{2}$$

Since $1 < N' < 10$, it follows that $10^0 < N' < 10^1$, and hence that $0 < \log N' < 1$. From Table I in the appendix, we can obtain a decimal approximation to the common logarithm of any number between 1 and 10, correct to four decimal places. Consequently, by referring to Eq. (2), we see that the common logarithm of any positive number not equal to 1 can be expressed approximately as an integer c plus a nonnegative decimal fraction $\log N'$. Since $1 = 10^0$, it follows that $\log 1 = 0$. Thus, for $\log 1$, the integer is zero and the decimal fraction is zero. We are now in position to state the following definition:

Characteristic

Mantissa

If the common logarithm of a positive number is expressed as an integer plus a nonnegative decimal fraction, the integer is called the *characteristic* of the logarithm, and the decimal fraction is the *mantissa*.

In the expression for $\log N$ in Eq. (2), the characteristic is the integer c, and the mantissa is the nonnegative number $\log N'$. Thus we have demonstrated that c is numerically equal to the number of digits between the reference position and the decimal point in N, and that it is positive or negative according as the decimal point is to the right or to the left of the reference position. Therefore, we have the following rule for determining the characteristic of the common logarithm of a positive number:

Value of the Characteristic

The characteristic of the common logarithm of a positive number N is numerically equal to the number of digits between the reference position and the decimal point in N and is positive or negative according as the decimal point is to the right or to the left of the reference position.

Example 1 The reference position in 236.78 is between the 2 and the 3. Hence there are two digits, 3 and 6, between the reference position and the decimal point. Since the decimal point is to the right of the reference position, the characteristic of the common logarithm of 236.78 is positive. It is 2.

Example 2 The characteristic of the common logarithm of 2.3678 is zero since the decimal point is in the reference position.

Example 3 The decimal point in .0023678 is three places to the left of the reference position. Therefore, the characteristic of $\log$ .0023678 is −3.

Since the position of the decimal point in the number N affects only the integer c in the scientific notation for N, and since the mantissa of $\log N$ is $\log N'$, the mantissa of $\log N$ depends only upon the sequence of digits in N.

If $N \geq 1$, the characteristic of log N is zero or a positive integer. Therefore, log N can be written as a single number. For example, if the mantissa of log 23,678 is .3743, then log 236.78 = 2 + .3743 = 2.3743; and log 2.3678 = 0 + .3743 = .3743.

In finding the logarithm of a positive number less than 1, however, we have a different situation. For example, log .0023678 = −3 + .3743 = −2.6257 = −2 − .6257. Now the decimal fraction −.6257 is negative and consequently is not a mantissa. We use the following device for dealing with such situations. If the characteristic of log N is −c, where $c > 0$, we express −c in the form $(10 - c) - 10$; then we write the mantissa to the right of $10 - c$. For example, we express log .0023678 in the form 7.3743 − 10 since the characteristic is −3, and $(10 - 3) - 10 = 7 - 10$. Similarly, log .23678 = 9.3743 − 10, log .023678 = 8.3743 − 10, and log .0000023678 = 4.3743 − 10.

13.7   GIVEN *N*, TO FIND LOG *N* BY TABLE AND BY CALCULATOR

The mantissa of the logarithm of a number does not depend upon the position of the decimal point since, as is pointed out earlier in this chapter, shifting the position of the decimal point is equivalent to multiplying by an integral power of 10 and, therefore, only the characteristic is affected.

Use of Tables

The mantissa is determined by the sequence of digits and can be found by use of a table such as Table I in the appendix. From this table we can find to four figures the mantissa of the logarithm of any three-digit number. To get the mantissa of the logarithm of such a number, we find the first two digits of the number in the column headed by N at the left of the page, and then move across the page to the column headed by the third digit. The desired mantissa is in the column headed by the third digit and in the line containing the first two digits.

In keeping with this procedure, to find log 327, we look in Table I across from 32 and under 7 and find 5145; so log 327 = 2.5145, since the characteristic is 2 and the decimal point that is a part of each mantissa is not printed. Furthermore, to get log .914, we look across from 91 and under 4 and find 9609; hence log .914 = 9.9609 − 10, since the characteristic is − 1 = 9 − 10.

log N from a Calculator

Given N, we can find log N by use of a variety of calculators. On certain calculators we put N on the display dial, punch the F button, and punch the *log* button. The value of log N then appears on the display dial. In order to find log 327 by use of a calculator, we put 327 on the display dial and then punch the F and *log* buttons. When this is done, 2.514548 appears on the display dial. Hence, rounding off to four decimal places, we see that log 327 = 2.5145 as found by use of tables. It is log N, not just the mantissa of log N, that appears on the display dial.

If $0 < N < 1$, then $\log N$ as obtained from a calculator is a negative number; but, if desirable, it may be written as a positive integer plus a positive fraction minus a positive integer. Thus, $\log .357$ is given as $-.447331$; this may be rounded off and then written as $-.4473 = 9.5527 - 10$ by adding $-.4473$ to $10 - 10$.

13.8 GIVEN LOG *N*, TO FIND *N*

Finding N if log N Is Given

If $\log N$ is given and the mantissa is in the table, we find the first two digits of N to the left of the mantissa, and the third digit above the mantissa. The position of the decimal point is determined by the value of the characteristic. Thus, if $\log N = 1.9212$, we first find 9212 in the body of the table. It is in the column headed by 4 and in line with 83 in the N column. Hence, the sequence of digits in N is 834. The characteristic of $\log N$ is 1. Hence, $N = 83.4$.

If $\log N$ is given and the mantissa is not in the table and we want N to only three digits, we find the mantissa in the table that is nearest to the given mantissa of $\log N$; we then use the corresponding value of N. If the mantissa of $\log N$ is midway between two entries, we use the value of N which corresponds to the larger entry. Accordingly, we see that if $\log N = 0.4822$, then the entry in the table that is nearest to it is 4829; the corresponding sequence of digits is 304. Furthermore, $N = 3.04$ since the characteristic is zero. If $\log N = 1.1804$, the mantissa is midway between 1790 and 1818; hence the sequence of digits used is 152, and $N = 15.2$ because the characteristic is 1.

The above procedure uses the tables to find N if $\log N$ is known. According to (13.1), the relationship between N and $\log N$ may be written

$$N = 10^{\log N}$$

The number N is sometimes referred to as the *antilog* of $\log N$.

If a calculator with a 10^x button is available, we can readily find N if $\log N$ is given. All we need to do is put the value of $\log N$ on the display dial and then punch the F and 10^x buttons. If this is done, the value of N appears on the display dial. For example, to find N if $\log N = 1.8623$, we display 1.8623, punch the F and 10^x buttons, and read 72.83025 on the display dial. Rounding off to three digits, we have $N = 72.8$.

Not all calculators have a 10^x button. If the available calculator does not have a 10^x button but does have an e^x button, the procedure for finding N involves one step in addition to those needed for a calculator with a 10^x button. For reasons not presented in this book, we must multiply $\log N$ by 2.3026, display this product, and then punch the F and e^x buttons so as to have N on the display dial. Thus, if $\log N = 1.8688$ and we want N but do not have a 10^x button, we display the 1.8688, multiply it by 2.3026, and get 4.3031 on the display dial; then we punch the F and e^x buttons and have $N = 73.9$ after rounding off.

Not all calculators carry the same number of digits in internal calculations; hence, different results may appear if we consider more than three or four digits.

13.9 INTERPOLATION

Quite often we need the value of the logarithm of a number that is not listed in the table or want the sequence of digits that corresponds to a mantissa value that is not in the table. Under such circumstances we resort to a procedure that is known as *linear interpolation.*

We shall illustrate the procedure by using it to get the mantissa of log 2537, and in the discussion we use $ml\ N$ to stand for the phrase, "the mantissa of log N."

Since

$$2570 < 2573 < 2580$$

then

$$ml\ 2570 < ml\ 2573 < ml\ 2580$$

Furthermore,

$$ml\ 2580 = ml\ 258 \qquad \text{and} \qquad ml\ 2570 = ml\ 257$$

We now let

$$ml\ 2573 = ml\ 2570 + c$$

and find the value of c by the following procedure:

$$10\left[{}_3\left[\begin{array}{l} ml\ 2580 = .4116 \\ ml\ 2573 = .4099 + c \\ ml\ 2570 = .4099 \end{array} \right]_c \right].0017$$

The numbers to the right and to the left of the brackets in this diagram are the differences of the numbers connected by the brackets. Now these differences are approximately proportional. Therefore,

$$\frac{c}{.0017} = \frac{3}{10}$$

and it follows that

$$c = \frac{3}{10}(.0017) = .0005$$

Therefore,

$$ml\ 2573 = .4099 + .0005 = .4104$$

Furthermore, since the decimal point in 2573 is three places to the right of the reference position, the characteristic of the logarithm is 3. Hence,

log 2573 = 3.4104

We use the same principle to obtain N if log N is given and ml N is not listed in the table. We illustrate the method by finding the value of N if log $N = 1.2869$. The mantissa .2869 is not in the table, but the two mantissas nearest to it are .2856 and .2878. Now, .2856 = ml 1930, and .2878 = ml 1940. We let n stand for the number composed of the four digits of N, then we find n, and finally we place the decimal point in the position indicated by the characteristic, thus obtaining N. Since

.2856 < .2869 < .2878

then

1930 < n < 1940

Now we let $n = 1930 + c$ and determine c as follows:

$$.0022\left[\begin{array}{l} .2878 = \text{ml } 1940 \\ .0013\left[\begin{array}{l}.2869 = \text{ml } 1930 + c \\ .2856 = \text{ml } 1930\end{array}\right]c\end{array}\right]10$$

Since the ratios of the differences to the right and left of the brackets are proportional, we have

$$\frac{c}{10} = \frac{.0013}{.0022}$$

$$c = \frac{.0013}{.0022}(10) = \frac{13}{22}(10) = 6 \qquad \textbf{to one digit}$$

Hence,

$n = 1930 + 6 = 1936$

Since the characteristic of log $N = 1.2869$ is 1, we place the decimal point to the right of 9 in 1936 and then get

$N = 19.36$

If a calculator is used, it is not necessary to interpolate. We just put all the digits of log N or of N on the display dial and proceed as in the cases discussed earlier. Thus, if $N = 27.563$, we find log N by displaying N, punching the F and log N buttons, and reading 1.440326 on the display dial. Hence, to four decimal places log 27.563 is 1.4403. Furthermore, if log $N = 2.8326$, we display it, punch the F and 10^x buttons, and find that $N = 680.171$. If the available calculator does not have a 10^x button, it is necessary to multiply by 2.3026 before punching the e^x button.

EXERCISE 13.3 Numbers and Logarithms of Numbers

Find the common logarithm of the number in each of Probs. 1 to 16.

1	38.7	**2**	5.96	**3**	809	**4**	73.8
5	731	**6**	9.92	**7**	40.7	**8**	29.9
9	.246	**10**	.0808	**11**	.00531	**12**	.483
13	.0101	**14**	.477	**15**	.0329	**16**	.674

Round off the number in each of Probs. 17 to 24 to four digits, and then find its common logarithm.

17	49.314	**18**	783.26	**19**	9.8247	**20**	34.712
21	.81153	**22**	.061795	**23**	.00024681	**24**	.59868

If log N is the number in each of Probs. 25 to 40, find N to three digits.

25	1.3263	**26**	2.8457	**27**	.6551	**28**	1.9562
29	.8476	**30**	1.4082	**31**	2.1399	**32**	.6232
33	$9.9149 - 10$	**34**	$8.5866 - 10$	**35**	$7.3729 - 10$	**36**	$8.9657 - 10$
37	$7.8640 - 10$	**38**	$9.9221 - 10$	**39**	$8.5745 - 10$	**40**	$9.2707 - 10$

Use interpolation to find the logarithm of the number in each of Probs. 41 to 48. If one is available, use a calculator to check the result.

41	45.23	**42**	781.6	**43**	2.037	**44**	5784
45	.3248	**46**	.06062	**47**	.8205	**48**	.009179

Use interpolation to find N to four digits if the number given in each of Probs. 49 to 56 is log N.

49	1.4431	**50**	3.1486	**51**	.5773	**52**	2.8715
53	$9.7137 - 10$	**54**	$8.2345 - 10$	**55**	$7.7777 - 10$	**56**	$9.3424 - 10$

13.10 LOGARITHMIC CALCULATIONS

As we stated previously, one of the most immediate and useful applications of logarithms is in the field of numerical computation. We shall explain presently the methods involved by means of several examples. Before considering special problems, however, we wish to call attention again to the fact that results obtained by the use of four-place tables are correct at most to four places. If the numbers in any computation problem contain only three places, the result is dependable to only three places. If a problem contains a mixture of three-place and four-place numbers, we cannot expect more than three places of the result to be correct, and so we round it off to three places. Hence, in the problems that follow, we shall not obtain any answer to more than four nonzero places, and sometimes not that many. Tables exist from which logarithms may be obtained to five, six, seven, and even more places. If results that are correct to more than four places are desired, longer tables should be used. The methods which we have presented may be applied to a table of any length.

We shall now present examples with explanations that illustrate the methods for using logarithms to obtain (1) products and quotients, (2) powers and roots, and (3) the solutions of miscellaneous computation problems.

In all computation problems we use the properties of logarithms developed in Sec. 13.5 to find the logarithm of the result. Then the value of the result can be obtained from the table.

PRODUCTS AND QUOTIENTS

Example 1 Find the value of $R = (8.56)(3.47)(198)$.

Solution Since R is equal to the product of three numbers, $\log R$ is equal to the sum of the logarithms of the three factors. Hence, we shall obtain the logarithm of each of the factors, add them together, and thus have $\log R$. Then we may use the table to get R. Before we turn to the table, it is advisable to make an outline, leaving blanks in which to enter the logarithms as they are found. It is also advisable to arrange the outline so that the logarithms to be added are in a column. We suggest the following plan:

$$\log R = \log (8.56)(3.47)(198)$$
$$= \log 8.56 + \log 3.47 + \log 198$$

$\log 8.56 =$
$\log 3.47 =$
$\underline{\log 198 =}$
$\quad \log R =$
$\qquad R = \qquad$ _____ enter sum here

Next we enter the characteristics in the blanks and have

$\log 8.56 = 0$
$\log 3.47 = 0$
$\underline{\log 198 = 2}$
$\quad \log R =$
$\qquad R =$

Now we turn to the tables, get the mantissas, and, as each is found, enter it in the proper place in the outline. Then we perform the addition and finally determine R by the method of Sec. 13.8. The completed solution then appears as

$$\log 8.56 = 0.9325$$
$$\log 3.47 = 0.5403$$
$$\underline{\log 198 = 2.2967}$$
$$\log R = 3.7695$$
$$R = 5.88(10^3)$$

Note   Each of the numbers in the problem contains only three digits. Hence, we can determine only three digits of R. Since the mantissa 7695 is between the two entries 7694 and 7701 and nearer the former than the latter, the first three digits of R are 588, corresponding to the matnissa .7694. The characteristic of $\log R$ is 3. Hence, the decimal point is three places to the right of the reference position. Therefore, we use scientific notation and multiply by 10^3 to place the decimal point.

   We have written the outline of the solution three times in order to show how it appears at the conclusion of each step. In practice, it is necessary to write the outline only once, since only the original blanks are required for computing all the operations.

Example 2   Use logarithms to find R, where

$$R = \frac{(337)(2.68)}{(521)(.763)}$$

Solution   In this problem R is a quotient in which the dividend and divisor are each the product of two numbers. Hence, we shall add the logarithms of the two numbers in the dividend, also add the logarithms of the two in the divisor, then subtract the latter sum from the former, and so obtain $\log R$. We suggest the following outline for the solution:

$$\log R = \log \frac{(337)(2.68)}{(521)(.763)}$$
$$= \log (337)(2.68) - \log (521)(.763)$$
$$= \log 337 + \log 2.68 - (\log 521 + \log .763)$$

$$\log 337 =$$
$$\underline{\log 2.68 =}$$
$$\log \text{dividend} = \qquad \text{——— enter sum here}$$

$$\log 521 =$$
$$\underline{\log .763 =}$$
$$\log \text{divisor} = \qquad \text{——— enter sum here}$$
$$\log R = \qquad \text{——— enter difference of the two sums here}$$
$$R =$$

After the characteristics are entered and the mantissas are found and listed in the proper places, the problem is completed as follows:

$$\begin{aligned}
\log 337 &= 2.5276 \\
\underline{\log 2.68 = 0.4281} \\
\log \text{ dividend} = \qquad\qquad 2.9557
\end{aligned}$$

$$\begin{aligned}
\log 521 &= 2.7168 \\
\underline{\log .763 = 9.8825 - 10} \\
\log \text{ divisor} = \qquad\qquad \cancel{1}2.5993 - \cancel{10} \\
\log R = \qquad\qquad 0.3564 \\
R = 2.27
\end{aligned}$$

Note   The logarithm of the divisor turned out to be $12.5993 - 10$. Hence, the characteristic is 2. Therefore, in the above outline, we strike out the 10 and the first digit in 12 before completing the solution.

Example 3   Use logarithms to evaluate

$$R = \frac{2.68}{33.2}$$

Solution $\log R = \log 2.68 - \log 33.2$; hence,

$$\begin{aligned}
\log 2.68 &= 0.4281 \\
\underline{\log 33.2 = 1.5211} \\
\log R &= -1.0930 = -1.0930 + 10 - 10 = 8.9070 - 10 \\
R &= .0807
\end{aligned}$$

In performing an indicated subtraction, as in $(7.3264 - 10) - (9.4631 - 10)$, we obtain a negative number, in this case -2.1367. Hence, we would add $10 - 10$ and have

$$-2.1367 = -2.1367 + 10 - 10 = 7.8633 - 10$$

Example 4   By use of logarithms, obtain the value of $R = (3.74)^5$.

Solution $\log R = \log (3.74)^5 = 5(\log 3.74) = 5(0.5729) = 2.8645$

Hence, $R = 732$.

Powers and Roots   We may also obtain the root of a number by means of logarithms. The method is illustrated in the following example:

Example 5   Evaluate $R = \sqrt[3]{62.3}$.

Solution   If we rewrite the problem in exponential form, we get

$$R = (62.3)^{1/3}$$

Hence,

$\log R = \frac{1}{3}\log 62.3 = \frac{1}{3}(1.7945) = 0.5982$

Therefore, $R = 3.96$.

In the application of logarithms to the problem of extracting a root of a decimal fraction, we use a device similar to that described in Example 3 in order to avoid a troublesome situation.

Example 6   By use of logarithms evaluate $R = \sqrt[6]{.0628}$.

Solution $\log R = \frac{1}{6}\log .0628$

$= \dfrac{8.7980 - 10}{6}$

If we perform the division indicated, we get

$\log R = 1.4663 - 1.6667$

$= -.2004 = -.2004 + 10 - 10 = 9.7996 - 10$

The last logarithm is in the customary form and, by referring to the table, we find that $R = 0.631$.

Miscellaneous Problems   Many computation problems require a combination of the processes of multiplication, division, raising to powers, and the extraction of roots. We now illustrate the general procedure for solving such problems.

Example 7   Use logarithms to find R if

$$R = \sqrt[5]{\dfrac{\sqrt{2.689}\,(3.478)}{(52.18)^2(51.67)}}$$

Solution   Since all the numbers in this problem contain four digits, we shall obtain the value of R to four places. Furthermore, we must use interpolation to obtain the mantissas. The steps in the solution are indicated in the following suggested outline:

$\log \sqrt{2.689} = \frac{1}{2}\log 2.689 = \frac{1}{2}($ $) =$
$\underline{\log 3.478 \hspace{8cm} =}$
$\hspace{4cm}\log \text{dividend} = \hspace{2cm}\underline{\hspace{1.5cm}}\text{ sum}$

$\log (52.18)^2 = 2\log 52.18 = 2($ $) =$
$\underline{\log 51.67 \hspace{8cm} =}$
$\hspace{4cm}\log \text{divisor} = \hspace{2cm}\underline{\hspace{1.5cm}}\text{ sum}$
$\hspace{4.5cm}\log R = \hspace{1.5cm}5\underline{\hspace{1.3cm}}\text{ difference}$
$\hspace{4.5cm}R =$

We now enter the characteristics in the proper places, then turn to the table, get the mantissas, enter each in the space left for it, and complete the solution. Then the outline appears as here:

$$\log \sqrt{2.689} = \tfrac{1}{2} \log 2.689 = \tfrac{1}{2}(0.4296) = 0.2148$$

$$\log 3.478 \qquad\qquad\qquad\qquad\qquad\qquad = 0.5413$$

$$\log \text{dividend} = \qquad\qquad 10.7561^{\dagger} - 10$$

$$\log (52.18)^2 = 2 \log 52.18 = 2(1.7175) = 3.4350$$

$$\log 51.67 \qquad\qquad\qquad\qquad\qquad\qquad = 1.7132$$

$$\log \text{divisor} = \qquad\qquad\qquad 5.1482$$

$$\begin{array}{r} 5\overline{)45.6079} - 50^{\ddagger} \end{array}$$

$$\log R = \qquad\qquad\qquad 9.1216 - 10$$

$$R = .1323$$

All the operations that have been illustrated in this section by use of logarithms can also be performed on a calculator. To *Products by Calculator* multiply M by N by use of a calculator, we display either number, depress the $\times$ button, then display the other number, press the $=$ button, and read the product on the display dial. A similar procedure is used to find the quotient of two numbers. Most calculators contain a square-root button. If the available one does have such a button, all that is required to get $\sqrt{N}$ is to display N, press the F button and the $\sqrt{\ }$ button, and read $\sqrt{N}$ on the display dial. The difficulties mentioned in Examples 3 and 6 do not arise if a calculator is used. A certain amount of skill and experience is required to perform the combination of operations called for in Example 7. If a calculator is available, the reader might enjoy verifying the results of Examples 1 to 7 by use of a calculator. Recall that both calculators and Table I use and give approximations; hence, the answers found by the two methods need not agree exactly.

EXERCISE 13.4 Calculations

Use logarithms and/or a calculator to perform the computations indicated in Probs. 1 to 52. Carry each to the justified number of significant digits.

1 $(3.15)(21.1)(1.76)$

2 $(71.4)(2.03)(.627)$

3 $(624)(.113)(1.07)$

4 $(809)(.102)(1.12)$

5 $(2.37)(48.4)(4.76)$

6 $(97.3)(5.89)(.473)$

7 $(34.8)(57.1)(12.3)$

8 $(70.7)(67.8)(3.27)$

9 $\dfrac{(29.6)(3.04)}{76.5}$

10 $\dfrac{(8.43)(96.5)}{60.7}$

11 $\dfrac{(308)(11.2)}{276}$

12 $\dfrac{(58.3)(38.5)}{358}$

13 $\dfrac{897}{(2.36)(47.1)}$

14 $\dfrac{43.7}{(38.4)(.808)}$

15 $\dfrac{98.5}{(37.2)(1.97)}$

†Note that we add $10 - 10$ here so that we can subtract 5.1482.
‡We add $40 - 40$ here so that we can divide by 5.

16 $\dfrac{2.39}{(5.83)(.877)}$

17 $\dfrac{28.7}{78.3}$

18 $\dfrac{405}{788}$

19 $\dfrac{37.1}{60.7}$

20 $\dfrac{4.87}{7.55}$

21 $(2.71^2)(3.62)^3$

22 $(7.96)^3(1.07)^4$

23 $(8.43)^3(1.04)^3$

24 $(5.43)^4(1.11)^3$

25 $(3.86)^{2/3}$

26 $(7.85)^{3/4}$

27 $(283)^{3/5}$

28 $(59.7)^{7/6}$

29 $\sqrt[3]{7.84}$

30 $\sqrt[4]{629}$

31 $\sqrt{(2.71)^3}$

32 $\sqrt[5]{(38.7)^3}$

33 $\sqrt[3]{.815}$

34 $\sqrt[4]{.666}$

35 $\sqrt[6]{.914}$

36 $\sqrt[7]{.403}$

37 $\sqrt{\dfrac{(3.26)^2(80.3)}{57.4}}$

38 $\sqrt{\dfrac{(\sqrt{38.1})(3.23)}{17.2}}$

39 $\sqrt[3]{\dfrac{78.4}{\sqrt{27.3}(1.88)^2}}$

40 $\sqrt[5]{\dfrac{(29.8)^2}{47.1\sqrt[3]{8.63}}}$

41 $(584.2)(78.13)(2.059)$

42 $(7.736)(87.43)(6.037)$

43 $(79.81)(8.07)(3.7623)$

44 $(5.974)(7.9468)(27.9)$

45 $\dfrac{(38.7)(69.6)}{(70.73)(23.4)}$

46 $\dfrac{(35.91)(1264)}{(407.63)(27.68)}$

47 $\dfrac{(983.4)(2.671)}{(67.2)(54.976)}$

48 $\dfrac{(5.916)(78.237)}{(49.71)(6.033)}$

49 $\sqrt{\dfrac{(4.963)(80.74)}{367.7}}$

50 $\sqrt{\dfrac{(803.6)(57.21)}{9843}}$

51 $\sqrt[3]{\dfrac{597.8}{(23.8)(37.3)}}$

52 $\sqrt[3]{\dfrac{865.7}{(32.14)(29.19)}}$

13.11   LOGARITHMIC AND EXPONENTIAL EQUATIONS

Exponential Equation
Logarithmic Equation

An equation is called an *exponential equation* if the variable appears in one or more exponents, and a *logarithmic equation* if the variable is part of a number whose logarithm is to be taken. Consequently, $2^x = 5$ and $x^{3-1} = 5^{x+y}$ are exponential equations, and

$$\log_2 x + \log_2(y - 1) = 2$$

is a logarithmic equation.

In general, exponential and logarithmic equations cannot be solved by the methods previously discussed. Many such equations, however, can be solved if we use the properties of logarithms. The following examples illustrate the procedure.

Example 1   Solve the equation

$$3^{x+4} = 5^{x+2}$$

Solution   We first equate the logarithms of the members of the equation and then proceed as indicated.

$$\log 3^{x+4} = \log 5^{x+2}$$
$$(x + 4) \log 3 = (x + 2) \log 5 \qquad \text{by (13.4)}$$
$$x \log 3 + 4 \log 3 = x \log 5 + 2 \log 5$$
$$x(\log 3 - \log 5) = 2 \log 5 - 4 \log 3 \qquad \begin{array}{l}\text{adding } -x \log 5 - 4 \log 3 \text{ to}\\ \text{each member}\end{array}$$
$$= \log 5^2 - \log 3^4$$
$$= \log 25 - \log 81 \qquad \text{by (13.4)}$$

Hence, $\qquad x = \dfrac{\log 25 - \log 81}{\log 3 - \log 5} \qquad \text{solving for } x$

$$= \frac{1.3979 - 1.9085}{0.4771 - 0.6990}$$

$$= \frac{-0.5106}{-0.2219}$$

$$= 2.301$$

Example 2 Solve $\log_6 (x + 3) + \log_6 (x - 2) = 1$.

Solution By applying the theorem for the logarithm of a product to the left member of the given equation, we have

$$\log_6 (x + 3)(x - 2) = 1$$

Hence,

$$(x + 3)(x - 2) = 6^1 \qquad \text{by (13.1)}$$
$$x^2 + x - 6 = 6 \qquad \text{performing indicated operations}$$
$$x^2 + x - 12 = 0$$
$$(x + 4)(x - 3) = 0$$

Solving this equation, we get $x = -4$ and $x = 3$, but we cannot admit -4 as a root since then $x + 3$ is negative, and we have not defined the logarithm of a negative number. Hence, the solution set is $\{3\}$.

Example 3 Solve the exponential equations

$$5^{x-2y} = 100 \tag{1}$$

and

$$3^{2x-y} = 10 \tag{2}$$

for x and y.

Solution If we equate the logarithms of the members of Eq. (1) and of Eq. (2), we get

$$(x - 2y) \log 5 = 2 \qquad \text{from (1) by (13.4)} \tag{3}$$

and

$$(2x - y) \log 3 = 1 \qquad \text{from (2)} \tag{4}$$

Therefore,

$$x - 2y = \frac{2}{\log 5} = 2.861 \tag{3'}$$

and

$$2x - y = \frac{1}{\log 3} = 2.096 \tag{4'}$$

Multiplying each member of (4') by 2 and subtracting from the corresponding member of (3'), we obtain

$$-3x = -1.331$$
$$x = .4437$$

Substituting this value for x in (3') and solving for y, we get

$$-2y = 2.861 - .4437$$
$$= 2.4173$$

Therefore,

$$y = -1.209$$

Hence, the solution set of the system is $\{.4437, -1.209\}$.

Example 4 Solve simultaneously the equations

$$5^{2x + 3y} = 120 \tag{1}$$

and

$$2^{3x + 5y} = 30 \tag{2}$$

Solution If we take the logarithm of each member of (1), we get

$$(2x + 3y) \log 5 = \log 120 \tag{3}$$

Similarly, from (2), we have

$$(3x + 5y) \log 2 = \log 30 \tag{4}$$

Therefore,

$$2x + 3y = \frac{\log 120}{\log 5} = 2.9746 \tag{3'}$$

and

$$3x + 5y = \frac{\log 30}{\log 2} = 4.9069 \tag{4'}$$

We shall now eliminate y between (3') and (4') by subtracting 3 times each member of (4') from 5 times the corresponding member of (3'). Thus, we obtain

$$x = 5(2.9746) - 3(4.9069) = .1523$$

If we put this in (3′), we have 2(.1523) + 3y = 2.9746; hence y = .8900, and the solution is (.1523, .8900).

13.12   THE GRAPHS OF $y = \log_b x$ AND $y = b^x$

The graph of $y = \log_b x$ is helpful in many situations. We shall make a table of corresponding values of x and $y = \log_b x$ with $b = 10$, plot the points thus determined, draw a smooth curve *Logarithmic* through them, and thereby have the graph of $y = \log_{10} x$. It is *Curve* called the *logarithmic curve*. If we arbitrarily assign the values .01, .1, .5, 1, 2, 5, 10, 16, 20, and 25 to x and use a table of logarithms to find each corresponding value of y, we get

x	.01	.1	.5	1	2	5	10	16	20	25
y	−2	−1	−.3	0	.3	.7	1	1.2	1.3	1.4

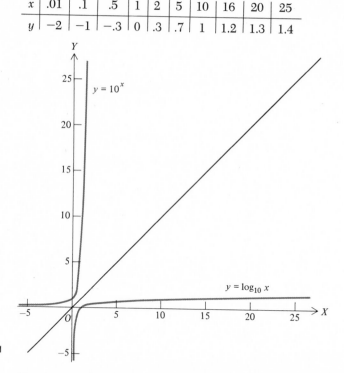

FIGURE 13.1

Consequently, the graph of $y = \log_{10} x$ is as shown in Fig. 13.1. The rate at which it rises as x increases is dependent on the value of the base b. The reader will see after working Probs. 33 to 36 of Exercise 13.5 that the larger the value of b, the slower the rate at which the curve rises; furthermore, the larger the value of b, the smaller the value of y for a given value of x (if $b > 1$).

We now make use of the definition of the logarithm of a number to change the equation $y = \log_b x$ to exponential form. We get $b^y = x$. This equation determines the same relation between x and y as is given by $y = \log_b x$; hence, the graphs of the two are the same. If, in $x = b^y$, we interchange x and y, we find that $y = b^x$ is the inverse of $y = \log_b x$. Consequently, the graph of $y = b^x$ is in the same position relative to the Y axis as that of $y = \log_b x$ relative to the X axis. Each can be obtained from the other by reflecting in the line determined by $y = x$. Both are shown in Fig. 13.1.

EXERCISE 13.5 $\text{Log}_b N$, $b \neq 10$, Exponential and Logarithmic Equations

In each of Probs. 1 to 24, find the value of x exactly or to four digits.

1 $2^{x+1} = 3^x$ **2** $3^{2x-1} = 5^x$ **3** $5^{2x-3} = 7^{x-1}$ **4** $7^{3x-1} = 10^{x+.7}$

5 $3^{3x+1} = 11^{4x-3}$ **6** $5^{3x-1} = 3^{3x}$ **7** $7^{2x-1} = 4^{3x-2}$ **8** $11^{5x+1} = 13^{2x+2}$

9 $\log_6 (x + 1) + \log_6 (x + 2) = 1$ **10** $\log_2 (x + 2) + \log_2 (x - 1) = 2$

11 $\log_{12} (x + 3) + \log_{12} (x - 1) = 1$ **12** $\log_3 (2x - 1) + \log_3 (x + 1) = 2$

13 $\log_2 (x - 3) + \log_2 (x + 4) = 3$ **14** $\log_7 (3x + 1) + \log_7 (2x + 3) = 2$

15 $\log_2 (x + 3) + \log_2 (3x + 7) = 3$ **16** $\log_5 (3x + 5) + \log_5 (2x + 5) = 2$

17 $\log_2 (x + 3) - \log_2 (2x - 3) = 1$ **18** $\log_7 (2x + 1) - \log_7 (3x - 1) = 0$

19 $\log_3 (4x + 5) - \log_3 (x + 2) = 1$ **20** $\log_2 (3x + 5) - \log_2 (x + 3) = 1$

21 $\log_2 (x^2 + x + 2) - \log_2 (x + 3) = 2$ **22** $\log_3 (x^2 + 3x + 2) - \log_3 (x + 1) = 1$

23 $\log_5 (x^2 + 3x + 5) - \log_5 (x + 4) = 1$ **24** $\log_2 (x^2 - 3x - 2) - \log_2 (x - 4) = 3$

Solve the following pairs of equations simultaneously.

25 $3^{2x+y} = 240$ **26** $2^{2x+y} = 15$ **27** $7^{x+y} = 50$ **28** $5^{x+2y} = 130$
        $3^{x-y} = 2$ $2^{3x+y} = 120$ $7^{2x-y} = 6$ $5^{2x+y} = 2$

29 $2^{2x+y} = 3$ **30** $3^{x+y} = 2$ **31** $5^{3x-y} = 6$ **32** $13^{4x-y} = 15$
        $7^{x+y} = 2$ $5^{3x+2y} = 600$ $11^{2x+y} = 2$ $7^{5x-y} = 50$

Sketch the graphs of the two curves defined by the equations in each of Probs. 33 to 36.

33 $y = \log_2 x;\ y = 2^x$ **34** $y = \log_5 x;\ y = 5^x$

35 $y = \log_{11} x;\ y = 11^x$ **36** $y = \log_{19} x;\ y = 19^x$

13.13   SUMMARY

This entire chapter is based on the definition

$$\log_b N = L \text{ if and only if } b^L = N, \quad \text{for } b > 0, b \neq 1 \quad (13.1)$$

It follows that $b^{\log_b L} = L$. The computation theorems are

$$\log_b MN = \log_b M + \log_b N \quad (13.2)$$

$$\log_b \frac{M}{N} = \log_b M - \log_b N \quad (13.3)$$

$$\log_b M^k = k \log_b M \quad (13.4)$$

$$\log_b \sqrt[r]{M} = \frac{1}{r} \log_b M \quad (13.4a)$$

We also make use of rounding off and scientific notation as discussed in Secs. 13.1 and 13.2, and give the definition of characteristic and mantissa in Sec. 13.7.

We point out at various times and places how the operations performed by use of logarithms can also be done with a calculator.

EXERCISE 13.6   Review

Round off the number in each of Probs. 1 to 4 to three significant digits, and express it in scientific notation.

1 786.291   **2** 786.928   **3** 5715   **4** 8025

Perform the operations indicated in Probs. 5 to 10. Then round off to the proper number of digits and express in scientific notation.

5 (78.2)(4.7)   **6** (8.418)(64)(3.72)   **7** 99.35 ÷ 1.46

8 58.37 ÷ 302   **9** 2.3 + 4.67 − 1.805   **10** 72.3 + 21.75 − 37.846

11 Put $3^4 = 81$ in logarithmic form.   **12** Express $\log_2 32 = 5$ in exponential form.

13 Solve $\log_b 81 = 4$ for b.   **14** Solve $\log_5 125 = L$ for L.

15 Find N if $\log_6 N = 3$.

Perform the operations indicated in Probs. 16 to 21 by use of logarithms and/or a calculator.

16 (4.71)(38.7)(.2375)   **17** (76.2)(58.1)(27.43)   **18** $\dfrac{(291)(81.4)}{48.3}$

19 $\dfrac{734}{(29.6)(49.2)}$ **20** $(3.74)^2(9.83)^3$ **21** $\sqrt[3]{\dfrac{(8.01)^2}{(7.37)^4}}$

22 Solve $3^{3x-1} = 15^x$.

23 Solve $7^{x+1} = 11^{x+.5}$.

24 Solve $\log_2 (3x - 1) - \log_2 (x + 1) = 1$.

25 Solve $5^{2x+y} = 23$ and $3^{4x-y} = 4$ simultaneously.

14

inequalities and systems of inequalities

We have studied equations in many previous chapters and shall study inequalities in this chapter. An equality may be considered as a statement that two expressions are equal, and an inequality as a statement that one quantity is larger than or equal to another. After some study of inequalities, we shall work with systems of inequalities and introduce linear programming.

14.1 LINEAR INEQUALITIES IN ONE VARIABLE

Inequality   A statement of the form $f(x) > g(x)$ or $f(x) \geq g(x)$ or $f(x) < g(x)$ or $f(x) \leq g(x)$ is called an *inequality*.

Linear Inequality   If $f(x)$ or $g(x)$ is a polynomial of degree 1 and the other is a constant or a polynomial of degree 1 with the coefficients of x in f and g different, then we have a *linear inequality*. Consequently, $3x - 1 > x + 3$ and $4x + 7 < 8x + 5$ are linear inequalities, but $3x + 4 < 3x + 9$ is not one.

Solution Set   The set of replacements for x for which an inequality is a true statement is called the *solution set* of the inequality. Thus, $x > 3$ is the solution set of the inequality $2x + 1 > 7$. This is often put in the form $\{x \mid 2x + 1 > 7\} = \{x \mid x > 3\}$.

The procedures for finding the solution set of an inequality are very similar to those for finding the solution set of an equation. The concept of equivalent inequalities is needed and used; hence, we shall now consider that subject. We begin by stating

Equivalent   that two inequalities are *equivalent* if they have the same solution set. We now give several theorems concerning equivalent inequalities.

Theorems on
Equivalent
Sets

If $f(x)$, $g(x)$, and $h(x)$ are real valued expressions, then each of the following is equivalent to $f(x) > g(x)$:

$$f(x) + h(x) > g(x) + h(x) \tag{14.1}$$
$$f(x) \cdot h(x) > g(x) \cdot h(x) \text{ for } \{x \mid h(x) > 0\} \tag{14.2}$$
$$f(x) \cdot h(x) < g(x) \cdot h(x) \text{ for } \{x \mid h(x) < 0\} \tag{14.3}$$

If k is a positive constant, then $f(x) > g(x)$ and $k \cdot f(x) > k \cdot g(x)$ are equivalent. $\hfill (14.2a)$

If c is a negative constant, then $f(x) > g(x)$ and $c \cdot f(x) < c \cdot g(x)$ are equivalent. $\hfill (14.3a)$

Similar statements are true for $f(x) < g(x)$.

Inequality 14.1 is sometimes put in this form: the same quantity may be added to each member of an inequality without affecting the direction or sense of the inequality. Also, (14.2) and (14.3) may be combined into: The sense of an inequality is unchanged by multiplying by a positive quantity but is changed by multiplying by a negative quantity.

The proofs of the above theorems are similar to the corresponding ones for equations and will not be given.

Example 1 Since $x^2 + 2 > 0$ for all real values of x, and $3x + 1 > 3x - 4$, it follows from (14.2) that $(x^2 + 2)(3x + 1) > (x^2 + 2)(3x - 4)$.

Example 2 Find the solution set of $6x - 1 > 2x + 7$.

Solution

$6x - 1 > 2x + 7$	the given inequality
$6x - 1 - 2x + 1 > 2x + 7 - 2x + 1$	by (14.1) with $h(x) = -2x + 1$
$4x > 8$	combining
$x > 2$	by (14.2a) with $k = \frac{1}{4}$

Consequently, the solution set is $\{x \mid x > 2\}$.

Example 3 To solve $-2x < 4$, we could write

$$-2x + 2x - 4 < 4 + 2x - 4$$
$$-4 < 2x$$
$$-2 < x \qquad \text{by (14.2)}$$

or we could write

$$-2x < 4$$
$$x > \frac{4}{-2} = -2 \qquad \text{by (14.3)}$$

Example 4 Solve $\frac{2}{3}x - 1 \leq \frac{7}{6}x - 3$.

Solution $\qquad$ $\frac{2}{3}x - 1 \leq \frac{7}{6}x - 3$ $\qquad$ the given inequality

$\frac{2}{3}x - 1 - \frac{7}{6}x + 1 \leq \frac{7}{6}x - 3 - \frac{7}{6}x + 1$ $\qquad$ by (14.1)

$-\frac{1}{2}x \leq -2$ $\qquad$ combining terms

$x \geq -2(-2)$ $\qquad$ by (14.3a)

$x \geq 4$

Therefore, the solution is $x \geq 4$.

EXERCISE 14.1   Linear Inequalities

Find the solution set of the inequality in each of Probs. 1 to 36.

1 $3x + 1 > 7$ $\qquad$ **2** $5x - 4 > 6$ $\qquad$ **3** $2x + 9 \geq 3$

4 $7x + 23 > -5$ $\qquad$ **5** $5x + 4 < 9$ $\qquad$ **6** $2x + 17 \leq 3$

7 $6x - 5 < 7$ $\qquad$ **8** $3x - 8 < 4$ $\qquad$ **9** $3x + 4 \geq x + 2$

10 $8x - 3 > 5x + 6$ $\qquad$ **11** $6x - 5 > 2x + 11$ $\qquad$ **12** $9x - 4 \geq 4x + 11$

13 $4x + 9 < x + 6$ $\qquad$ **14** $8x + 3 < 3x + 8$ $\qquad$ **15** $7x - 4 \leq 3x + 2$

16 $4x + 5 < 2x - 3$ $\qquad$ **17** $2x + 5 > 5x - 1$ $\qquad$ **18** $3x - 7 \geq 7x + 5$

19 $5x + 3 > 7x - 3$ $\qquad$ **20** $4x - 3 > 7x + 6$ $\qquad$ **21** $2x - 7 \leq 5x + 2$

22 $x + 5 < 5x - 3$ $\qquad$ **23** $3x + 2 < 6x - 4$ $\qquad$ **24** $5x + 7 \leq 9x - 5$

25 $\frac{2}{3}x + 3 > \frac{4}{3}x - 1$ $\qquad$ **26** $\frac{3}{4}x - 2 < \frac{1}{4}x + 2$ $\qquad$ **27** $\frac{3}{4}x - 4 \leq \frac{1}{3}x + 1$

28 $\frac{4}{5}x - 5 > \frac{3}{4}x - 4$

29 If $a > b$, prove that $2a > a + b$.

30 If $2a \leq 3b$, prove that $4a \leq 2a + 3b$.

31 If $a < b$, prove that $5a + 3b < 3a + 5b$.

32 If $a \geq 3b$, prove that $3a - b \geq a + 5b$.

14.2   NONLINEAR INEQUALITIES

There are many types of nonlinear inequalities, just as there are many types of nonlinear equations. We shall consider those that are made up of the product and quotient of linear factors or can be changed to that form. In solving nonlinear inequalities, we need to make use of the fact that a product or quotient that contains an odd number of negative factors is negative, and other products and quotients of linear factors are positive provided no zero factor is involved.

Example 1 $\quad$ Find the solution set of $(2x - 1)/(3x + 5) < 0$.

Solution $\quad$ The left number of the given inequality is the quotient of two linear factors; hence, it is negative if either factor is negative and the other is positive. Consequently, we solve

$$2x - 1 > 0 \quad \text{and} \quad 3x + 5 < 0 \tag{1}$$

simultaneously and solve

$$2x - 1 < 0 \quad \text{and} \quad 3x + 5 > 0 \tag{2}$$

simultaneously. If we add 1 to each member of the first inequality in (1), and -5 to each member of the second, we get $2x > 1$ and $3x < -5$; hence, $x > \frac{1}{2}$ and $x < -\frac{5}{3}$. Therefore the part of the solution set of the given equation that comes from (1) is

$$\{x \mid x > \tfrac{1}{2}\} \cap \{x \mid x < -\tfrac{5}{3}\} = \varnothing \tag{3}$$

Now, solving the two inequalities in (2) simultaneously, we get $x < \frac{1}{2}$ and $x > -\frac{5}{3}$. Consequently, the part of the solution set of the given inequality that comes from the pair of inequalities in (2) is

$$\{x \mid x < \tfrac{1}{2}\} \cap \{x \mid x > -\tfrac{5}{3}\} = \{x \mid -\tfrac{5}{3} < x < \tfrac{1}{2}\} \tag{4}$$

Finally, the complete solution set of the given inequality is the union of the sets given in (3) and (4); hence, it is

$$\{x \mid -\tfrac{5}{3} < x < \tfrac{1}{2}\} \cup \varnothing = \{x \mid -\tfrac{5}{3} < x < \tfrac{1}{2}\}$$

Example 2 Find the solution set of $(2x - 7)(3x + 5) > 0$.

Solution Since the left member of the given inequality is the product of two linear factors, it is positive if both factors are positive or if both are negative. Therefore, we get a part of the desired solution by simultaneously solving

$$2x - 7 > 0 \quad \text{and} \quad 3x + 5 > 0 \tag{5}$$

Thus, we have

$$2x > 7 \quad \text{and} \quad 3x > -5 \qquad \textbf{by (14.1)}$$
$$x > \tfrac{7}{2} \quad \text{and} \quad x > -\tfrac{5}{3} \qquad \textbf{by (14.2)}$$

Noting that $x > -\frac{5}{3}$ automatically if $x > \frac{7}{2}$, we see that a part of the desired solution is

$$\{x \mid x > \tfrac{7}{2}\} \tag{5'}$$

To find the other part of the desired solution we put

$$2x - 7 < 0 \quad \text{and} \quad 3x + 5 < 0 \tag{6}$$

and solve simultaneously. Thus, adding 7 to each member of the first of these inequalities, and -5 to each member of the second, gives

$$2x < 7 \quad \text{and} \quad 3x < -5$$

Now, dividing these inequalities by 2 and 3, respectively, we get $x < \frac{7}{2}$ and $x < -\frac{5}{3}$. Therefore,

$\{x \mid x < -\frac{5}{3}\}$ (6′)

is the part of the desired solution set that comes from both factors being negative. Consequently, by (5′) and (6′),

$\{x \mid x > \frac{7}{2}\} \cup \{x \mid x < -\frac{5}{3}\}$

is the solution set of the given inequality.

Graphical solution   The solution can also be found graphically by sketching the graph of $y = (2x - 7)(3x + 5)$ and seeing where it is above the X axis. Clearly $y = 0$ if $x = \frac{7}{2}$ or $x = -\frac{5}{3}$, and the graph is as shown in Fig. 14.1 (since $y = -35$ for $x = 0$). From this figure, we see that $y > 0$ for $x < -\frac{5}{3}$ and for $x > \frac{7}{2}$, as was found in the algebraic solution.

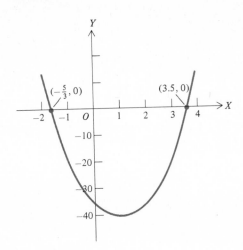

FIGURE 14.1

Example 3   Find the solution set of $(2x - 5)(3x - 4)(4x + 1) > 0$.

Solution   The left member of the given inequality is the product of three linear factors; hence, it is positive if all three factors are positive or if any two are negative and the third is positive.

   If all three factors are positive, we have

$2x - 5 > 0 \qquad 3x - 4 > 0 \qquad 4x + 1 > 0$ (7)

If we add 5, 4, and −1, respectively, to these three inequalities, we obtain

$2x > 5 \qquad 3x > 4 \qquad 4x > -1$

Each member of each inequality is now divided by the coefficient of x in the inequality to get

$x > \frac{5}{2} \qquad x > \frac{4}{3} \qquad x > -\frac{1}{4}$

If $x > \frac{5}{2}$, it is automatically greater than $\frac{4}{3}$ and $-\frac{1}{4}$. Consequently, the part of the solution set of the given inequality that comes from all three factors being positive is

$$\{x \mid x > \tfrac{5}{2}\} \tag{7'}$$

Another part of the solution set is the common part of the solutions of the inequalities obtained by requiring the first factor to be positive and the other two negative. Thus, we have

$$
\begin{array}{llll}
2x - 5 > 0 & 3x - 4 < 0 & 4x + 1 < 0 & \tag{8} \\
\quad 2x > 5 & \quad 3x < 4 & \quad 4x < -1 & \text{by (14.1)} \\
\quad x > \tfrac{5}{2} & \quad x < \tfrac{4}{3} & \quad x < -\tfrac{1}{4} & \text{by (14.2)}
\end{array}
$$

The common part or intersection of these three sets is $\varnothing$, since a number greater than $\tfrac{5}{2}$ cannot be simultaneously less than $-\tfrac{1}{4}$.

A third part of the solution is obtained by requiring the second factor to be positive and the other two negative. Thus, we have

$$
\begin{array}{llll}
2x - 5 < 0 & 3x - 4 > 0 & 4x + 1 < 0 & \tag{9} \\
\quad 2x < 5 & \quad 3x > 4 & \quad 4x < -1 & \text{by (14.1)} \\
\quad x < \tfrac{5}{2} & \quad x > \tfrac{4}{3} & \quad x < -\tfrac{1}{4}
\end{array}
$$

The intersection of these three sets is also $\varnothing$, since a number cannot be simultaneously greater than $\tfrac{4}{3}$ and less than $-\tfrac{1}{4}$.

The final part of the solution set of the given inequality is obtained by requiring that the last factor be positive and the other two negative. Thus, we have

$$
\begin{array}{llll}
2x - 5 < 0 & 3x - 4 < 0 & 4x + 1 > 0 & \tag{10} \\
\quad 2x < 5 & \quad 3x < 4 & \quad 4x > -1 & \text{by (14.1)} \\
\quad x < \tfrac{5}{2} & \quad x < \tfrac{4}{3} & \quad x > -\tfrac{1}{4} & \text{by (14.2)}
\end{array}
$$

Since a number is automatically less than $\tfrac{5}{2}$ if it is less than $\tfrac{4}{3}$, the intersection of the solutions of (10) is

$$\{x \mid x < \tfrac{4}{3}\} \cap \{x \mid x > -\tfrac{1}{4}\} = \{x \mid -\tfrac{1}{4} < x < \tfrac{4}{3}\} \tag{10'}$$

Finally, the solution set of the given inequality is the union of the sets given by (7') and (10'), since the inequalities (8) and (9) lead to $\varnothing$. Hence, the desired set is

$$\{x \mid x > \tfrac{5}{2}\} \cup \{x \mid -\tfrac{1}{4} < x < \tfrac{4}{3}\}$$

Graphical solution The given inequality can be solved graphically by sketching the graph of $y = (2x - 5)(3x - 4)(4x + 1)$ and finding the values of x for which $y > 0$ as required by the problem.

To sketch the graph, note that $y = 0$ if $2x - 5 = 0$ or $3x - 4 = 0$ or $4x + 1 = 0$. The corresponding values of x are $\tfrac{5}{2}$, $\tfrac{4}{3}$, and $-\tfrac{1}{4}$. The graph may be drawn quickly by choosing a value of x smaller than $-\tfrac{1}{4}$, say $x = -1$; a value of x between $-\tfrac{1}{4}$ and $\tfrac{4}{3}$, say $x = 0$; a value of x between $\tfrac{4}{3}$ and $\tfrac{5}{2}$, say 2; and a value of x

greater than $\frac{5}{2}$, say $x = 3$. For each x, we find the corresponding y by using $y = (2x - 5)(3x - 4)(4x + 1)$. This gives the table of values

x	-1	$-\frac{1}{4}$	0	$\frac{4}{3}$	2	$\frac{5}{2}$	3
y	-147	0	20	0	-18	0	65

We see from Fig. 14.2 that $y > 0$ if $x > 2.5$ or if x is between $-\frac{1}{4}$ and $\frac{4}{3}$, as obtained in the algebraic solution.

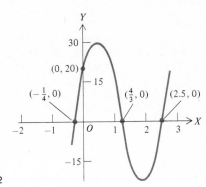

FIGURE 14.2

It is a simple matter to check this solution by assigning a value to x in each interval of the solution and substituting this value for x in the given inequality. We shall use $x = 3$ since it is greater than $\frac{5}{2}$, and use $x = 0$ since it is between $-\frac{1}{4}$ and $\frac{4}{3}$. For $x = 3$, the given inequality becomes $[(2)(3)-5][(3)(3) - 4][(4)(3) + 1)] = (1)(5)(13) > 0$ as it should be. Similarly, if $x = 0$, the given inequality becomes $(-5)(-4)(1) > 0$.

EXERCISE 14.2   Nonlinear Inequalities

Find the solution set of the inequality in each of Probs. 1 to 28.

1 $(x - 2)(x + 3) > 0$ **2** $(x - 1)(x - 4) > 0$ **3** $(x + 1)(x + 4) > 0$

4 $(x + 2)(x - 3) > 0$ **5** $x^2 - 5x + 4 < 0$ **6** $x^2 - 2x - 3 < 0$

7 $x^2 + x - 6 < 0$ **8** $x^2 + 6x + 8 < 0$ **9** $(2x + 5)/(x - 1) > 0$

10 $(3x + 1)/(x + 3) > 0$ **11** $(2x - 1)/(3x + 2) > 0$ **12** $(4x + 3)/(3x - 4) > 0$

13 $(5x - 8)/(8x + 5) < 0$ **14** $(2x + 7)/(3x - 4) < 0$ **15** $(7x + 3)/(3x + 7) < 0$

16 $(8x - 1)/(2x - 9) < 0$

17 $(x - 1)(2x - 3)(3x + 5) > 0$ **18** $(x + 2)(3x - 4)(5x + 9) > 0$

19 $(2x + 5)(3x - 1)(5x + 7) > 0$ **20** $(7x - 9)(9x + 7)(3x + 7) > 0$

21 $\dfrac{(5x + 3)(3x - 8)}{4x + 9} < 0$ **22** $\dfrac{(3x + 2)(5x - 6)}{2x - 7} < 0$

23 $\dfrac{3x + 5}{(2x + 3)(7x - 8)} < 0$ **24** $\dfrac{5x - 7}{(3x - 5)(2x + 3)} < 0$

25 $\dfrac{2x^2 + 5x - 3}{2x - 3} > 0$ **26** $\dfrac{6x^2 + 11x - 10}{x + 4} > 0$

27 $\dfrac{2x - 7}{2x^2 - 3x - 2} < 0$ **28** $\dfrac{4x + 5}{3x^2 - 13x + 12} < 0$

14.3   LINEAR INEQUALITIES THAT INVOLVE ABSOLUTE VALUES

If we use the definition of the absolute value of a number, we see that an inequality of the type

Solution of
$|ax + b| < c$

$$|ax + b| < c \qquad c > 0 \qquad\qquad (1)$$

requires that $ax + b$ be between c and $-c$; hence, if a replacement for x satisfies *both $ax + b < c$ and $ax + b > -c$*, it will satisfy (1). Therefore,

---

The solution set of (1) is the intersection

$$\{x \mid ax + b < c\} \cap \{x \mid ax + b > -c\} = \{x \mid -c < ax + b < c\}$$

Example 1   Solve $|3x - 4| < 5$.

Solution 1   This inequality is satisfied if x satisfies both $3x - 4 < 5$ and $3x - 4 > -5$. So, the solution is $\{x \mid 3x - 4 < 5\} \cap \{x \mid 3x - 4 > -5\}$. Now, adding 4 to each member of each of these inequalities and dividing by 3, we see that

$$\{x \mid x < 3\} \cap \{x \mid x > -\tfrac{1}{3}\} = \{x \mid -\tfrac{1}{3} < x < 3\}$$

is the desired solution.

Solution 2   If we divide each member of $|3x - 4| < 5$ by 3, we have $|x - \tfrac{4}{3}| < \tfrac{5}{3}$. Hence, the distance between x and $\tfrac{4}{3}$ is less than $\tfrac{5}{3}$, as shown in Fig. 14.3.

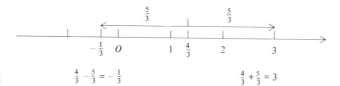

FIGURE 14.3 $\tfrac{4}{3} - \tfrac{5}{3} = -\tfrac{1}{3}$ $\tfrac{4}{3} + \tfrac{5}{3} = 3$

Example 2   Solve $|-2x + 7| < 9$.

Solution   This inequality is satisfied if both $-2x + 7 < 9$ and $-2x + 7 > -9$ are satisfied, and hence if both $-2x < 2$ and $-2x > -16$ are satisfied. Now, dividing by -2 in each inequality and changing the sense of each inequality in keeping with (14.3*a*) we find that $\{x \mid -1 < x < 8\}$ is the solution set.

If we apply the definition of absolute value to

$$|ax + b| > c \qquad c > 0 \tag{2}$$

we find that (2) is satisfied if *either* $ax + b > c$ *or* $ax + b < -c$. Consequently

---

Solution of $|ax + b| > c$

The solution set of (2) is the union of

$\{x \mid ax + b > c\}$ and $\{x \mid ax + b < -c\}$

Example 3 Solve $|3x + 2| > 4$.

Solution 1 If we apply the definition of the absolute value of a number to the given equation, we see that it is satisfied if either $3x + 2 > 4$ or $3x + 2 < -4$; now, solving this pair of inequalities, we find that $x < -2$ and $x > \frac{2}{3}$. Therefore, the desired solution set is

$$\{x \mid x < -2\} \cup \{x \mid x > \tfrac{2}{3}\}$$

Solution 2 If we divide each member of the given inequality by 3, we get $|x + \frac{2}{3}| > \frac{4}{3}$. Since $|x + \frac{2}{3}| = |x - (-\frac{2}{3})|$, the distance of x from $-\frac{2}{3}$ must be greater than $\frac{4}{3}$, as shown in Fig. 14.4.

FIGURE 14.4

Example 4 Solve $|-5x + 7| > 2$.

Solution If we apply the definition of absolute value to the given inequality, we see that it is satisfied if either $-5x + 7 > 2$ or $-5x + 7 < -2$; now, adding -7 to each member of each of these and dividing by -5, we find that they are satisfied by $x < 1$ and $x > \frac{9}{5}$, respectively. Therefore, the solution set of the given inequality is $\{x \mid x < 1\} \cup \{x \mid x > \frac{9}{5}\}$.

EXERCISE 14.3 Inequalities and Absolute Values

Solve each of the following inequalities.

1 $|2x - 3| < 5$ **2** $|3x - 2| < 7$ **3** $|5x - 1| < 9$

4 $|7x - 9| < 5$ **5** $|4x + 7| < 7$ **6** $|6x + 5| < 17$

7 $|3x + 2| < 8$ **8** $|5x + 1| < 11$ **9** $|-7x + 3| < 4$

10 $|-2x - 5| < 7$ **11** $|-3x + 8| < 2$ **12** $|-4x + 9| < 13$

13 $|3x + 2| > 14$ **14** $|2x + 3| > 7$ **15** $|2x + 5| > 11$

16 $|5x + 6| > 16$ **17** $|4x - 1| > 7$ **18** $|3x - 5| > 7$

19 $|7x - 4| > 10$ **20** $|9x - 11| > 2$ **21** $|-5x + 3| > 8$

22 $|-4x + 9| > 1$ **23** $|-2x - 5| > 1$ **24** $|-3x - 8| > 10$

25 $|2x - 9| < 3$ **26** $|-2x + 9| < 3$ **27** $|3x + 5| > 7$

28 $|-3x - 5| > 7$ **29** Verify Eq. (1).

14.4 LINEAR INEQUALITIES IN TWO VARIABLES

We shall begin by finding how to indicate graphically the solution set of a linear inequality in two variables. Any such linear inequality is equivalent to one of the forms $y > ax + b$, $y < ax + b$, $x > a$, or $x < a$. At times we may include the equality sign, as in $x \geq a$, and still call the result an inequality, in spite of the fact that it is a combination of an equation and an inequality.

In order to find the solution set of $y \geq ax + b$, we first draw the line represented by $y = ax + b$. Then, if $P(x, y)$ is on this line and if $y' > y$, it follows that $P(x, y')$ is above the line. Furthermore, since $y' > y$ and $y = ax + b$, it follows that $y' > ax + b$. Consequently, *the solution set of $y \geq ax + b$ is all points on*

Solution of
$y \geq ax + b$
and of
$y \leq ax + b$

and above the graph of $y = ax + b$. It can be shown similarly that *the solution set of $y \leq ax + b$ is all points on and below the line represented by $y = ax + b$.*

In keeping with the above discussion, the graph of $y \leq .3x - 1$ consists of the points on and below the line whose equation is $y = .3x - 1$, as indicated in Fig. 14.5 by not having that marked out.

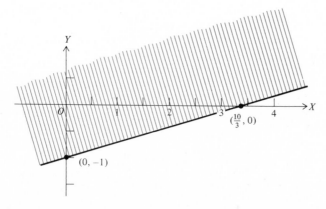

FIGURE 14.5

Simultaneous
Solution

We shall now consider how to determine the region in which $P(x, y)$ lies if (x, y) satisfies each of a set of two or more inequalities. In order to do this, we begin by drawing each related line, then indicate each region in which $P(x, y)$ must lie, and finally determine the intersection of these regions.

Example   Find the region determined by $x \leq 1$, $y \geq 0$, $y \leq 3x - 1$, and $y \leq .5x + 1$, and determine its vertices.

Solution   As indicated above, we begin by sketching the lines represented by $x = 1$, $y = 0$, $y = 3x - 1$, and $y = .5x + 1$, as shown in Fig. 14.6. We then indicate the regions determined by the four inequalities and note the intersection of these regions. We find the coordinates of the vertices of this region by solving simultaneously the equations of the lines that intersect at the vertices. Thus, we solve $y = 3x - 1$ and $y = .5x + 1$ simultaneously to find the coordinates of D. They are $x = .8$ and $y = 1.4$, and we write $D(.8, 1.4)$. Similarly, the other vertices are $A(\frac{1}{3}, 0)$ $B(1, 0)$ and $C(1, 1.5)$.

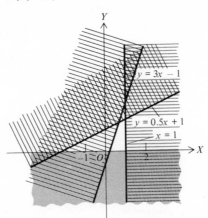

FIGURE 14.6

If all points of the line segment PQ are in a region whenever
Convex Region P and Q are in it, the region is called a *convex region*. The
Convex Polygon   boundary is called a *convex polygon* if the region is bounded by line segments. Thus, the region $ABCD$ in Fig. 14.7a is a convex region, and that in Fig. 14.7b is not convex. A circle and its interior form a convex region.

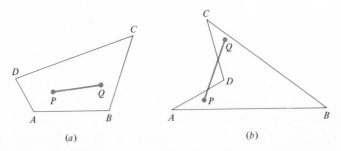

FIGURE 14.7   (a)   (b)

EXERCISE 14.4 Convex Polygons

Use arrows to indicate the half plane determined by the inequality in each of Probs. 1 to 4.

1 $x \geq 2$ **2** $y \leq -1$ **3** $3x + y \leq 6$ **4** $2x - y \leq 4$

Show the convex region determined by the set of inequalities in each of Probs. 5 to 16.

5 $x \geq 0, y \geq 0, 2x + y \leq 2$

6 $x \geq 1, y \leq -1, 2x - 3y \leq 6$

7 $x \geq 2, y \leq 1, 2x - 3y \leq 6$

8 $x \leq 3, y \geq 2, y - x \leq 0$

9 $x - 2y \leq 4, 2x - y \leq 6, 3x + y \geq 3$

10 $-2x + y \leq 6, 3x - y \geq 3, 4x - y \leq 8$

11 $-3x + 2y \geq 6, 2x - 5y \leq 10, x + 4y \leq 4$

12 $4x + 3y \leq 12, 2x - y \geq 8, -x + 5y \geq 10$

13 $x + y \leq 3, x - 2y \geq 0, x + y \geq 0, x + 2y + 3 \geq 0$

14 $3x + y \geq 7, 5x - y - 17 \leq 0, 4x - y \geq 1$

15 $5x + y \leq 12, y \leq -x, x - 3y + 4 \leq 0$

16 $x + 3y \leq 7, 3x - 5y - 7 \leq 0, x - 3y + 5 \geq 0, 3x + y + 5 \geq 0$

Show that the polygon with vertices at the points in each of Probs. 17 to 20 is not convex.

17 $(-1, -1), (3, 1), (1, 1), (2, 4)$

18 $(1, -2), (5, 0), (2, 1), (1, 3)$

19 $(3, -2), (6, -1), (5, 3), (3, 0), (0, 0)$

20 $(0, 3), (1, -1), (5, 0), (0, 4), (-2, 1)$

Find the vertices of the polygonal region determined by the inequalities in each of Probs. 21 to 28.

21 $-2x + y + 4 \geq 0, 5x + y - 3 \geq 0, x + 3y - 9 \leq 0$

22 $3x - 2y - 10 \leq 0, 5x + y - 8 \geq 0, 2x + 3y - 11 \leq 0$

23 $-2x + y + 5 \geq 0, 3x + 5y - 1 \geq 0, x + 6y - 9 \leq 0$

24 $5x - 2y - 11 \leq 0, -x + 5y - 7 \leq 0, 4x + 3y + 5 \geq 0$

25 $3x - 2y - 4 \leq 0, x + y - 3 \leq 0, x - y + 3 \geq 0, 2x + 3y + 6 \geq 0$

26 $x \leq 3, x - 3y \leq 12, 2x + y + 4 \geq 0, 2x - 5y + 4 \geq 0$

27 $-x + 2y + 2 \geq 0, x - y \leq 2, x + 3y \leq 6, -2x + y \leq 2, x + y + 1 \geq 0$

28 $y \geq 0, y \geq 2x - 6, x + 2y \leq 8, x - y + 1 \geq 0, x \geq 0$

14.5   LINEAR PROGRAMMING

Maximum   If there is a largest value of a function, it is called the *maximum*.
Minimum   The smallest value of a function, if there is one, is called the
Extreme   *minimum*.   The two are often referred to as the *extremes*. In
keeping with these definitions, the minimum value of $y = f(x) =$
$5x + 2$, for $-1 \leq x \leq 3$, occurs for $x = -1$ and is -3; furthermore,
the maximum occurs for $x = 3$ and is 17.

If a problem involves two variables x and y and conditions
that restrict (x, y) to a region S that is determined by a set of
linear inequalities, then the determination of (x, y) such that a

Linear
Programming
Constraints
Feasible
Solutions
Objective
Function

given linear combination of x and y is an extreme is called *linear programming*. The inequalities that determine the region S are called *constraints*; the region S is called the set of *feasible solutions*; and the linear function that is to be an extreme is called the *objective function*.

We shall now state without proof a theorem on linear programming and then give two examples.

Theorem on
Extremes

If S is a convex polygon, if $f(x, y) = ax + by + c$ has domain S, and if B is the polygonal boundary of S, then f has a maximum and a minimum and each occurs at a vertex of B.

Example 1 Find the maximum and minimum of $f(x, y) = 3x + 5y - 4$ for the region with vertices at $(3, 1)$, $(2, 4)$, $(0, 2)$, and $(1, 1)$.

Solution The function f is linear, and the region with vertices at the given points is convex, as is seen in Fig. 14.8; hence, by the theorem, the maximum and minimum values occur at vertices. Consequently, we shall evaluate $f(x, y)$ at each vertex. We find that $f(3, 1) = 3(3) + 5(1) - 4 = 10$; $f(2, 4) = 22$; $f(0, 2) = 6$; and $f(1, 1) = 4$. Consequently, the maximum is 22 and it occurs at $(2, 4)$, whereas the minimum is 4 and it occurs at $(1, 1)$.

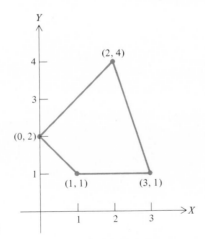

FIGURE 14.8

Example 2 A shoe manufacturer makes boots, men's shoes, and women's shoes and can produce 600 pairs in a unit of time. The manufacturer has standing orders for 150 pairs of men's shoes and 240 pairs of women's shoes and can sell at most 100 pairs of boots and 300 pairs of men's shoes. How many pairs of each type should be produced to make a maximum profit, provided the profit is $4 per pair of boots, $2.50 per pair of men's shoes, and $1.75 per pair of women's shoes?

Solution   If we use x to represent the number of pairs of boots made, and y the number of pairs of men's shoes made, then $600 - x - y$ is the number of pairs of women's shoes manufactured. Therefore, the profit in dollars is $4x$ for boots, $2.5y$ for men's shoes, and $1.75(600 - x - y)$ for women's shoes; hence, the entire profit is

$$f(x, y) = 4x + 2.5y + 1.75(600 - x - y)$$

This objective function is subject to the following constraints as imposed by the statement of the problem:

$$y \geq 150$$ since there is a standing order for 150 pairs of men's shoes

$$600 - x - y \geq 240$$ since there is a standing order for 240 pairs of women's shoes

$$x \leq 100$$ since not more than 100 pairs of boots can be sold

$$y \leq 300$$ since not more than 300 pairs of men's shoes can be sold

$$x \geq 0, y \geq 0, 600 - x - y \geq 0$$ since all types are for sale

These constraints are shown in graphical form in Fig. 14.9. The graph of $600 - x - y \geq 0$ is not shown since this inequality is automatically satisfied if $600 - x - y \geq 240$.

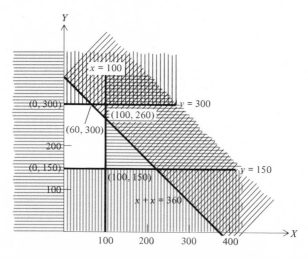

FIGURE 14.9

After the region of feasible solutions is found, we obtain the coordinates of the vertices by simultaneously solving each pair of equations which determine lines that intersect at a vertex. Thus, we solve $y = 300$ and $x + y = 360$ simultaneously to obtain coordinates of the vertex on the top at right. In this way we find that the vertices are at $(60, 300)$, $(0, 300)$, $(0, 150)$, $(100, 150)$, and $(100, 260)$, shown in Fig. 14.9. We must evaluate the objective function $f(x, y) = 4x + 2.5y + 1.75(600 - x - y)$ for the coordinates of each vertex in order to find the extremes. Evaluating at $(60, 300)$, we find that

$$f(60, 300) = 4(60) + 2.5(300) + 1.75(600 - 60 - 300)$$
$$= 240 + 750 + 1.75(240) = 240 + 750 + 420$$
$$= 1410$$

Similarly $f(0, 300) = 1275$; $f(0, 150) = 1162.50$; $f(100, 150) = 1387.50$; and $f(100, 260) = 1470$. Therefore, the maximum profit is obtained if 100 pairs of boots, 260 pairs of men's shoes, and $600 - 100 - 260 = 240$ pairs of women's shoes are manufactured.

Example 3 Repeat Example 2, except that the profit on a pair of women's shoes is $2.75 instead of $1.75.

Solution Because of the change in assumed profit, the total profit is now given by

$$g(x, y) = 4x + 2.5y + 2.75(600 - x - y)$$

This must be evaluated for the values of x and y at each vertex of the region of feasible solutions so as to find the maximum profit and the number of pairs of each type of footwear required for this profit. By substitution, we find that $g(0, 300) = 4(0) + 2.5(300) + 2.75(600 - 0 - 300) = 1575$; $g(0, 150) = 1612.50$; $g(100, 150) = 1737.50$; $g(100, 260) = 1710$; and $g(60, 300) = 1650$. Therefore, the maximum profit is obtained if 100 pairs of boots, 150 pairs of men's shoes, and $600 - 100 - 150 = 350$ pairs of women's shoes are manufactured.

EXERCISE 14.5   Linear Programming

In each of Probs. 1 to 8, find the maximum and minimum of f for the convex polygonal region with the specified vertices.

1   $f(x, y) = 7x + 2y - 4; (2, 0), (4, 7), (1, 3)$

2   $f(x, y) = 4x + 3y - 2; (1, 1), (2, 4), (3, 2)$

3   $f(x, y) = 8x - 3y - 7; (2, 1), (5, 2), (3, 4)$

4   $f(x, y) = 6x + 5y - 3; (1, 0), (5, 1), (6, 4)$

5   $f(x, y) = 5x + 6y - 8; (1, 5), (5, 1), (3, 3), (2, 1)$

6   $f(x, y) = 9x + 3y - 4; (4, 3), (11, 2), (7, 6), (8, 1)$

7   $f(x, y) = 8x + 7y - 9; (1, 4), (5, 7), (9, 2), (2, 2)$

8   $f(x, y) = 7x + 5y - 7; (3, 5), (7, 1), (9, 3), (8, 6)$

In each of Probs. 9 to 16, find the extremes of f for the convex polygonal region determined by the given inequalities. Find where each extreme occurs.

9 $-4x + 3y + 13 \geq 0,\ x + 3y - 7 \geq 0,\ x - 2y + 3 \geq 0;\ f(x, y) = 2x + y - 1$

10 $2x + 3y \geq 5,\ 3x - y \leq -9,\ -x + 2y \leq 3;\ f(x, y) = 4x + 5y - 2$

11 $3x + 8y \geq 24,\ 4x + 3y \leq 32,\ x - 5y + 15 \geq 0;\ f(x, y) = 5x + 4y + 2$

12 $3x + 2y \geq 9,\ x - 2y \leq 3,\ x + 6y \leq 19;\ f(x, y) = 6x + 5y - 3$

13 $4x + y \geq 4,\ x - 4y \leq 1,\ x + y \leq 6,\ x + 3y \leq 12;\ f(x, y) = 3x + 5y - 4$

14 $x \geq 0,\ y \geq 0,\ -2x + y + 14 \geq 0,\ x + 8y \leq 24;\ f(x, y) = 4x + 7y - 5$

15 $5x + y \geq 5,\ 2x + y \leq 13,\ y \geq 1,\ 2x + 5y \leq 25;\ f(x, y) = 6x + 3y + 7$

16 $-5x + 4y \leq -1,\ 2x + y \leq 3,\ -x + y \geq -3,\ x - 3y \leq 9;\ f(x, y) = 7x + 5y - 2$

17 A tailor has 160 units of material and 210 units of labor available for use in a given period of time. He can make suits of type A or type B or both. How many of each type should he make so as to have a maximum profit if he clears $10 on each suit of type A and $4 on each suit of type B? What is this maximum profit? Assume that a suit of type A requires 2 units of material and 14 of labor, whereas a suit of type B uses 4 units of material and 2 of labor.

18 Repeat Prob. 17 except that the profit on each suit of type A is $12.

19 Repeat Prob. 17 except that 184 units of labor are available.

20 Repeat Prob. 17 except that the profit on each suit of type A is $11, and on each suit of type B is $10.

21 A farmer raises wheat and corn. The wheat requires 9 units of insecticide and 6 units of fertilizer per acre, whereas the corn needs 7 units of each. How many acres of each should she plant to make the maximum profit if the profit on an acre of wheat is $32, and on an acre of corn is $29? Assume that she has all the needed land, 715 units of insecticide, and 640 units of fertilizer. What is the profit?

22 Repeat Prob. 21 except that the wheat requires 8 units of insecicide and 6 units of fertilizer per acre.

23 Repeat Prob. 21 except that the profit is $38 per acre on wheat and $31 per acre on corn.

24 Repeat Prob. 21 except that the profit per acre is $29 for wheat and $32 for corn.

25 A truck farmer has 47 acres available for planting pepper, lettuce, and tomatoes. He thinks he can make a profit of $300 per acre on pepper, $400 on lettuce, and $350 on tomatoes. He cannot take care of more than 20 acres of pepper, more than 26 acres of lettuce, or more than 17 acres of tomatoes. How many acres of each should he grow so as to make a maximum profit?

26 Repeat Prob. 25 except that the profit per acre on tomatoes is $425.

27 Repeat Prob. 25 except that the farmer cannot take care of more than 22 acres of lettuce.

28 Repeat Prob. 25 except that the farmer anticipates a profit of $500 per acre on pepper.

14.6   SUMMARY

This chapter deals with linear inequalities. We first treat linear inequalities in one variable in Sec. 14.1. We give five theorems that enable us to solve such inequalities. The theorems are concerned with adding a constant to each member and multiplying each member by a positive number and by a negative number. We then discuss nonlinear inequalities that are made up of the product or quotient of two or more linear factors. The solution of such an inequality is based on our earlier study of graphs and inequalities and the fact that the product or quotient of an odd number of negative numbers is negative, whereas other products and quotients of positive or negative numbers are positive. We then solve inequalities that involve absolute values by showing that such a situation is similar to solving two linear inequalities that do not involve absolute values. Finally, we present linear programming as the simultaneous solution of a set of inequalities in two variables.

EXERCISE 14.6   Review

Solve the inequality in each of Probs. 1 to 16.

1 $2x + 5 > 9$ **2** $3x - 2 < 4$ **3** $|2x + 5| > 9$

4 $|3x - 2| < 4$ **5** $-5x + 1 < 11$ **6** $|-5x + 1| < 11$

7 $(x - 1)(2x + 3) > 0$ **8** $\dfrac{x - 1}{2x + 3} > 0$ **9** $\dfrac{2x - 5}{x + 3} > 0$

10 $(2x - 5)(x + 3) < 0$ **11** $(3x + 2)(x - 4) < 0$ **12** $(3x + 7)(x - 2) < 0$

13 $(x - 3)(x - 1)(2x + 3) > 0$ **14** $(4x + 9)(x + 1)(2x - 3) < 0$

15 $\dfrac{2x^2 - 5x + 2}{x - 3} < 0$ **16** $\dfrac{3x + 4}{2x^2 - x - 3} > 0$

Draw the convex polygons with vertices as given in Probs. 17 and 18, and with sides as given in Probs. 19 and 20.

17 $(2, 1), (8 - 3), (10, 6)$ **18** $(-2, 4), (-5, -7), (9, -2), (10, 6)$

19 $2x + 7y \geq 3, 5x + 2y \leq 7, 3x - 5y \geq -11$

20 $x + 6y \geq -14, 3x + 5y \leq 0, -7x + 3y \leq -8, 2x - 5y \leq 26$

21 Show that the polygon with vertices at $(1, 5), (5, 1), (4, -3)$, and $(6, 2)$ is not convex.

22 Show that the intersection of two convex regions is also convex.

Find the vertices of the polygonal region determined by the inequalities in each of Probs. 23 and 24.

23 $3x + y \leq 11, -5x + y \leq -5, x - y \leq 1$

24 $4x + y \geq 11, -x + 3y \geq -6, 5x - 2y \leq 30, x - 3y + 7 \geq 0$

25 Find the extreme of $f(x, y) = 5x + 3y - 2$ in the region with vertices at $(2, 3)$, $(3, -1)$, $(6, 0)$, and $(8, 5)$.

26 Find the extremes of $f(x, y) = 13x + 7y - 4$ in the polygonal region determined by $x + y \leq 13$, $-x + y \leq 1$, $x + 3y \geq 7$, and $3x - 5y \leq 7$.

27 A builder has 269 units of lumber and 208 units of brick to use along with other materials in building houses. He can build an A type house by using 9 units of lumber and 16 units of brick, and he will make a profit of $3200 on each such house. He can also build a type B house by using 14 units of lumber and 8 units of brick, and he will make $2900 on each of these. How many of each type should he build to have a maximum profit?

28 Repeat Prob. 27 except that the profit is $4000 on an A type house and $1800 on a B type.

29 Repeat Prob. 27 except that 193 units of lumber are available.

30 Show that $(2x - 1)(3x + 7) < 0$ has the same solutions as $(2x - 1)/(3x + 7) < 0$.

Tables

TABLE I   COMMON LOGARITHMS

N	0	1	2	3	4	5	6	7	8	9
10	0000	0043	0086	0128	0170	0212	0253	0294	0334	0374
11	0414	0453	0492	0531	0569	0607	0645	0682	0719	0755
12	0792	0828	0864	0899	0934	0969	1004	1038	1072	1106
13	1139	1173	1206	1239	1271	1303	1335	1367	1399	1430
14	1461	1492	1523	1553	1584	1614	1644	1673	1703	1732
15	1761	1790	1818	1847	1875	1903	1931	1959	1987	2014
16	2041	2068	2095	2122	2148	2175	2201	2227	2253	2279
17	2304	2330	2355	2380	2405	2430	2455	2480	2504	2529
18	2553	2577	2601	2625	2648	2672	2695	2718	2742	2765
19	2788	2810	2833	2856	2878	2900	2923	2945	2967	2989
20	3010	3032	3054	3075	3096	3118	3139	3160	3181	3201
21	3222	3243	3263	3284	3304	3324	3345	3365	3385	3404
22	3424	3444	3464	3483	3502	3522	3541	3560	3579	3598
23	3617	3636	3655	3674	3692	3711	3729	3747	3766	3784
24	3802	3820	3838	3856	3874	3892	3909	3927	3945	3962
25	3979	3997	4014	4031	4048	4065	4082	4099	4116	4133
26	4150	4166	4183	4200	4216	4232	4249	4265	4281	4298
27	4314	4330	4346	4362	4378	4393	4409	4425	4440	4456
28	4472	4487	4502	4518	4533	4548	4564	4579	4594	4609
29	4624	4639	4654	4669	4683	4698	4713	4728	4742	4757
30	4771	4786	4800	4814	4829	4843	4857	4871	4886	4900
31	4914	4928	4942	4955	4969	4983	4997	5011	5024	5038
32	5051	5065	5079	5092	5105	5119	5132	5145	5159	5172
33	5185	5198	5211	5224	5237	5250	5263	5276	5289	5302
34	5315	5328	5340	5353	5366	5378	5391	5403	5416	5428
35	5441	5453	5465	5478	5490	5502	5514	5527	5539	5551
36	5563	5575	5587	5599	5611	5623	5635	5647	5658	5670
37	5682	5694	5705	5717	5729	5740	5752	5763	5775	5786
38	5798	5809	5821	5832	5843	5855	5866	5877	5888	5899
39	5911	5922	5933	5944	5955	5966	5977	5988	5999	6010
40	6021	6031	6042	6053	6064	6075	6085	6096	6107	6117
41	6128	6138	6149	6160	6170	6180	6191	6201	6212	6222
42	6232	6243	6253	6263	6274	6284	6294	6304	6314	6325
43	6335	6345	6355	6365	6375	6385	6395	6405	6415	6425
44	6435	6444	6454	6464	6474	6484	6493	6503	6513	6522
45	6532	6542	6551	6561	6571	6580	6590	6599	6609	6618
46	6628	6637	6646	6656	6665	6675	6684	6693	6702	6712
47	6721	6730	6739	6749	6758	6767	6776	6785	6794	6803
48	6812	6821	6830	6839	6848	6857	6866	6875	6884	6893
49	6902	6911	6920	6928	6937	6946	6955	6964	6972	6981
50	6990	6998	7007	7016	7024	7033	7042	7050	7059	7067
51	7076	7084	7093	7101	7110	7118	7126	7135	7143	7152
52	7160	7168	7177	7185	7193	7202	7210	7218	7226	7235
53	7243	7251	7259	7267	7275	7284	7292	7300	7308	7316
54	7324	7332	7340	7348	7356	7364	7372	7380	7388	7396
N	0	1	2	3	4	5	6	7	8	9

TABLE I   COMMON LOGARITHMS   (continued)

N	0	1	2	3	4	5	6	7	8	9
55	7404	7412	7419	7427	7435	7443	7451	7459	7466	7474
56	7482	7490	7497	7505	7513	7520	7528	7536	7543	7551
57	7559	7566	7574	7582	7589	7597	7604	7612	7619	7627
58	7634	7642	7649	7657	7664	7672	7679	7686	7694	7701
59	7709	7716	7723	7731	7738	7745	7752	7760	7767	7774
60	7782	7789	7796	7803	7810	7818	7825	7832	7839	7846
61	7853	7860	7868	7875	7882	7889	7896	7903	7910	7917
62	7924	7931	7938	7945	7952	7959	7966	7973	7980	7987
63	7993	8000	8007	8014	8021	8028	8035	8041	8048	8055
64	8062	8069	8075	8082	8089	8096	8102	8109	8116	8122
65	8129	8136	8142	8149	8156	8162	8169	8176	8182	8189
66	8195	8202	8209	8215	8222	8228	8235	8241	8248	8254
67	8261	8267	8274	8280	8287	8293	8299	8306	8312	8319
68	8325	8331	8338	8344	8351	8357	8363	8370	8376	8382
69	8388	8395	8401	8407	8414	8420	8426	8432	8439	8445
70	8451	8457	8463	8470	8476	8482	8488	8494	8500	8506
71	8513	8519	8525	8531	8537	8543	8549	8555	8561	8567
72	8573	8579	8585	8591	8597	8603	8609	8615	8621	8627
73	8633	8639	8645	8651	8657	8663	8669	8675	8681	8686
74	8692	8698	8704	8710	8716	8722	8727	8733	8739	8745
75	8751	8756	8762	8768	8774	8779	8785	8791	8797	8802
76	8808	8814	8820	8825	8831	8837	8842	8848	8854	8859
77	8865	8871	8876	8882	8887	8893	8899	8904	8910	8915
78	8921	8927	8932	8938	8943	8949	8954	8960	8965	8971
79	8976	8982	8987	8993	8998	9004	9009	9015	9020	9025
80	9031	9036	9042	9047	9053	9058	9063	9069	9074	9079
81	9085	9090	9096	9101	9106	9112	9117	9122	9128	9133
82	9138	9143	9149	9154	9159	9165	9170	9175	9180	9186
83	9191	9196	9201	9206	9212	9217	9222	9227	9232	9238
84	9243	9248	9253	9258	9263	9269	9274	9279	9284	9289
85	9294	9299	9304	9309	9315	9320	9325	9330	9335	9340
86	9345	9350	9355	9360	9365	9370	9375	9380	9385	9390
87	9395	9400	9405	9410	9415	9420	9425	9430	9435	9440
88	9445	9450	9455	9460	9465	9469	9474	9479	9484	9489
89	9494	9499	9504	9509	9513	9518	9523	9528	9533	9538
90	9542	9547	9552	9557	9562	9566	9571	9576	9581	9586
91	9590	9595	9600	9605	9609	9614	9619	9624	9628	9633
92	9638	9643	9647	9652	9657	9661	9666	9671	9675	9680
93	9685	9689	9694	9699	9703	9708	9713	9717	9722	9727
94	9731	9736	9741	9745	9750	9754	9759	9763	9768	9773
95	9777	9782	9786	9791	9795	9800	9805	9809	9814	9818
96	9823	9827	9832	9836	9841	9845	9850	9854	9859	9863
97	9868	9872	9877	9881	9886	9890	9894	9899	9903	9908
98	9912	9917	9921	9926	9930	9934	9939	9943	9948	9952
99	9956	9961	9965	9969	9974	9978	9983	9987	9991	9996
	0	1	2	3	4	5	6	7	8	9

TABLE II POWERS AND ROOTS

No.	Sq.	Sq. Root	Cube	Cube Root	No.	Sq.	Sq. Root	Cube	Cube Root
1	1	1.000	1	1.000	51	2,601	7.141	132,651	3.708
2	4	1.414	8	1.260	52	2,704	7.211	140,608	3.733
3	9	1.732	27	1.442	53	2,809	7.280	148,877	3.756
4	16	2.000	64	1.587	54	2,916	7.348	157,464	3.780
5	25	2.236	125	1.710	55	3,025	7.416	166,375	3.803
6	36	2.449	216	1.817	56	3,136	7.483	175,616	3.826
7	49	2.646	343	1.913	57	3,249	7.550	185,193	3.849
8	64	2.828	512	2.000	58	3,364	7.616	195,112	3.871
9	81	3.000	729	2.080	59	3,481	7.681	205,379	3.893
10	100	3.162	1,000	2.154	60	3,600	7.746	216,000	3.915
11	121	3.317	1,331	2.224	61	3,721	7.810	226,981	3.936
12	144	3.464	1,728	2.289	62	3,844	7.874	238,328	3.958
13	169	3.606	2,197	2.351	63	3,969	7.937	250,047	3.979
14	196	3.742	2,744	2.410	64	4,096	8.000	262,144	4.000
15	225	3.873	3,375	2.466	65	4,225	8.062	274,625	4.021
16	256	4.000	4,096	2.520	66	4,356	8.124	287,496	4.041
17	289	4.123	4,913	2.571	67	4,489	8.185	300,763	4.062
18	324	4.243	5,832	2.621	68	4,624	8.246	314,432	4.082
19	361	4.359	6,859	2.668	69	4,761	8.307	328,509	4.102
20	400	4.472	8,000	2.714	70	4,900	8.367	343,000	4.121
21	441	4.583	9,261	2.759	71	5,041	8.426	357,911	4.141
22	484	4.690	10,648	2.802	72	5,184	8.485	373,248	4.160
23	529	4.796	12,167	2.844	73	5,329	8.544	389,017	4.179
24	576	4.899	13,824	2.884	74	5,476	8.602	405,224	4.193
25	625	5.000	15,625	2.924	75	5,625	8.660	421,875	4.217
26	676	5.099	17,576	2.962	76	5,776	8.718	438,976	4.236
27	729	5.196	19,683	3.000	77	5,929	8.775	456,533	4.254
28	784	5.291	21,952	3.037	78	6,084	8.832	474,552	4.273
29	841	5.385	24,389	3.072	79	6,241	8.888	493,039	4.291
30	900	5.477	27,000	3.107	80	6,400	8.944	512,000	4.309
31	961	5.568	29,791	3.141	81	6,561	9.000	531,441	4.327
32	1,024	5.657	32,768	3.175	82	6,724	9.055	551,368	4.344
33	1,089	5.745	35,937	3.208	83	6,889	9.110	571,787	4.362
34	1,156	5.831	39,304	3.240	84	7,056	9.165	592,704	4.380
35	1,225	5.916	42,875	3.271	85	7,225	9.220	614,125	4.397
36	1,296	6.000	46,656	3.302	86	7,396	9.274	636,056	4.414
37	1,369	6.083	50,653	3.332	87	7,569	9.327	658,503	4.431
38	1,444	6.164	54,872	3.362	88	7,744	9.381	681,472	4.448
39	1,521	6.245	59,319	3.391	89	7,921	9.434	704,969	4.465
40	1,600	6.325	64,000	3.420	90	8,100	9.487	729,000	4.481
41	1,681	6.403	68,921	3.448	91	8,281	9.539	753,571	4.498
42	1,764	6.481	74,088	3.476	92	8.464	9.592	778,688	4.514
43	1,849	6.557	79,507	3.503	93	8,649	9.644	804,357	4.531
44	1,936	6.633	85,184	3.530	94	8,836	9.695	830,584	4.547
45	2,025	6.708	91,125	3.557	95	9,025	9.747	857,375	4.563
46	2,116	6.782	97,336	3.583	96	9,216	9.798	884,736	4.579
47	2,209	6.856	103,823	3.609	97	9,409	9.849	912,673	4.595
48	2,304	6.928	110,592	3.634	98	9,604	9.899	941,192	4.610
49	2,401	7.000	117,649	3.659	99	9,801	9.950	970,299	4.626
50	2,500	7.071	125,000	3.684	100	10,000	10.000	1,000,000	4.642

ANSWERS

EXERCISE 1.1

1 A, {2, 4, 8}, {3, 7}   **2** B, $\varnothing$, $\varnothing$

3 A, B, $\{x \mid x \in A$ and does not play a trumpet$\}$

5 $\{x \mid x$ is in a French class *or* in an algebra class *or* in both$\}$,
$\{x \mid x$ is in a French class *and* in an algebra class$\}$,
$\{x \mid x \in C$ but is not in an algebra class$\}$

6 $\{x \mid x$ has black hair or likes strawberry ice cream or both$\}$
$\{x \mid x \in M$ and likes strawberry ice cream$\}$
$\{x \mid x \in M$ but does not like strawberry ice cream$\}$

7 A, B, $\{x \mid x$ is a male student with red hair$\}$

9 A, {1, 5, 7, 11}, {5, 7}   **10** $A - C$, {6}   **11** {1, 2, 4, 5, 6, 7, 8, 10, 11}

EXERCISE 1.2

1 14	**2** 49	**3** 83	**5** 55	**6** 73
7 1241	**9** 4	**10** 4	**11** −4	**13** 43
14 161	**15** 85	**17** 36	**18** 36	**19** 360
21 378	**22** 396	**23** 1426	**25** 19	**26** 9.5
27 $2.333 \cdots$	**29** .8	**30** $.555 \cdots$	**31** $1.555 \cdots$	**33** Rational
34 Real	**35** Irrational	**37** True	**38** True	**39** False
41 False	**42** True	**43** False		

EXERCISE 1.3

1 Commutative law for multiplication   **2** Associative law for addition

3 Closure law for addition   **5** Multiplicative identity

6 Additive inverse   **7** Additive identity

9 True   **10** True   **11** True   **13** True   **14** True   **15** False

21 3   **22** 4   **23** 22

EXERCISE 1.4

1 32   **2** -45 **3** -56 **5** 32   **6** -54 **7** 100

9 64   **10** 729   **11** 2304   **13** 18   **14** -40 **15** -81

17 3^9 **18** -3^9 **19** 5^{14} **21** -5^5 **22** 11^3 **23** -1

25 5^{24} **26** 3^{40} **27** 5^9 **29** $-2^{12}(3^2)$ **30** $5^2(3^4)$ **31** $2^6 \cdot 3 \cdot 5$

EXERCISE 1.5

9 $\frac{5}{7}$ **10** $\frac{13}{3}$ **11** $\frac{14}{5}$ **13** $\frac{6}{11}$ **14** $\frac{13}{7}$ **15** $\frac{1}{4}$

17 $\frac{5}{6}$ **18** $\frac{8}{15}$ **19** $\frac{11}{28}$ **21** $\frac{11}{12}$ **22** $\frac{23}{40}$ **23** $\frac{37}{42}$

25 $\frac{1}{12}$ **26** $\frac{9}{20}$ **27** $-1/6$ **29** $-8/9$ **30** $\frac{73}{24}$ **31** $-103/28$

33 $\frac{1}{6}$ **34** $\frac{3}{2}$ **35** $\frac{3}{5}$ **37** 1   **38** $\frac{1}{2}$ **39** $\frac{4}{7}$

41 $\frac{4}{9}$ **42** $\frac{2}{3}$ **43** $\frac{1}{6}$

EXERCISE 1.6

1 $\{2, 4, 10\}$ **2** $\{1, 3, 5, 6, 7, 8, 9\}$ **3** $\{6, 8\}$ **17** Yes

18 Yes   **19** Yes   **22** 6   **23** 5

25 $.454545\cdots$ **26** $.46666\cdots$ **27** 128   **29** 16

30 36   **31** 27   **33** 3^{20} **34** 5^8

37 $\frac{21}{40}$ **38** $\frac{7}{12}$ **39** $\frac{8}{5}$ **41** $\frac{5}{2}$

42 $\frac{9}{7}$

EXERCISE 2.1

1 Binomial, binomial   **2** None, polynomial   **3** Trinomial, none

5 Monomial, 12, 3   **6** Trinomial, 3, 2   **7** Monomial, 8, 0

9 $7a$ **10** $12y$ **11** $9xy$ **13** $10ab + 2ac$ **14** $7x + 5y$

15 $-4xy + 7xz + 4yz$ **17** $-4a - 2b + c$ **18** $4ab - 4ac + 8bc$

19 $6x - 2xy + 11z$ **21** $-x + 2y - 2w$ **22** $a - 5b + c + 2d$

23 $9a + 4b + c - 7d$ **25** $-3x + 8y$ **26** $5x + 2y$

27 $-7a - 20b$ **29** $-20x + 30y$ **30** $38x - 93y$

31 $27a - 13$ **33** $-6x + 4y + 4z$ **34** $-3a - 5b + 6c$

35 $12d - 6e - 6f$ **37** $b + 8c - 6d$ **38** $x^2 + 8x - 6$

39 $-3p + 2q - 4r$ **41** $2x^3 - 2x^2 + 9x - 2$ **42** $11x^4 + 5x^2 - 4x - 1$

43 $3a + 2b - 5c + 2d$ **45** $5x + y + 2z$ **46** $2a - 2e$

47 $-g + e + m$ **49** -589 **50** 63   **51** 40   **53** $-3x - 2y - z$

54 $5a - b - 3d$ **55** $3x^2 + 3x - 8$ **57** $3a^2 + 6b^2$

58 $-x + 2y^2 - z^3$ **59** $-x^3 + 7x^2y + xy^2 - 7y^3$

EXERCISE 2.2

1 $10x^5y^6$

2 $-6x^3b^5$

3 $18x^2y^7z^4$

5 $5x^5 + 10x^3$

6 $-12x^9 + 6x^6$

7 $16x^2y^3 + 32x^2y^4$

9 $12x^3y^3z^3 - 18x^3y^4z^2$

10 $-60x^3y^2z^4 + 40x^3y^2z^2$

11 $6x^3y^3z^5 - 8x^2y^4z^7$

13 $x^7y^3z^2$

14 $-15a^7b^4c^7$

15 $12p^6q^8r^4$

17 $x^3 + 4x^2 - 2x + 3$

18 $2x^3 - 3x^2 - 5x + 1$

19 $3x^4 + 2x^3 - x^2 + 2x - 4$

21 $17x$

22 $-2x^3 + 36x^2 - 23x$

23 $12x^4 - 5x^3 - 8x^2$

25 $2x^2 - x - 15$

26 $12b^2 - b - 1$

27 $2a^2 - 5ab + 2b^2$

29 $x^2 - y^2$

30 $x^2 + 2xy + y^2$

31 $x^2 - 2xy + y^2$

33 $x^3 + 3x^2y + 3xy^2 + y^3$

34 $x^3 - 3x^2y + 3xy^2 - y^3$

35 $x^3 + y^3$

37 $x^3 - x^2y - 3xy^2 + 2y^3$

38 $2a^3 + 5a^2b - 14ab^2 - 8b^3$

39 $x^3 + x^2 - 7x + 5$

41 $6x^4 - 5x^3y - 7x^2y^2 + 8xy^3 - 2y^4$

42 $6x^4 + x^3y - 6x^2y^2 + 5xy^3 - 6y^4$

43 $15x^4 + 4x^3y + 4x^2y^2 + 7xy^3 - 2y^4$

45 $10x^4 + 17x^3 + 11x^2 - x - 2$

46 $12x^4 + 11x^3 - 3x - 2$

47 $-2x^4 + 5x^3 + 8x^2 - 11x - 12$

49 $8a^2 + 8ab - 6ac - 6b^2 + 7bc - 2c^2$

50 $a^5 + 1$

51 $x^5 + 32$

EXERCISE 2.3

1 $x^2 + 2x - 3$

2 $x^2 - 2x - 8$

3 $x^2 + 5x + 6$

5 $6x^2 + 5x - 6$

6 $3x^2 + 11x - 20$

7 $-8x^2 + 14x - 3$

9 $6x^2 - xy - y^2$

10 $x^2 - 2xy - 8y^2$

11 $6x^2 + xy - 15y^2$

13 $72k^2 - 67km + 15m^2$

14 $12c^2 - 52cd + 55d^2$

15 $12a^2 - 19ab - 70b^2$

17 $a^2 + 4ab + 4b^2$

18 $9h^2 - 6hk + k^2$

19 $4x^2 + 12x + 9$

21 $25m^2 + 20mn + 4n^2$

22 $16x^2 - 40x + 25$

23 $100x^2 + 100x + 25$

25 1225

26 2025

27 5625

29 2304

30 2809

31 1681

33 $x^2 - 9$

34 $a^2 - 16$

35 $4x^2 - 1$

37 $9x^2 - 4y^2$

38 $25a^2 - 4b^2$

39 $16c^2 - 49d^2$

41 9996

42 3596

43 2491

45 851

46 1575

47 6384

49 $a^3 + 6a^2b + 12ab^2 + 8b^3$

50 $8a^3 - 12a^2b + 6ab^2 - b^3$

51 $x^3 + 9x^2y + 27xy^2 + 27y^3$

53 $125x^3 + 150x^2 + 60x + 8$

54 $27a^3 - 108a^2 + 144a - 64$

55 $8x^3 - 36x^2 + 54x - 27$

57 $x^2 + y^2 + 1 + 2xy - 2x - 2y$

58 $4x^2 + y^2 + 9 + 4xy + 12x + 6y$

59 $a^2 + 4b^2 + 9c^2 - 4ab + 6ac - 12bc$

61 $4x^2 + 4xy + y^2 - z^2$

62 $x^2 - 6xy + 9y^2 - 4z^2$

63 $a^2 - b^2 + 2bc - c^2$

65 $a^2 + b^2 + c^2 + d^2 + 2ab - 2ac + 2ad - 2bc + 2bd - 2cd$

66 $x^2 + y^2 + z^2 + 1 - 2xy + 2xz - 2x - 2yz + 2y - 2z$

67 $4x^2 + y^2 + a^2 + 9 + 4xy - 4xa + 12x - 2ya + 6y - 6a$

EXERCISE 2.4

1 $2 \cdot 3^2$ **2** $2^3 \cdot 5$ **3** $2 \cdot 3 \cdot 5$ **5** $2(x + y)$ **6** $3(x - a)$ **7** $7(x + t)$

9 $4(x + 2y - 4t)$ **10** $3(2x^2 + x + 5)$ **11** $2(3xy - xz + 4yz)$

13 $x(x + y)$ **14** $a(a^2 - 2)$ **15** $xy(x - y)$

17 $x^n(x^n - 1)$ **18** $a^{n-1}(a^3 - 1)$ **19** $3ab(a - 4b + 3)$

21 $(a - b)(3 - x)$ **22** $(x + 2y)(4a - b)$ **23** $(a + 3b)(2x + 3y)$

25 $(a + b)(x + y)$ **26** $(2a - b)(x - y)$ **27** $(a - 3b)(2x + y)$

29 $(a - 1)(x + y - z)$ **30** $3(x + y - 1)(a - 2b)$ **31** $(2a + 1)(x - 2y - 3z)$

33 $(2x + 5)(x - y)$ **34** $(x - 4)(3x + y)$ **35** $(a + b)(2 - a)$

37 $(x - 1)(x + 3)$ **38** $(2x - 1)(x + 2)$ **39** $(1 - x)(2 + 3x)$

41 $(x + 1)(x^2 + 1)$ **42** $(x - 2)(x^2 - 3)$ **43** $x(x - 1)(x^2 + 2)$

EXERCISE 2.5

1 Yes, PS   **2** No   **3** Yes   **5** Yes   **6** No

7 No   **9** No   **10** Yes, PS   **11** Yes   **13** Yes, PS

14 Yes   **15** Yes   **17** $(x - 3)(x + 5)$

18 $(x + 2)(x + 6)$ **19** $(x - 4)(x - 8)$ **21** $(y - 1)(2y + 3)$

22 $(y + 1)(3y - 4)$ **23** $(y + 2)(2y + 5)$ **25** $(2a + 1)(4a - 3)$

26 $(2a + 7)^2$ **27** $(5a - 2)(2a - 5)$ **29** $(6b + 7)(b + 3)$

30 $(8p - 1)^2$ **31** $(9r - 5)(6r - 13)$ **33** $(x - 2y)^2$

34 $(x - 2y)(2x - y)$ **35** $(2a - 5b)^2$ **37** $(4y - 3z)(4y - 5z)$

38 $(6x + 5y)(5x - 3y)$ **39** $(10a + 3b)(3a - 4b)$ **41** $(6x - 5y)^2$

42 $(8a + 5b)(12a - 11b)$ **43** $(9c + 8d)^2$ **45** $(x + y + 1)^2$

46 $(a - b + 1)^2$ **47** $(2x + y - 2)^2$ **49** $(y - 7)(y - 6)$

50 $(b - 4)(b + 3)$ **51** $(a + b + 2)(a + b - 1)$ **53** $(x - y)(2x + y - 1)$

54 $(2x + y)(x + y + 3)$ **55** $(a - 3b)(a - 3b + c)$ **57** $3(2a + 3b)(2a + b - 1)$

58 $4(x - y)(x - y + z)$ **59** $x(x + 2y)(x + 2y - 2)$

EXERCISE 2.6

9 $(a - 2)(a + 2)$ **10** $(b - 6)(b + 6)$

11 $(x - 2y)(x + 2y)$ **13** $(5x - 6y)(5x + 6y)$

14 $(11a - 17b)(11a + 17b)$ **15** $(3x^2 - 8y)(3x^2 + 8y)$

17 $(11h^6 - 2t^3)(11h^6 + 2t^3)$ **18** $(10x^{50} - 8y^{32})(10x^{50} + 8y^{32})$

19 $(8x^4 - 5z^5)(8x^4 + 5z^5)$ **21** $(a - 2)(a^2 + 2a + 4)$

22 $(b + 3)(b^2 - 3b + 9)$ **23** $(2a - 1)(4a^2 + 2a + 1)$

25 $(2x - 3y)(4x^2 + 6xy + 9y^2)$ **26** $(4a + 3b)(16a^2 - 12ab + 9b^2)$

27 $(5y + 2x)(25y^2 - 10yx + 4x^2)$ **29** $(x^2 + 1)(x^4 - x^2 + 1)$

30 $(y^3 + 2)(y^6 - 2y^3 + 4)$ **31** $(b^5 - 2c^3)(b^{10} + 2c^3b^5 + 4c^6)$

33 $(3a^9 + 6b^{72})(9a^{18} - 18a^9b^{72} + 36b^{144})$ **34** $(7x^3 + 3y^4)(49x^6 - 21x^3y^4 + 9y^8)$

35 $(8r^8 - 3s^2)(64r^{16} + 24r^8s^2 + 9s^4)$ **37** $(x - y)(x + y)(x^2 + y^2)$

38 $(9x^4 + 1)(3x^2 - 1)(3x^2 + 1)$

39 $(a + b)(a^2 - ab + b^2)(a^6 - a^3b^3 + b^6)$

41 $(x + y + 3)(x + y - 3)$

42 $(a + 2b + 4)(a + 2b - 4)$

43 $(x - 3 - y)(x - 3 + y)$

45 $(x + y + z)(x^2 + 2xy + y^2 - xz - yz + z^2)$

46 $(x + y - 2a)(x^2 + 2xy + y^2 + 2ax + 2ay + 4a^2)$

47 $(x + y + a + b)(x + y - a - b)$

49 $2(x^2 - 2)(x^2 + 2)$

50 $a(a - 2b)(a + 2b)$

51 $3c^2(2c - d)(2c + d)$

53 $2(a - 2b)(a^2 + 2ab + 4b^2)$

54 $3(3 + a^2)(9 - 3a^2 + a^4)$

55 $a(a - b)(a^2 + ab + b^2)$

57 $(a - b)(a + b + 1)$

58 $(x + 2y)(x - 2y - 1)$

59 $(2x - y)(2x + y + 2)$

61 $(a^2 - 1 + a)(a^2 - 1 - a)$

62 $(x - 2)(x + 2)(x - 1)(x + 1)$

63 $(y^4 + 3 - y^2)(y^4 + 3 + y^2)$

65 $(a^2 - ab + b^2)(a^2 + ab + b^2)$

66 $(x^2 + 2xy - 2y^2)(x^2 - 2xy - 2y^2)$

67 $(3m^2 + mn - n^2)(3m^2 - mn - n^2)$

EXERCISE 2.7

1 x^4

2 a^5

3 $-z^3$

5 $4a$

6 $3x^2$

7 $-8ac^2$

9 $-5d^3$

10 $-4r^2s^3t^2$

11 $6a^7b^3c^3$

13 $3a^4 + 4a^2$

14 $3b^4 - 2b^2$

15 $-4x^4 + 3$

17 $3x^3 - 2x^2y + y^2$

18 $3a^8b^4 - 2a^4b^2 - 1$

19 $-4r^6s^3 + 5r^3s^2t + 7t^2$

37 $28, 4$

38 $83, 6$

39 $429, 29$

41 $3a - 4, 2$

42 $2x + 5, 8$

43 $3b + 4d, 7d^2$

45 $2x^2 + x - 3, 5$

46 $7x^2 + 4x + 3, -5$

47 $2x^2 + 4xy - 7y^2, 0$

49 $3x - 1, 0$

50 $3t^2 + t + 2, 17$

51 $8y + 3z, 6z^3 + 2z^2$

53 (a) $3x + 4y, 7y^2$; (b) $11y - 11x, 28x^2$

54 (a) $4x + 5y, 6y^2$; (b) $3y - \frac{8}{3}x, \frac{8}{3}x^2$

55 (a) $4x^2 + xy - y^2, 2y^4$; (b) $y^2 + 3yx + 2x^2, -6yx^3 + 4x^4$

EXERCISE 2.8

1 $8, 4, 4, 6$

2 $4x + 6y$

3 $-27x - 2y$

5 $-4x^3 + 3x^2 + 10x + 2$

6 $-7a + 7b + 2ab$

7 $-27x^3y^5z + 12x^3y^8z^3$

9 $15x^3 + 16x^2 - 6x + 2$

10 $14x^3 - 27x^2y - 22xy^2 + 5y^3$

11 $16a^2 + 26ab + 52ac - 86bc - 35b^2 - 48c^2$

13 $9x^2 - 12xy + 4y^2$

14 $27x^3 - 54x^2y + 36xy^2 - 8y^3$

15 $81x^4 - 216x^3y + 216x^2y^2 - 96xy^3 + 16y^4$

17 4836

18 $x^4 - 6x^3 + 11x^2 - 6x + 1$

19 $3x(6xy - 2yz^2 + 3z^2)$

21 $(6x - 5)(a - 2b)$

22 $(a - 4b)(2x + 7y - 3)$

23 $2(a - b)(4x - y)$

25 $(2x^2 + 3)(x - 5)$

26 $(4x - 5)(6x + 5)$

27 $(6x - 5y)(4x + 3y)$

29 $(3x + y - 1)^2$

30 $x(x + y)(x + y - 2)$

31 $2x(4x - 7y)(4x + 7y)$

33 $(4a^3 - b^4)(16a^6 + 4a^3b^4 + b^8)$

34 $(x^3 - 2)(x^6 + 2x^3 + 4)$

35 $(x^4 - 3)(x^4 + 3)$

37 $(x^2 + 2y^2 + xy)(x^2 + 2y^2 - xy)$

38 $x + 2, -4$

39 $5x + 3y, -3y^2$

41 $x^2 + 4, -3$

EXERCISE 3.1

1 $\frac{5}{7}$

2 $\frac{1}{15}$

3 $\frac{8}{13}$

5 $\frac{x}{y^2}$

6 $\frac{ac}{b}$

7 $\frac{y^2}{xz^2}$

9 $\frac{4}{x + y}$

10 x^2

11 $\frac{5xy}{1 + 2y}$

13 $\frac{x + 1}{2x + 1}$

14 $\frac{x - 2y}{2x + y}$

15 $\frac{a - 2b}{2a - b}$

17 $\frac{2a + b}{3a - b}$

18 $\frac{2a + 3b}{3a + 2b}$

19 $\frac{1}{2a + 1}$

21 $6x^2 + 3xy$

22 $4a^2b + 10ab^2$

23 $-a - b$

25 5

26 2

27 $x + 2$

29 False

30 True

31 True

33 False

34 False

35 False

37 $\frac{1}{3}$

38 $\frac{3}{4}$

39 $\frac{12}{25}$

41 $\frac{5}{6b^5}$

42 $\frac{3y^6}{7x^6}$

43 $\frac{x^7y^{10}}{6}$

45 x^6

46 $\frac{x}{y^5z}$

47 $\frac{a^4}{b^2}$

49 $\frac{x(x - 2y)}{6(x + 2y)}$

50 $\frac{a(a - c)}{c(a + c)}$

51 $\frac{x(x - 3y)}{16(x + y)}$

53 b

54 $\frac{-y}{x}$

55 $\frac{xy}{(x - y)(x^3 + y^3)}$

57 $\frac{x + y}{x - y}$

58 1

59 $\frac{x^2 + xy + y^2}{x}$

61 $a + 3b$

62 -1

63 $\frac{x - 3}{x + 2}$

EXERCISE 3.2

1 $2x^2yz^4$

2 abc

3 $72x^2yz$

5 $(x - y)^2(x + y)$

6 $(x^2 + y^2)(x + y)^2(x - y)^2$

7 $(c + d)^3(c - 3d)(c - d)^2$

9 $(x - y)(x + y)^3(x + 2y)^2$

10 $(a - 2b)^2(a + 2b)^2(a - b)$

11 $(a + 2b)(a - b)(a + b)(a - 2b)$

13 $\frac{17}{6}$

14 $\frac{-1}{2}$

15 $\frac{23}{40}$

17 $\frac{a + 12}{9}$

18 $\frac{114b - 327}{175}$

19 $\frac{8x - 11y}{36}$

21 $\frac{5x^2 + 9y^2 - 4z^2}{30xyz}$

22 $\frac{12c^2 + 10d^2 - 7e^2}{28cde}$

23 $\frac{7}{6r^2}$

25 $\frac{3y^2}{2x(x + y)}$

26 $\frac{a}{b}$

27 $\frac{-w^2}{3z(z - w)}$

29 $\frac{3a + 2b}{a + b}$

30 $\frac{-a}{a - b}$

31 $\frac{x}{x - 2y}$

33 $\frac{2a^2}{b(a^2 - b^2)}$

34 $\dfrac{42c^2}{(3c-2d)(c+2d)(2c+d)}$ **35** $\dfrac{12t^2}{(r-2t)(r+t)(r+2t)}$

37 $\dfrac{9r^2}{(p+r)(2p-r)(p-2r)}$ **38** $\dfrac{6}{(h+1)(h-1)(2h+1)}$ **39** $\dfrac{6a^2}{(a-b)(a+b)(2a+b)}$

41 $\dfrac{1}{c-2d}$ **42** $\dfrac{2}{x+2y}$ **43** 2

45 $\dfrac{2r-t}{r+2t}$ **46** $-\dfrac{w+z}{2w+z}$ **47** $-\dfrac{2a+b}{a+2b}$

49 $\dfrac{1}{x}$ **50** $\dfrac{c}{d(2c-3d)}$ **51** $\dfrac{3}{b}$

53 $\dfrac{4(x-2y)}{3(x^2-y^2)}$ **54** $\dfrac{a-b}{(3a-b)(a+b)}$ **55** $\dfrac{c+d}{(c-d)(3c+2d)}$

57 $\dfrac{w+z}{w-z}$ **58** $\dfrac{1}{3h+k}$ **59** $\dfrac{3(r+7s)}{(3r-2s)(2r-5s)}$

EXERCISE 3.3

1 $\frac{15}{11}$ **2** 7 **3** $\frac{22}{9}$ **5** $\dfrac{-5}{4}$

6 $\frac{23}{51}$ **7** $\dfrac{-3}{14}$ **9** $\dfrac{x}{2x-1}$ **10** $\dfrac{a+1}{a}$

11 $2x-1$ **13** $3x^2+1$ **14** $\dfrac{d-8}{5}$ **15** $\dfrac{2a-2b}{3a+2b^2}$

17 $2h+3$ **18** $\dfrac{a+3b}{2}$ **19** $\dfrac{2(x-y)}{x+y}$ **21** $(a-2)(a-1)$

22 $\dfrac{a+b}{3a-b}$ **23** $\dfrac{x}{x+y}$ **25** $\dfrac{x+2}{x-1}$ **26** $\dfrac{x-1}{x+3}$

27 $\dfrac{x-2}{2x-1}$ **29** $\dfrac{x^2+xy+y^2}{x+y}$ **30** $\dfrac{c-2d}{2c+3d}$ **31** $\dfrac{3b-a}{2(a+2b)}$

33 $\dfrac{-1}{w+1}$ **34** $\dfrac{-a-3}{2(a+1)}$ **35** $\dfrac{2(x-y)}{x+y}$ **37** $1-a$

38 $3-a$ **39** $\dfrac{x+y}{x-y}$

EXERCISE 3.4

1 $\dfrac{x}{yz^2}$ **2** $\dfrac{2x-y}{x-4y}$ **3** $\dfrac{a-2b}{a+b}$ **5** False

6 True **7** $\dfrac{2ab^2}{9c^5}$ **9** $\dfrac{3x-2y}{3x-y}$ **10** $\dfrac{1}{x}$

11 $12x^2y$ **13** $x(x-1)(x+2)^2(x+3)$ **14** $\dfrac{13x^2-25x+36}{105}$ **15** $\dfrac{8c-5b+3a}{abc}$

17 $\dfrac{50x^2-53xy+11y^2}{(x-y)(2x-y)(3x-y)}$ **18** $\dfrac{-4(a+17)}{(2a+5)(3a-1)(a-3)}$

19 $\dfrac{-2b(30a + b)}{(4a + b)(a + 3b)(2a + b)}$

21 $\dfrac{x^2 - 2}{x^2 + 2}$

22 $\dfrac{5(8a - b)}{15a + 2b^2}$

23 $\dfrac{4a - 15b}{4a - 20b - 3}$

25 $\dfrac{x^2 + 9xy - 3y^2}{5x^2 - 2xy - 39y^2}$

EXERCISE 4.1

1 $\{2, 4, 6, 8\}$ **2** $\{1, 2, 6, 10\}$ **3** $\{\text{Alaska, Hawaii}\}$

5 $\{\text{Eisenhower, Kennedy, Johnson, Nixon, Ford, Carter}\}$

6 $\{\text{Neil Armstrong}\}$ **7** $\{\text{Oriole, Cardinal}\}$ **9** $\{2, 5\}$

10 $\{\{2, 5\}\}$ **11** $\{285\}$ **13** Conditional

14 Conditonal   **15** Identity   **17** Conditional

18 Identity   **19** Conditional   **29** Yes

30 No   **31** Yes   **33** Yes

34 No   **35** Yes   **37** No

38 No   **39** No

EXERCISE 4.2

1 $\dfrac{-5}{2}$ **2** $\frac{2}{3}$ **3** 7 **5** 3 **6** -1 **7** -4

9 4 **10** -5 **11** $\dfrac{-7}{5}$ **13** $\dfrac{-13}{4}$ **14** -34 **15** $\dfrac{-21}{2}$

17 4 **18** 3 **19** -2 **21** $-\frac{5}{3}$ **22** $\frac{2}{3}$ **23** $\frac{3}{2}$

25 3 **26** -5 **27** -1 **29** 4 **30** 6 **31** $\frac{1}{2}$

33 12 **34** 18 **35** 36 **37** a^2 **38** $2ab$ **39** $\dfrac{b^2}{a}$

41 $8, -2$ **42** $8, 2$ **43** $-8, 4$ **45** $3, -2$ **46** $2, -7/2$ **47** $2, -2/3$

EXERCISE 4.3

1 4 **2** $\frac{5}{2}$ **3** -11 **5** -1 **6** 3 **7** 9

9 5 **10** 8 **11** 4 **13** 2 **14** -2 **15** 7

17 -2 **18** -4 **19** -3 **21** 4 **22** 3 **23** 7

25 1 **26** 3 **27** 3 **29** 5 **30** -4 **31** -7

33 6 **34** 8 **35** 5 **45** 0 **46** 0 **47** 1

49 $\dfrac{-5}{2}$ **50** $\dfrac{-11}{2}$ **51** $\frac{7}{3}$

EXERCISE 4.4

1 23, 24, 25 **2** 5, 6, 9 **3** $36, $54 **5** $173, $386 **6** 140 hours

7 Lucy, 8.8 pounds; Dean, 12.2 pounds

9 Sue, 4 hours; Jack, 6 hours; Joe, 11 hours **10** 35 miles per hour

11 $104 **13** 4 **14** $100 **15** 8 hours **17** 8 **18** 16,000

19 1000 **21** 12, 24, 6 **22** 50 quarters, 65 dimes, 80 nickels

23 73°F **25** 3 miles **26** $6\frac{1}{2}$ hours

27 50 miles per hour, freeway; 15 miles per hour, suburbs

29 $6\frac{6}{7}$ hours **30** $1\frac{1}{5}$ hours **31** $4\frac{4}{9}$ hours **33** 7 hours

34 4 hours **35** 11:00 A.M. **37** 6 hours **38** 2 hours

39 6 **41** 0.4 gallons **42** 25 milliliters

43 30 miles per hour, 45 miles per hour **45** 112 miles per hour

46 25 miles per hour **47** $50

EXERCISE 4.5

1 Water **2** Conditional equation **3** True **6** Yes **7** Yes

9 5 **10** 7 **11** 12 **13** a/b **14** $1, -5/3$

15 5 **17** 3 **18** 7 **19** 4 **22** $2x + 3$

23 $\dfrac{x + 3}{2}$ **25** 6 **26** 18 hours

27 2.4 liters of 7% solution and 3.6 liters of 12% solution

EXERCISE 5.1

1 32 **2** 729 **3** 125 **5** a^8 **6** a^2

7 a^5 **9** 8 **10** 27 **11** 5 **13** b^3

14 b^5 **15** b^5 **17** 729 **18** 729 **19** 16

21 a^6 **22** a^8 **23** a^8 **25** $6a^5$ **26** $10a^7$

27 $12a^7$ **29** $4a^2$ **30** $3a^3$ **31** $5a^4$ **33** $4a^6$

34 $27a^6$ **35** 1 **37** a^4b^6 **38** a^9b^6 **39** $a^{12}b^3$

41 a^6/b^9 **42** a^8/b^2 **43** $1/b^5$ **45** $4xy^2$ **46** $5xy^2$

47 $3a^2b^3$ **49** $12x^5y^7$ **50** $6x^5y^4$ **51** $30x^2y^5$ **53** abc^2d

54 $9a^2bcd^3$ **55** $6xy^3z^2w$ **57** $9b^4/a^4$ **58** $125a^{15}b^6$

59 $16a^4/b^4$ **61** $16a^4b^2/c^2$ **62** $45xy^4/14z^4$ **63** $27b^3/8$

65 $9x^2y^8/4$ **66** x^{12}/y^{12} **67** a^{10}/y^{10} **69** a^{4x-1}

70 a^{x+1} **71** b^{3x} **73** a^{x+3} **74** a^{2x+2}

75 c^{1-2y} **77** x^{2a}/y^2 **78** $a^{3b+9}c^3$ **79** $a^{s+4}b^{s+3}$

EXERCISE 5.2

1 $\frac{1}{8}$	2 $\frac{1}{9}$	3 $\frac{1}{64}$	5 $\frac{1}{27}$	6 $\frac{1}{256}$
7 5	9 4	10 $\frac{1}{25}$	11 27	13 $\frac{1}{9}$
14 $\frac{1}{9}$	15 1	17 $\frac{4}{81}$	18 $\frac{9}{16}$	19 $\frac{1}{1024}$
21 $\frac{4}{3}$	22 5184	23 $\frac{9}{256}$	25 a^2b^{-3}	26 a^3b^{-4}
27 $a^{-2}b^{-1}$	29 $2^{-1}ab^{-1}$	30 $a^{-3}b^{-3}$	31 $6v^3sr^{-2}$	33 $3/a^4$
34 $1/2a^2$	35 $1/8x$	37 x	38 $1/x^4$	39 $1/a$
41 y^4/x^2z	42 y^2z^3/x	43 dq/p^4	45 $16e^3/3th$	46 $ha/2m$
47 $e^2at^3/2$	49 a^8e^4/t^2	50 c^3a^3/r^3	51 h^3/a^2t	53 t^6/a^6
54 $m^2/9a^4$	55 $36h^2/e^2$	57 $(1-x^2)/x$	58 $(1+x^4)/x^2$	

59 $\dfrac{a^2+b^2}{ab}$ 61 $\dfrac{1}{y-x}$ 62 $\dfrac{b+a}{ab}$ 63 $\dfrac{1}{b+a}$

65 $\dfrac{-(x-5)(x-1)}{(x+1)^4}$ 66 $\dfrac{(x-2)^2(x+13)}{(x+3)^3}$ 67 $\dfrac{-(3x+2)}{(x+2)^5}$

69 $\dfrac{2(3x+2)(3x-5)}{(2x-1)^2}$ 70 $\dfrac{(6x+31)(3x-1)^2}{(2x+3)^3}$ 71 $\dfrac{32x+46}{(4x-7)^4(2x+5)^2}$

EXERCISE 5.3

1 3	2 4	3 -2	5 4	6 32
7 27	9 125	10 8	11 9	13 $\frac{1}{4}$
14 $\frac{1}{125}$	15 $\frac{1}{4}$	17 0.3	18 0.4	19 0.2
21 $6xy$	22 $12xy$	23 $10x$	25 $32ab^{11/6}$	26 $18a^{14/15}b^{13/15}$
27 $40a^{13/20}b^{23/20}$	29 $9x^{4/15}y^{1/20}/8$	30 $2x^{8/15}y^{5/12}/x$	31 $x^{20/21}y^{9/10}/3x$	
33 $3x^{4/5}y^{3/4}$	34 $2x^{2/3}y^{2/5}/3$	35 $x^{31/42}y^{1/60}/3$	37 $4y^{2/5}x^{2/3}$	
38 $81y/x^2$	39 $25x^{8/5}y^{2/3}/y^2$	41 $3\sqrt{3}\,b^{4/5}/b$	42 $16x^{1/5}y^3/x^2$	
43 $8\sqrt{2}\,a^{1/5}b^{4/3}/a$	45 $16x^{4/3}y^{2/3}/81y$	46 $4x^{2/3}y^{2/3}/9x$	47 $5x^{1/4}y^{1/3}/4y$	
49 a	50 $1/a^x$	51 a^{x-y}	53 $x-2x^{1/2}y^{1/2}+y$	
54 $x-y$	55 $x-y$	57 $1/(b-a)$	58 $(b-a)/ab$	

59 $1/(a^2+b^2)^2$ 61 $\dfrac{(9x+4)(x+1)^{1/2}}{x+1}$ 62 $\dfrac{(5x-1)(x-2)^{2/3}}{x-2}$

63 $\dfrac{(14x-11)(2x-3)^{2/5}}{2x-3}$ 65 $\dfrac{(5x-7)(2x-1)^{1/2}(x-2)^{2/3}}{(2x-1)(x-2)}$

66 $\dfrac{(7x+1)(4x-3)^{1/4}(3x+4)^{1/3}}{(4x-3)(3x+4)}$ 67 $\dfrac{(9x-13)(5x-3)^{2/5}(2x-5)^{1/2}}{(5x-3)(2x-5)}$

EXERCISE 5.4

1 11	2 6	3 8	5 4	6 3
7 2	9 $3\sqrt{2}$	10 $5\sqrt{3}$	11 $3\sqrt[3]{4}$	13 18
14 70	15 60	17 $4\sqrt{15}$	18 $12\sqrt{14}$	19 $14\sqrt{3}$

21 $\frac{3}{2}$ **22** $\frac{5}{7}$ **23** $\frac{2}{3}$ **25** $\dfrac{\sqrt[3]{9}}{6}$ **26** $\frac{2}{3}$

27 $2\sqrt[3]{6}/3$ **29** $3xy\sqrt{2y}$ **30** $5x^2y^2\sqrt{3x}$ **31** $3xy\sqrt[3]{4y^2}$

33 $2xy\sqrt[4]{2y^3}$ **34** $3xy\sqrt[4]{2xy^2}$ **35** $2xy\sqrt[5]{3xy^4}$ **37** $0.4ab\sqrt{b}$

38 $3a\sqrt{b}/b$ **39** $3ab\sqrt{b}/4$ **41** $10y\sqrt{6x}$ **42** $30x^2y\sqrt{x}$

43 $8y\sqrt{6y}/x^2$ **45** $\sqrt{x}/4x^2y$ **46** $x^3/3y^4$ **47** $5\sqrt{xy}/3xy$

49 $y\sqrt{x}/2$ **50** $70\sqrt{3y}/3xy$ **51** $0.3y\sqrt[3]{x^2}$ **53** $\sqrt{5}$

54 $\sqrt{6}$ **55** $\sqrt{5}$ **57** $2\sqrt{2}$ **58** $\sqrt[4]{8}$

59 $\sqrt{2}$ **61** $x\sqrt{3xy}$ **62** $y\sqrt{5xy}$ **63** $y\sqrt{3x}$

65 $\sqrt[4]{8x^2y}$ **66** $\sqrt[3]{4xy^2}$ **67** $y\sqrt{2xy}$ **69** y/a

70 1 **71** $\sqrt[4]{1715}/7$

EXERCISE 5.5

1 -1 **2** 4 **3** 3 **5** $-5\sqrt{6}$

6 $11 - 5\sqrt{10}$ **7** $-1 + 5\sqrt{35}$ **9** 31 **10** -3

11 $7 - 5\sqrt{3}$ **13** $2\sqrt{6}$ **14** $7 - 6\sqrt{2}$ **15** $-1 + 2\sqrt{15}$

17 $-7 - 4\sqrt{3}$ **18** $-8 + 3\sqrt{7}$ **19** $(19 - 8\sqrt{3})/13$ **21** $(4 - 3\sqrt{15})/17$

22 $(-29 + 7\sqrt{35})/23$ **23** $3 + 2\sqrt{3}$

25 $(3 + 3\sqrt{5} + 5\sqrt{3} + \sqrt{15})/12$ **26** $(3 + 3\sqrt{2} - 2\sqrt{3} - \sqrt{6})/3$

27 $(7\sqrt{2} + 2\sqrt{7} - 2\sqrt{21} - 2\sqrt{6})/5$ **29** $6\sqrt{3}$

30 $2\sqrt{2}$ **31** $3\sqrt{5}$ **33** 0

34 $2\sqrt[3]{3}$ **35** $7\sqrt[3]{5}$ **37** $5(\sqrt{3} + \sqrt[3]{3})$

38 $8(\sqrt{2} + \sqrt[3]{2})$ **39** $2\sqrt{5} + 11\sqrt[3]{2}$ **41** $3\sqrt{3} + 2\sqrt[3]{3}$

42 $6\sqrt{2} + 3\sqrt[3]{2}$ **43** $2\sqrt{2}$ **45** $2\sqrt{2y}(x - y)$

46 $\sqrt{3xy}(11x + 2y)$ **47** $-\sqrt{5x}(4x + 3y)$ **49** $5x^2y^3\sqrt{xy}$

50 $5x^2y\sqrt{2x} - x^2y\sqrt{2xy}$ **51** $2x^2\sqrt[3]{3xy^2}(1 - 6xy^2)$ **53** $1.5xy\sqrt{6}$

54 $5\sqrt{3y}/6x$ **55** $(x^2 - 1)(\sqrt{3x} + \sqrt[3]{3x^2})/x^2$

EXERCISE 5.6

1 a^8 **2** a^9 **3** a^4 **5** a^6

6 1 **7** $6a^5$ **9** $16a^6$ **10** a^9/b^6

11 $8x^4y^4$ **13** $a^{2b-4}c^{2b+10}$ **14** a^{3x}/b^3 **15** $\frac{1}{16}$

17 $\frac{1}{5}$ **18** 1 **19** $10xy^2z^2$ **21** $\dfrac{(5x - 1)(x - 1)}{(x + 1)^4}$

22 $\dfrac{(3x + 2)\sqrt{(x + 3)(2x - 1)}}{(x + 3)(2x - 1)}$ **23** $6x^{7/12}y^{5/12}$ **25** $b^{4/3}/a^{9/2}$

26 a^x **27** $3xy^2\sqrt[4]{2xy}$ **29** 2

30 $2\sqrt{15} - 10\sqrt{3} - 3 + 3\sqrt{5}$ **31** $\dfrac{3\sqrt{35} - 17}{2}$ **33** $\dfrac{3\sqrt{2} - 4}{2}$

EXERCISE 6.1

1 $3, -3$
2 $4, -4$
3 $6, -6$
5 $1/7, -1/7$
6 $1/8, -1/8$
7 $.1, -.1$
9 $3/2, -3/2$
10 $4/3, -4/3$
11 $5/4, -5/4$
13 $2, 3$
14 $3, 4$
15 $1, 5$
17 $-1, -3$
18 $-2, -3$
19 $-1, -4$
21 $2, -3$
22 $1, -5$
23 $2, 4$
25 $1/2, -1/3$
26 $1/2, -1/4$
27 $1/3, -1/2$
29 $3/2, -2/3$
30 $2/5, -1/2$
31 $1/4, -4/3$
33 $3/5, -5/3$
34 $-2/3, -3/4$
35 $3/2, -1/3$
37 $1/2, -2/5$
38 $1/5, -3/2$
39 $4/3, 3/4$
41 $3/4, -1/4$
42 $-3/5, -5/3$
43 $5/3, -2/3$
45 $2/7, -3/2$
46 $5/6, -5/2$
47 $1/3, -5/4$
49 $4d/3, -3d/2$
50 $3d/4, -d/2$
51 $3d/2, -2d$
53 $-2/3a, 5/2a$
54 $3/2a, -4/3a$
55 $2/5a, -3/2a$

EXERCISE 6.2

1 $2, 3$
2 $1, 4$
3 $3, 4$
5 $-3, -2$
6 $-5, -1$
7 $-3, -1$
9 $-1, 2$
10 $-2, 3$
11 $-2, 5$
13 $-1/2, 2$
14 $-3/2, 1$
15 $-1, 4/3$
17 $1/4, 1$
18 $-2/3, 4$
19 $-2, 4/3$
21 $1/2, 2/3$
22 $-3/2, -1/3$
23 $-4/3, 3/4$
25 $-5/2, 2/5$
26 $-7/4, 1/2$
27 $-2/3, 3/5$
29 $2 \pm \sqrt{3}$
30 $3 \pm \sqrt{2}$
31 $5 \pm \sqrt{3}$
33 $(1 \pm \sqrt{3})/2$
34 $(-2 \pm \sqrt{2})/3$
35 $(1 \pm \sqrt{3})/5$
37 $1 \pm i$
38 $2 \pm i$
39 $3 \pm 2i$
41 $(1 \pm i)/2$
42 $(2 \pm i)/2$
43 $(4 \pm 2i)/3$
45 $a, -5a$
46 $2m, -3m$
47 $-b, 2c$
49 $-b/a, 2b/a$
50 $-3a/b, 2a/b$
51 $(a + 2b)/b, 1$

EXERCISE 6.3

1 $-3, 1$
2 $-1, 4$
3 $-4, 2$
5 $-3/2, 1$
6 $-2, 2/3$
7 $-3, 1/3$
9 $1/3, 3/2$
10 $-2/3, 1/4$
11 $-7/5, 1/2$
13 $1 \pm \sqrt{2}$
14 $2 \pm \sqrt{3}$
15 $-3 \pm \sqrt{5}$
17 $1 \pm \sqrt{3}$
18 $-2 \pm \sqrt{2}$
19 $3 \pm \sqrt{7}$
21 $(2 \pm \sqrt{3})/3$
22 $(3 \pm \sqrt{2})/2$
23 $(-3 \pm \sqrt{5})/4$
25 $1 \pm i$
26 $2 \pm i$
27 $3 \pm 2i$
29 $-3 \pm 2i$
30 $-2 \pm 3i$
31 $-4 \pm 2i$
33 $1 \pm \sqrt{3}i$
34 $-2 \pm \sqrt{2}i$
35 $-3 \pm \sqrt{5}i$
37 $(-3 \pm \sqrt{2}i)/2$
38 $(-2 \pm \sqrt{3}i)/3$
39 $(1 \pm \sqrt{6}i)/3$
41 $a \pm b$
42 $-2a \pm b$
43 $r \pm 2s$
45 $(c + 2d)/3, 3d$
46 $(a + 2b)/2, (a - 2b)/3$
47 $b/d, -2b/5d$
49 $3/(2m + n), -2/(m + n)$
50 $-1/(3m - n), 1/(m - n)$
51 $-2/(a - 2b), 3/(2a + b)$

EXERCISE 6.4

1 $\pm 1, \pm 3$ **2** $\pm 2, \pm i$ **3** $\pm 2, \pm 2i$

5 $1, 2, -1 \pm i\sqrt{3}, (-1 \pm i\sqrt{3})/2$

6 $1, -3, \dfrac{-1 \pm i\sqrt{3}}{2}, \dfrac{3 \pm 3i\sqrt{3}}{2}$

7 $-3, -1, \dfrac{1 \pm i\sqrt{3}}{2}, \dfrac{3 \pm 3i\sqrt{3}}{2}$

9 $\pm 2, \pm 1, \pm 2i, \pm i$

10 $\pm 3, \pm 1, \pm 3i, \pm i$

11 $\pm\sqrt{2}, \pm 1/2, \pm\sqrt{2}i, \pm i/2$

13 $1/2, 1/3$

14 $1, -1/6$

15 $-3, -1/4$

17 $1, -1/2, \dfrac{1}{1 \pm i\sqrt{3}}, \dfrac{2}{-1 \pm i\sqrt{3}}$

18 $-1, 1/3, \dfrac{2}{-3 \pm 3i\sqrt{3}}, \dfrac{2}{1 \pm i\sqrt{3}}$

19 $1/2, 2, \dfrac{4}{-1 \pm i\sqrt{3}}, \dfrac{1}{-1 \pm i\sqrt{3}}$

21 $\pm 1/2, \pm i$

22 $\pm 1/3, \pm i$

23 $\pm 2, \pm 1/3$

25 $\pm 2, \pm 1$

26 $\pm 1, \pm 2i$

27 $\pm 3i, \pm i$

29 $-1, -3/2, (-5 \pm\sqrt{33})/4$

30 $(-1 \pm\sqrt{13})/6, (-1 \pm\sqrt{37})/6$

31 $2/5, -1, (-3 \pm\sqrt{29})/10$

33 $3, -1/2, 3/2, 1$

34 $-1, -3, -2 \pm\sqrt{7}$

35 $-2, -1, (-3 \pm\sqrt{17})/2$

37 $0, -2, -3 \pm\sqrt{5}$

38 $\pm\sqrt{5}, (1 \pm\sqrt{29})/2$

39 $0, 3.5, (1 \pm i\sqrt{23})/4$

41 $1/4, 1$

42 $4, -4/13$ **43** $-6, 2/3$

45 $-1, 3$ **46** $11, 6$

47 $5/3, 2/3$ **49** $-2, 1$

50 $-1, 11$ **51** $29/3$

EXERCISE 6.5

1 5 **2** 5 **3** $\varnothing$ **5** 3 **6** 7 **7** 1

9 5 **10** -1 **11** 5 **13** $-3, -2$ **14** 2 **15** $0, 3$

17 4 **18** -1 **19** -2 **21** 1 **22** -3 **23** 2

25 $5, -2$ **26** $4, -3$ **27** $-1, 2$ **29** 9 **30** $6, 0$ **31** $1/2, 1$

33 9 **34** $5, 21$ **35** $1, 5$ **37** 3 **38** 4 **39** 2

41 $0, -1/3$ **42** 2 **43** $3/2, 1$ **45** 2 **46** 5 **47** 4

EXERCISE 6.6

1 4, 4, rational and equal

2 6, 9, rational and equal

3 -2, 1, rational and equal

5 4/3, 4/9, rational and equal

6 $-3/2$, 9/16, rational and equal

7 -5, 25/4, rational and equal

9 0, 4, conjugate complex

10 0, 9, conjugate complex

11 0, 16, conjugate complex

13 4/3, 5/9, conjugate complex

14 3, 5/2, conjugate complex

15 -1, 13/16, conjugate complex

17 7/2, 3, rational and unequal

18 5/3, 2/3, rational and unequal

19 −5/3, −4, rational and unequal

21 7/6, 1/3, rational and unequal

22 11/15, 2/15, rational and unequal

23 1/15, −2/5, rational and unequal

25 2, −1, irrational and unequal

26 4, 1, irrational and unequal

27 −6, 6, irrational and unequal

29 4/3, 2/9, irrational and unequal

30 3, 1, irrational and unequal

31 −2/5, −1/5, irrational and unequal

33 $x^2 - 4x + 4 = 0$

34 $x^2 - 6x + 9 = 0$

35 $x^2 + 10x + 25 = 0$

37 $6x^2 - x - 2 = 0$

38 $20x^2 - 23x + 6 = 0$

39 $10x^2 - x - 3 = 0$

41 $x^2 - 2x + 5 = 0$

42 $x^2 - 4x + 13 = 0$

43 $4x^2 - 12x + 11 = 0$

45 $x^2 - x - 1 = 0$

46 $4x^2 - 12x + 7 = 0$

47 $49x^2 + 42x + 4 = 0$

EXERCISE 6.7

1 5, 6

2 8, −7

3 8, 10; −6, −8

5 5, 10

6 3/2, 2/3

7 5, 10; −10, −5

9 50 by 50 yards

10 15 by 20 feet

11 10 feet

13 200 miles per hour

14 180 miles per hour

15 35 miles per hour

17 4, 6

18 5 hours, 7.5 hours

19 10, 15

21 4 by 12 feet

22 28 by 21 feet

23 1 inch

25 3 by 24 feet

26 30, 40

27 4

29 6, 8

30 50, 60

31 45 cubic feet per minute

EXERCISE 6.8

1 2, 5

2 1, −3

3 4, 3/2

5 −3/2, −4/3

6 2/3, −3/4

7 2, −3/2

9 $1 \pm \sqrt{3}$

10 $(-3 \pm \sqrt{2})/2$

11 $(3 \pm 2i)/2$

13 3, −2/3

14 3/2, −5/3

15 $(3 \pm \sqrt{6})/2$

17 $(1 \pm 3i)/3$

18 $(-3 \pm 2i)/4$

19 $-2b/a, 3b/a$

21 $\pm\sqrt{3}/2, \pm 2i$

22 4, 3/2, 1, −3/2

23 1/7, 3

25 1, −5/11

26 2

27 5/4, −3/8, rational, unequal

29 −4/3, 7/9, conjugate complex

30 1/2, −5/16, irrational, unequal

31 10 days, 15 days

EXERCISE 7.1

1 No, more than one second member for some first member

2 Yes, no first number has more than one second member

3 Yes, no first member has more than one second member

5 {(2, 3), (4, 7), (8, 13), (16, 21)}

6 {(b, l), (a, a), (s, b)}, yes

7 {(s, c), (h, l), (e, i), (e, f), (r, f)}; there are two second elements i and f for the first element e

9 Yes, no first element has more than one second element

10 Yes, no first element has more than one second element

11 No, some first elements have two or more second elements

13 −3, 1, 7   **14** 1, −9, −29   **15** 0, −2, 10

17 $R = \{3, 7, 11, 15\}$ **18** $R = \{-5, -2, 1, 7\}$ **19** $R = \{3, \sqrt{7}, 2, 0\}$

21 $\{(-1, -5), (0, -2), (1, 1), (2, 4)\}$, yes

22 $\{(-5, -9), (-3, -5), (-1, -1)\}$, yes

23 $\{(-2, 1), (-1, -2), (0, -3), (1, -2), (2, 1)\}$, yes

25 $\{(-1, 6), (0, 4), (2, 6), (4, 16)\}$, yes   **26** $\{(-1, 3), (0, 5), (1, 5), (2, 9)\}$, yes

27 $\{(-3, -11), (-1, -1), (0, 4), (1, 9), (2, 14)\}$, yes   **29** 3   **30** 2   **31** $2x + h - 2$

33 $f \cup g = \{(-3, -5), (-1, -3), (-1, -1), (3, 1), (3, 7), (5, 3), (5, 11)\}$; $f \cap g = \{(-3, -5)\}$

34 $f \cup g = \{(-4, -3), (-4, -11), (-2, -1), (-2, -7), (1, 2), (1, -1), (4, 5)\}$; $f \cap g = \{(4, 5)\}$

35 $f \cup g = \{(-2, 5), (-1, -2), (-1, 0), (0, -3), (0, -1), (1, 2)\}$, $f \cap g = \{(-2, 5), (1, 2)\}$

EXERCISE 7.2

2 (*a*) The bisector of the second and fourth quadrants; (*b*) the bisector of the fourth quadrant; (*c*) the *Y* axis; (*d*) the line three units above and parallel to the *X* axis

3 (*a*) The bisector of the first and third quadrants; (*b*) the bisector of the first quadrant; (*c*) the line parallel to and four units to the right of the *Y* axis; (*d*) the line parallel to and two units below the *X* axis

25 $x = -2, y = 6$ **26** $x = 2.5, y = -5$ **27** $x = -4, y = -4$

EXERCISE 7.3

29 $x = y + 1, y = x - 1$, all numbers ≥ 3

30 $x = -y + 3, y = -x + 3$, all nonnegative numbers

31 $x = 2y - 5, y = \frac{1}{2}x + 2.5$, all numbers ≤ 0

33 $x = (y + 1)/(y - 1), y = (x + 1)/(x - 1)$, all numbers ≤ 3 and > 1

34 . $x = (2y - 3)/(y + 2), y = (2x + 3)/(-x + 2)$, all numbers < 2 and ≥ -1.5

35 $x = (y + 3)/y, y = 3/(x - 1)$, all numbers > 1

37 $x = 2 + \sqrt{y - 1}, y = x^2 - 4x + 5$, all numbers ≥ 2

38 $x = 5 - \sqrt{y + 3}, y = x^2 - 10x + 22$, all numbers ≤ 5

39 $x = 3 - \sqrt{2y + 5}, y = 0.5x^2 - 3x + 2$, all numbers $\leq 3 - \sqrt{3}$

41 $x^2 = y^2 - 4$, a relation, all numbers $< \sqrt{21}$ and $> -\sqrt{21}$

42 $x^2 = 4 - y^2$, a relation, all numbers ≤ 2 and > -2

43 $x^2 = 4y - 12$, a function, all real numbers

EXERCISE 7.4

1 Yes, there is exactly one *y* for each *x*

2 Yes, there is exactly one *y* for each *x*

3 (1, 2), (3, 4), (5, 6), (7, 8), (9, 10), (11, 12)

5 Yes, there is exactly one *y* for each value of *x*

6 No, there is more than one y for some values of x

7 $\{-3, -1, 1, 3\}$ **9** $\{(-2, -1)\}$

23 $y = (x + 1)/2$, all nonnegative numbers

25 $y = \sqrt{4 - x^2}$, all numbers < 2 and ≥ 0 **26** $y = x^2 - 4x + 7$, numbers ≥ 2

EXERCISE 8.1

1 Independent   **2** Independent   **3** Dependent   **5** Dependent

6 Dependent   **7** Inconsistent   **9** Independent   **10** Dependent

11 Inconsistent   **21** $(4, 5)$ **22** $(2, -1)$ **23** $(-4, 2)$

25 $(1, 0)$ **26** $(-6, -4)$ **27** $(4, 1)$ **29** $(1, 7)$

30 $(-2, -5)$ **31** $(-4, 4)$ **33** $(3/2, 7/2)$ **34** $(5/2, 1)$

35 $(1, 5/2)$ **37** $(1/2, 1/2)$ **38** $(1, 3/2)$ **39** $(1, -1)$

EXERCISE 8.2

1 $(4, 3)$ **2** $(3, 2)$ **3** $(2, 1)$ **5** $(2, -1)$ **6** $(-3, 2)$

7 $(-3, 4)$ **9** $(4, 2)$ **10** $(8, -3)$ **11** $(3, 5)$ **13** $(8, -11)$

14 $(4, 5)$ **15** $(3, -7)$ **17** $(8, 5)$ **18** $(1, 1)$ **19** $(2, 1)$

41 $(1/2, 1/3)$ **42** $(1/4, -1/5)$ **43** $(1/4, 1)$

EXERCISE 8.3

1 $(4, 2, 3)$ **2** $(1, 5, 4)$ **3** $(-1, 2, 3)$ **5** $(3, 1, 3)$

6 $(4, -1, 1)$ **7** $(6, 1, -2)$ **9** $(6, 4, 3)$ **10** $(-3, -8, 6)$

11 $(7, -5, -3)$ **13** $(1, -2, 3)$ **14** $(2, 1, -3)$ **15** $(3, -4, 2)$

17 $(5, -6, 4)$ **18** $(8, 5, 6)$ **19** $(2, 4, 1)$ **21** $(3/2, 2/5, 4/3)$

22 $(7/8, 5/6, 3/4)$ **23** $(-3/4, 4/3, 5/2)$ **25** $(2, 1, 3)$ **26** $(-2, 3, 1)$

27 $(2/3, 3/4, 1/2)$ **29** $(4, 5, -7)$ **30** $(4, 3, 5)$ **31** $(2, 3, 6)$

33 $(a, b, a - b)$ **34** $(a + b, a - b, a^2 - b^2)$ **35** $(a, a + b, a + c)$

EXERCISE 8.4

1 48, 24   **2** 18, 6   **3** 36, 24   **5** 100, 150   **6** 20, 15

7 40 at \$35,000; 60 at \$30,000   **9** 2 miles per hour, 4 miles per hour

10 300 miles per hour, 20 miles per hour   **11** 200 miles per hour, 20 miles per hour

13 27 miles by bus, 800 miles by plane

14 Horseback, $2\frac{1}{2}$ hours; car, $\frac{1}{2}$ hours; plane, 1 hour

15 Morning, 150 miles; afternoon, 130 miles

17 300 by 200 feet   **18** 30 by 80 feet, 60 by 80 feet

19 20,000 square yards, 10,000 square yards

21 15 sedans, 20 sports cars, 10 station wagons

22 100 bales, $17,500

23 A $38.50, B $25.25

25 8 hours, 6 hours

26 Older girl, 4 hours; younger girl 5 hours

27 8 pounds of the $3.50 grade, 12 pounds of the $3 grade

29 Sue, $4\frac{1}{2}$ hours; Bill, $3\frac{3}{5}$ hours; Tom, 6 hours

30 24, 12, 16

31 5, 600, 4

EXERCISE 8.5

1 1	**2** -19	**3** 36	**5** 34
6 49	**7** -2	**9** -13	**10** 2
11 -10	**13** 5	**14** -17	**15** 39
17 -2	**18** 2	**19** 2	**29** $(4, 1)$
30 $(2, -1)$	**31** $(-3, 2)$	**33** $(4, -1)$	**34** $(-4, 3)$
35 $(-2, 3)$	**37** $(-1, 5)$	**38** $(-2, 4)$	**39** $(-4, 5)$
41 $(1/2, 3/5)$	**42** $(4/3, 1/4)$	**43** $(1/4, 2/3)$	**45** $(3/4, -1/4)$
46 $(-3/4, 1/7)$	**48** $(5/3, 3/5)$		

EXERCISE 8.6

1 9	**2** 26	**3** -15	**5** 0	**6** -26
7 0	**9** -9	**10** -26	**11** 15	**13** 9
14 0	**15** 26	**17** 80	**18** -3	**19** -95
29 $(1, 2, 1)$	**30** $(1, 2, 2)$	**31** $(1, 1, 2)$	**33** $(2, 3, 2)$	**34** $(2, 1, 2)$
35 $(0, 0, 3)$	**37** $(0, 2, 1)$	**38** $(2, 2, 1)$	**39** $(3, 0, 1)$	
41 $(1/3, 1/4, 1/3)$	**42** $(1/4, 1/2, 1/3)$	**43** $(2/3, 1/2, 1/3)$	**45** $(3/8, 4/5, 1/2)$	
46 $(2/3, 5/3, 3/2)$	**47** $(2/3, 1/7, 1/7)$	**49** 32	**50** 0 **51** 15/2	

EXERCISE 8.7

1 $4/6 = -14/(-21) = 6/9$

2 $9/(-12) = 12/(-16) \neq -18/30$

3 $(4, -1)$	**5** $(3, 5)$	**6** $(0, 3)$	**7** $(3, 7)$	**9** $(-4, 7)$
10 $(11, 8)$	**11** $(5, 1)$	**13** $(8, 7)$	**14** $(12, 9)$	**15** $(3, 8)$
17 $(-3, 10)$	**18** $(5, -4)$	**19** $(1, 2, 4)$	**21** $(1, 2, 5)$	**22** $(1/3, 1/4, 1)$
23 $(2/3, 1/4, -1)$		**25** $(4, 1)$		**31** $120 and $100 per month

33 700 $5 tickets, 70 $10 tickets, and 32 $25 tickets **34** 12

EXERCISE 9.2

1 $(1, 2), (9, -6)$

2 $(12, 5), (12, -5)$

3 $(1, 2), (1, -2), (4, 4), (4, -4)$

5 $(9, 3), (9, -3), (\frac{1}{4}, \frac{1}{2}), (\frac{1}{4}, -\frac{1}{2})$

6 $(1, 3), (1, -3), (-1, 3), (-1, -3)$

7 $(4, 6), (4, -6), (-4, 6), (-4, -6)$

9 $(1.5, 1.5), (0, -.5)$

10 $(3.5, 1.5), (0, 3.5)$

11 $(2, 3), (-2, -3)$

13 $(2.5, .5), (-4, -2)$

14 $(-.5, 1), (0, 1)$

15 $(1, 4), (-1, -4)$

17 Two   **18** Two   **19** Two   **21** Two   **22** None

23 Two   **25** Four   **26** Four   **27** Four

EXERCISE 9.3

1 $(2, 1), (19/4, -9/2)$

2 $(1, 3), (6/25, -4/5)$

3 $(1, 1), (17/16, 5/4)$

5 $(1, 1/3), (-5/9, -19/27)$

6 $(1, -1), (-17/13, -3/13)$

7 $(2, -2), (-2/3, -14/3)$

9 $(b, a), (-b, -a)$

10 $(3b, -b/a), (-b, 3b/a)$

11 $(-b/m, 0), (0, b)$

13 $(2, 1), (-1, -2)$

14 $(2, 3), (-3/2, -4)$

15 $(-1, 3), (-3, 1)$

17 $(2, 14), (-2, 10), (2i, -12 + 2i), (-2i, -12 - 2i)$

18 $(0, 1), (-2, -1), (1 + \sqrt{3}, \sqrt{3}), (1 - \sqrt{3}, -\sqrt{3})$

19 $(2, 1), (-1, -2)$

21 $(2, 1/2), (0, -1/2), (-3 + \sqrt{3}i, \sqrt{3}i/2), (-3 - \sqrt{3}i, -\sqrt{3}i/2)$

22 $(5, 1), (-3, -1), (14 + 4\sqrt{14}, \sqrt{14}), (14 - 4\sqrt{14}, -\sqrt{14})$

23 $(10, 2), (-2, -2), (5, 3/2), (-4, -3/2)$

25 $(2, 1), (-2, -1), (\sqrt{3}, 2\sqrt{3}/3), (-\sqrt{3}, -2\sqrt{3}/3)$

26 $(3, 1), (-3, -1), (\sqrt{6}/2, \sqrt{6}), (-\sqrt{6}/2, -\sqrt{6})$

27 $(2, 1), (-2, -1), (\sqrt{2}/2, 2\sqrt{2}), (-\sqrt{2}/2, -2\sqrt{2})$

29 $(1, 2), (-1, -2), (-2i, i), (2i, -i)$ **30** $(3/2, 1), (-3/2, -1), (1/2, 3) (-1/2, -3)$

31 $(1, 2), (-1, -2), (2\sqrt{2}, \sqrt{2}/2), (-2\sqrt{2}, -\sqrt{2}/2)$

EXERCISE 9.4

1 $(1, 0), (-1, 0)$

2 $(1, 1), (1, -1), (-1, 1), (-1, -1)$

3 $(2, 1), (2, -1), (-2, 1), (-2, -1)$

5 $(2, 1), (-2, 1), (2, -1), (-2, -1)$

6 $(1, 2), (1, -2), (-1, 2), (-1, -2)$

7 $(2, 3), (2, -3), (-2, 3), (-2, -3)$

9 $(1, 1), (1, -1), (-1, 1), (-1, -1)$

10 $(3, 0) (-3, 0)$

11 $(2, 3), (2, -3), (-2, 3), (-2, -3)$

13 $(2, 1), (2, -1), (-1/3, \sqrt{86}/3), (-1/3, -\sqrt{86}/3)$

14 $(2, 3), (-2, 3), (i\sqrt{11}/2, -3/2), (-i\sqrt{11}/2, -3/2)$

15 $(2, 1), (2, -1), (-11/3, \sqrt{26}/3), (-11/3, -\sqrt{26}/3)$

17 $(1, 2i), (1, -2i)$

18 $(3, 2), (3, -2), (2, 1), (2, -1)$

19 $(1, 1), (1, -1), (-1/2, \sqrt{29}i/2), (-1/2, -\sqrt{29}i/2)$

21 $(2, 5), (-2, 5), (\sqrt{61.19}/2, -43/10), (-\sqrt{61.19}/2, -43/10)$

22 $(2, 3), (-2, 3), (0, -1)$

23 $(2, 3), (2, -3), (-11/5, 3i\sqrt{3}/5), (-11/5, -3i\sqrt{3}/5)$

25 $(2, -2.5), (3, -1)$

27 $(1, -3), (-2.8, -134/7)$

30 $(-3, 2), (1, -2)$

26 $(1, 2), (-1/6, -3/2)$

29 $(1, 2), (-1/6, -3/2)$

31 $(2, 3), (-2, -1)$

EXERCISE 9.5

1 $(2, 5), (-2, 5)$

3 $(1, 3), (1, -3), (-1, 3), (-1, -3)$

6 $(4, 3), (-4, 3)$

9 $(10, 2), (-2, -2), (5, 3/2), (-4, 3/2)$

10 $(1, -2), (-1, -4), (\sqrt{7}, 3 + \sqrt{7}), (-\sqrt{7}, 3 - \sqrt{7})$

11 $(-16 - 2i\sqrt{5}, i\sqrt{5}), (-16 + 2i\sqrt{5}, -i\sqrt{5}), (2 + 2\sqrt{3}, \sqrt{3}), (2 - 2\sqrt{3}, -\sqrt{3})$

13 $(2, 3), (2, -3), (-2, 3), (-2, -3)$

15 $(2, 3), (-11/5, 3i\sqrt{3}/5), (-11/5, -3i\sqrt{3}/5), (2, -3)$

17 $(-3, 2), (1, -2)$

2 $(1, 6), (-4, 3)$

5 $(2, 3), (2, -3), (-2, 3), (-2, -3)$

7 $(0, 5), (-30/7, -55/7)$

14 $(\sqrt{5}, 1), (\sqrt{5}, -1), (-\sqrt{5}, 1), (-\sqrt{5}, -1)$

18 $(1, 2), (-1/6, -3/2)$

EXERCISE 10.1

1 $\frac{4}{1}$

6 $\frac{4}{1}$

11 $\frac{1}{4}$

15 3 hot dogs/boy

21 1.5

27 ± 5

34 27

41 $x = 15, y = 3$

46 12

2 $\frac{5}{1}$

7 $\frac{5}{1}$

13 47 miles/hour

17 .32

22 12

29 10

35 4.5

42 $x = 14, y = 7$

47 28 sacks

3 $\frac{8}{5}$

9 $\frac{33}{1}$

14 $1360/month

18 $\frac{8}{15}$

23 $\frac{8}{5}$

30 1

37 ± 3

43 $x = 10, y = 6$

5 $\frac{7}{2}$

10 $\frac{5}{8}$

19 $\frac{5}{12}$

25 2

31 21

38 ± 4

45 5 seconds

26 3

33 4

39 ± 18

EXERCISE 10.2

1 (a) $p = kq$ (b) $a = k/b$ (c) $x = kyz$ (d) $u = kw^2/v$

2 6

9 2.54 centimeters

13 1640.4 yards

17 453.6 grams

21 68.324 pounds

25 3.7852 liters

29 25°C

33 77°F

37 90°, 86°

41 5 square inches

3 30

10 15.24 centimeters

14 1609.3 meters

18 28.35 grams

22 .002204 pounds

26 1.057 quarts

30 −50/9°C

34 104°F

38 70°, 68°

42 7 grams

5 2

11 30.48 centimeters

15 109.4 yards

19 1360.8 grams

23 1.7357 pounds

27 14.19 liters

31 −40°C

35 212°F

39 40°, 41°

43 $121.50

6 120

7 112

45 2 liters

46 The second part requires $\frac{4}{5}$ as much as the first

47 205 cubic feet   **49** 3.5 inches

50 The safe load of the first is two-thirds that of the second

51 1.8 pounds

53 The second jet has four times the power of the first

54 10 inches   **55** 15 pounds   **57** 670

58 300 rpm   **59** 1/200 seconds

EXERCISE 11.1

1 1, 3, 5, 7, 9, 11   **2** 7, 5, 3, 1, −1   **3** 12, 9, 6, 3, 0, −3, −6

5 2, 6, 10, 14, 18   **6** 7, 6, 5, 4, 3, 2   **7** 4, 6, 8, 10, 12, 14, 16

9 $a_n = 13, s = 49$ **10** $a_n = 23, s = 100$ **11** $a_n = 4, s = 39$

13 $d = -5, s = 60$ **14** $d = 2, s = 80$ **15** $d = 3, s = 119$

17 $n = 8, s = -24$ **18** $n = 6, s = 156$ **19** $a_1 = 51, s = 156$

21 $n = 7, d = 6$ **22** $n = 7, d = -3$ **23** $a_1 = 17, d = -2$

25 $a_n = 13, d = 3$ **26** $a_n = 9, d = 1$ **27** $a_n = 2, d = -3$

29 $a_1 = 1, n = 7$ **30** $a_1 = 2, n = 8$ **31** $a_1 = 9, n = 6$

33 459   **34** 735   **35** 585

37 $84   **38** $19,700, $226,000   **39** 73.5

41 147   **42** All x **43** $\frac{3}{7}$

EXERCISE 11.2

1 1, 2, 4, 8, 16, 32   **2** 7, −14, 28, −56, 112

3 12, −36, 108, −324, 972, −2916   **5** 2, 6, 18, 54, 162

6 7, ±7, 7, ±7, 7, ±7   **7** ±3, 6, ±12, 24, ±48, 96, ±192

9 $a_n = 32. s = 62$ **10** $a_n = 486, s = 728$ **11** $a_n = 256, s = 341$

13 $n = 7, s = 43$ **14** $n = 6, s = -63$ **15** $n = 7, s = 21.5$

17 $r = \frac{1}{3}, s = 363$ **18** $r = -\frac{1}{4}, s = 3276.75$ **19** $a_1 = 256, s = 511$

21 $r = 2, n = 5$ **22** $r = 3, n = 6$ **23** $a_n = 256. n = 5$

25 $a_1 = 1, n = 7$ **26** $a_1 = 3, n = 6$ **27** $n = 7, a_n = 32$

29 $a_1 = 243, r = \frac{1}{3}$ **30** $a_1 = 4096, r = -\frac{1}{4}$ **31** $a_1 = 256, r = \frac{1}{2}$

33 Twentieth, twenty-first   **34** Two   **35** 531, 441n

37 2, 5, 8; 3   **38** 4, 8, 16; 4   **39** 2, $\frac{1}{2}$

EXERCISE 11.3

1 4, 8/3, 16/9, 32/27; 12   **2** 4, 4/3, 4/9, 4/27; 6   **3** 6, 12/5, 24/25, 48/125; 10

5 4   **6** 8   **7** 11   **9** $\frac{3}{2}$ **10** 12

11 7   **13** 8   **14** 8   **15** $\frac{81}{2}, \frac{81}{10}$ **17** $\frac{1}{3}$

18 $\frac{7}{9}$ **19** $\frac{5}{9}$ **21** $\frac{25}{33}$ **22** $\frac{4}{11}$ **23** $\frac{3}{11}$

25 $\frac{790}{37}$ **26** $\frac{1033}{333}$ **27** $\frac{1537}{666}$ **29** 1   **30** $\frac{1}{2}$

31 $\frac{1}{6}$ **33** 2   **34** $\frac{3}{4}$ **35** 4   **37** 33 feet

38 128 square inches   **39** $32\sqrt{2}(\sqrt{2}+1)$ inches   **41** $4800

42 80 feet   **43** $n < \frac{7}{3}$ **45** $1/x$

46 $\dfrac{1}{x-3}$ **47** $\dfrac{1}{2(x-1)}$

EXERCISE 11.4

1 Arithmetic progression, 17, 21   **2** Arithmetic progression, -10, -14

3 Arithmetic progression, 22, 27   **5** Geometric progression, 162, 486

6 Geometric progression, 48, 96   **7** Geometric progression, 16, -32

9 Geometric progression, 4, -2 **10** Geometric progression, 3, -1

11 Geometric progression, 1/5, 1/25   **13** Harmonic progression, 1/13, 1/16

14 Harmonic progression, 1/23, 1/28   **15** Harmonic progression, 6/11, 6/13

17 5, 7   **18** 4, 7, 10   **19** 5, 8, 11, 14

21 4 or -4 **22** 9, 27   **23** $\pm 1, 4, \pm 16$

25 $\pm 64, 32, \pm 16, 8, \pm 4$ **26** $\pm 54, 18, \pm 6$ **27** 625, 125, 25, 5

29 $\frac{1}{7}$ **30** $\frac{1}{16}$ **31** $\frac{1}{5}$

EXERCISE 11.5

1 2, 7/3, 8/3, 3, 10/3   **2** 9/16, 3/8, 1/4, 1/6, 1/9, 2/27

3 3/7, 1/3, 3/11, 3/13   **5** 9, 24

6 1, 0   **7** $-1, 1, 1/3, 1/5, 1/7, 1/9$ **9** $r = 2/3$, $a_1 = 3$, $a_n = 32/81$, $s = 665/81$ **10** 2

11 $-1/7, 2$ **13** 1125   **14** 126   **15** 9837   **17** 5/11   **23** $\pm 4, 8, \pm 16$

25 2/19, 1/17, 2/49   **26** Arithmetic progression, 15/4, 9/2

27 Geometric progression, 12, 24

EXERCISE 12.1

1 $a^5 + 5a^4b + 10a^3b^2 + 10a^2b^3 + 5ab^4 + b^5$

2 $b^7 + 7b^6c + 21b^5c^2 + 35b^4c^3 + 35b^3c^4 + 21b^2c^5 + 7bc^6 + c^7$

3 $c^4 + 4c^3d + 6c^2d^2 + 4cd^3 + d^4$

5 $b^7 - 7b^6d + 21b^5d^2 - 35b^4d^3 + 35b^3d^4 - 21b^2d^5 + 7bd^6 - d^7$

6 $a^4 - 4a^3c + 6a^2c^2 - 4ac^3 + c^4$

7 $a^6 - 6a^5b + 15a^4b^2 - 20a^3b^3 + 15a^2b^4 - 6ab^5 + b^6$

9 $16a^4 + 32a^3x + 24a^2x^2 + 8ax^3 + x^4$

10 $729b^6 + 1458b^5c + 1215b^4c^2 + 540b^3c^3 + 135b^2c^4 + 18bc^5 + c^6$

11 $32a^5 + 80a^4b + 80a^3b^2 + 40a^2b^3 + 10ab^4 + b^5$

13 $a^6 - 12a^5b + 60a^4b^2 - 160a^3b^3 + 240a^2b^4 - 192ab^5 + 64b^6$

14 $b^5 - 15b^4c + 90b^3c^2 - 270b^2c^3 + 405bc^4 - 243c^5$

15 $c^3 - 12c^2a + 48ca^2 - 64a^3$

17 $8a^3 + 36a^2b + 54ab^2 + 27b^3$

18 $27a^3 + 108a^2b + 144ab^2 + 64b^3$

19 $16a^4 + 160a^3b + 600a^2b^2 + 1000ab^3 + 625b^4$

21 $x^8 - 12x^6y + 54x^4y^2 - 108x^2y^3 + 81y^4$

22 $a^{18} - 12a^{15}y + 60a^{12}y^2 - 160a^9y^3 + 240a^6y^4 - 192a^3y^5 + 64y^6$

23 $b^{10} - 20b^8y + 160b^6y^2 - 640b^4y^3 + 1280b^2y^4 - 1024y^5$

25 $32a^5 + 400a^4y^2 + 2000a^3y^4 + 5000a^2y^6 + 6250ay^8 + 3125y^{10}$

26 $81a^4 + 432a^3y^3 + 864a^2y^6 + 768ay^9 + 256y^{12}$

27 $1024b^5 + 3840b^4y^4 + 5760b^3y^8 + 4320b^2y^{12} + 1620by^{16} + 243y^{20}$

29 $64x^{12} - 576x^{10}y^3 + 2160x^8y^6 - 4320x^6y^9 + 4860x^4y^{12} - 2916x^2y^{15} + 729y^{18}$

30 $243a^{15} - 1620a^{12}b^2 + 4320a^9b^4 - 5760a^6b^6 + 3840a^3b^8 - 1024b^{10}$

31 $64x^6 - 240x^4y^3 + 300x^2y^6 - 125y^9$

33 $x^5 + 10x^3 + 40x + 80x^{-1} + 80x^{-3} + 32x^{-5}$

34 $x^4 - 12x^2 + 54 - 108x^{-2} + 81x^{-4}$

35 $16x^4 - \frac{32}{3}x^2 + \frac{8}{3} - \frac{8}{27}x^{-2} + \frac{1}{81}x^{-4}$

37 $a^8 - 12a^5 + 54a^2 - 108a^{-1} + 81a^{-4}$

38 $b^{15} - 10b^{11} + 40b^7 - 80b^3 + 80b^{-1} - 32b^{-5}$

39 $x^6 - 6x + 12x^{-4} - 8x^{-9}$ **41** 1.125508   **42** 1.27625   **43** 1.50036

45 0.90392   **46** 0.78272   **47** 0.8145

49 $a^{10} + 20a^9b + 180a^8b^2 + 960a^7b^3$

50 $4096a^{12} - 24{,}576a^{11}b + 67{,}584a^{10}b^2 - 112{,}640a^9b^3$

51 $x^9 - 27x^8y + 324x^7y^2 - 2268x^6y^3$ **53** $x^{28} - 14x^{25} + 91x^{22} - 364x^{19}$

54 $x^{27} + 9x^{22} + 36x^{17} + 84x^{12}$ **55** $x^{16} - 4x^8 + 6 - 4x^{-8}$

57 $672x^6y^3$ **58** $1001(2^{10})x^{10}y^4$ **59** $-77(3^7)(2^6)x^6a^5$ **61** $-40(3^8)x^6y^7$

62 $560x^{12}y^3$ **63** $84a^2x^{10}$ **65** $5670a^4y^4$ **66** $-160a^3b^3$

67 $600x^2y^2$ **69** $2000x^3y^2$ **70** $5103x^2y^5$ **71** $4860x^4y^4$

EXERCISE 13.1

1 $3.284(10^3)$ **2** $7.315(10^2)$ **3** $6.85713(10^5)$ **5** $5.984(10^{-1})$

6 $4.862(10^{-3})$ **7** $1.405(10^{-4})$ **9** $8.47(10^3)$ **10** $8.470(10^3)$

11 $2.1000(10^4)$ **13** $6.480(10^3)$ **14** $6.48(10^3)$ **15** $2.786(10^6)$

17 48.46; 48.5   **18** 873.8; 874   **19** 27.16; 27.2

21 697,800; 698,000   **22** 89,410; 89,400   **23** 500,800; 501,000

25 8969; 8970   **26** 407.6; 408   **27** 237,600; 238,000

29 36.35; 36.4   **30** 36.35; 36.3   **31** 80.26; 80.3

33 9.7	**34** 6.2	**35** $1.8(10^4)$	**37** $2.2(10^4)$	**38** $2.2(10)$
39 44.8	**41** 3.0	**42** 2.1	**43** 21	**45** $5.0(10)$
46 .25	**47** 3.5	**49** 8.2	**50** 10.9	**51** 10.83
53 1.7	**54** 10.07	**55** 25.3		

EXERCISE 13.2

1 $\log_2 8 = 3$	**2** $\log_3 9 = 2$	**3** $\log_5 625 = 4$	**5** $\log_5 \frac{1}{125} = -3$
6 $\log_7 \frac{1}{49} = -2$	**7** $\log_2 \frac{1}{16} = -4$	**9** $\log_8 4 = \frac{2}{3}$	**10** $\log_4 8 = \frac{3}{2}$
11 $\log_{16} 8 = \frac{3}{4}$	**13** $\log_{1/2} 8 = -3$	**14** $\log_{1/4} 16 = -2$	**15** $\log_{1/3} 243 = -5$
17 $3^4 = 81$	**18** $2^3 = 8$	**19** $5^2 = 25$	**21** $2^{-4} = \frac{1}{16}$
22 $5^{-3} = \frac{1}{125}$	**23** $7^{-2} = \frac{1}{49}$	**25** $8^{2/3} = 4$	**26** $36^{3/2} = 216$
27 $32^{3/5} = 8$	**29** $(\frac{1}{3})^{-2} = 9$	**30** $(\frac{1}{4})^{-3} = 64$	**31** $(\frac{1}{2})^{-4} = 16$

33 3	**34** 4	**35** 2	**37** $\frac{3}{2}$	**38** $\frac{3}{4}$	**39** $\frac{3}{2}$
41 8	**42** 25	**43** 81	**45** 9	**46** 8	**47** 27
49 5	**50** 3	**51** 2	**53** 8	**54** 9	**55** 64

57 1	**58** .7781	**59** 1.1761	**61** $-.2219$	**62** $-.1761$
63 .8751	**65** .9030	**66** .9542	**67** 2.7960	**69** 1.1582
70 3.9263	**71** -1.9209			

EXERCISE 13.3

1 1.5877	**2** .7752	**3** 2.9079	**5** 2.8639
6 .9965	**7** 1.6096	**9** $9.3909 - 10$	**10** $8.9074 - 10$
11 $7.7251 - 10$	**13** $8.0043 - 10$	**14** $9.6785 - 10$	**15** $8.5172 - 10$
17 1.6929	**18** 2.8939	**19** .9923	**21** $9.9093 - 10$
22 $8.7910 - 10$	**23** $6.3923 - 10$	**25** 21.2	**26** 701
27 4.52	**29** 7.04	**30** 25.6	**31** 138
33 .822	**34** .0386	**35** .00236	**37** .00731
38 .836	**39** .0375	**41** 1.6554	**42** 2.8930
43 .3090	**45** $9.5116 - 10$	**46** $8.7826 - 10$	**47** $9.9141 - 10$
49 27.74	**50** 1408	**51** 3.778	**53** .5173
54 .01716	**55** .005994		

EXERCISE 13.4

1 117	**2** 90.9	**3** 75.4	**5** 546	**6** 271
7 $2.44(10^4)$	**9** 1.18	**10** 13.4	**11** 12.5	**13** 8.07
14 1.41	**15** 1.34	**17** .367	**18** .514	**19** .611
21 348	**22** 661	**23** 674	**25** 2.46	**26** 4.69

27	29.6	29	1.99	30	5.01	31	4.46	33	.934
34	.903	35	.985	37	3.86	38	1.08	39	1.62
41	$9.398(10^4)$	42	$4.083(10^3)$	43	$2.42(10^3)$	45	1.63	46	4.023
47	.711	49	1.044	50	2.161	51	.876		

EXERCISE 13.5

1 $\dfrac{\log 2}{\log 1.5} = 1.710$

2 $\dfrac{\log 3}{\log 1.8} = 1.869$

3 $\dfrac{\log \dfrac{125}{7}}{\log \dfrac{25}{7}} = 2.264$

5 $\dfrac{\log 3 + 3 \log 11}{4 \log 11 - 3 \log 3} = 1.317$

6 $\dfrac{\log 5}{\log 125 - \log 27} = 1.050$

7 $\dfrac{\log 16 - \log 7}{\log 64 - \log 49} = 3.096$

9	1	10	2	11	3	13	4	14	2	15	-1
17	3	18	2	19	1	21	$5, -2$	22	1	23	$-3, 5$

25 $x = 1.873, y = 1.242$

26 $x = 3, y = -2.093$

27 $x = .9771, y = 1.033$

29 $x = 1.229, y = -.8726$

30 $x = 2.714, y = -2.083$

31 $x = .2805, y = -.2719$

EXERCISE 13.6

1	$7.86(10^2)$	2	$7.87(10^2)$	3	$5.72(10^3)$	5	$3.7(10^2)$
6	$2.0(10^3)$	7	$6.80(10)$	9	5.2	10	$5.62(10)$
11	$\log_3 81 = 4$	13	3	14	3	15	216
17	$1.21(10^5)$	18	490	19	.504	21	.279
22	1.869	23	1.653	25	$x = .535, y = .878$		

EXERCISE 14.1

1	$x > 2$	2	$x > 2$	3	$x \geq -3$	5	$x < 1$	6	$x \leq -7$
7	$x < 2$	9	$x \geq -1$	10	$x > 3$	11	$x > 4$	13	$x < -1$
14	$x < 1$	15	$x \leq 1.5$	17	$x < 2$	18	$x \leq -3$	19	$x < 3$
21	$x \geq -3$	22	$x > 2$	23	$x > 2$	25	$x < 6$	26	$x < 8$
27	$x \leq 12$								

EXERCISE 14.2

1 $\{x \mid x < -3\} \cup \{x \mid x > 2\}$

2 $\{x \mid x < 1\} \cup \{x \mid x > 4\}$

3 $\{x \mid x < -4\} \cup \{x \mid x > -1\}$

5 $\{x \mid x > 1\} \cap \{x \mid x < 4\}$, that is, $\{x \mid 1 < x < 4\}$

6 $\{x \mid -1 < x < 3\}$

7 $\{x \mid -3 < x < 2\}$

9 $\{x \mid x < -\frac{5}{2}\} \cup \{x \mid x > 1\}$

10 $\{x \mid x < -3\} \cup \{x \mid x > -\frac{1}{3}\}$

11 $\{x \mid x < -\frac{2}{3}\} \cup \{x \mid x > \frac{1}{2}\}$

13 $\{x \mid -\frac{5}{8} < x < \frac{8}{5}\}$

14 $\{x \mid -\frac{7}{2} < x < \frac{4}{3}\}$

15 $\{x \mid -\frac{7}{3} < x < -\frac{3}{7}\}$

17 $\{x \mid x > \frac{3}{2}\} \cup \{x \mid -\frac{5}{3} < x < 1\}$

18 $\{x \mid x > \frac{4}{3}\} \cup \{x \mid -2 < x < -\frac{9}{5}\}$

19 $\{x \mid x > \frac{1}{3}\} \cup \{x \mid -\frac{5}{2} < x < -\frac{7}{5}\}$

21 $\{x \mid x < -\frac{9}{4}\} \cup \{x \mid -\frac{3}{5} < x < \frac{8}{3}\}$

22 $\{x \mid x < -\frac{2}{3}\} \cup \{x \mid \frac{6}{5} < x < \frac{7}{2}\}$

23 $\{x \mid x > \frac{2}{3}\} \cup \{x \mid -4 < x < -\frac{5}{2}\}$

25 $\{x \mid x > \frac{3}{2}\} \cup \{x \mid -3 < x < \frac{1}{3}\}$

26 $\{x \mid x < -\frac{5}{3}\} \cup \{x \mid -\frac{3}{2} < x < \frac{8}{7}\}$

27 $\{x \mid x < -\frac{1}{2}\} \cup \{x \mid 2 < x < \frac{7}{2}\}$

EXERCISE 14.3

1 $-1 < x < 4$

2 $-\frac{5}{3} < x < 3$

3 $-\frac{8}{5} < x < 2$

5 $-\frac{7}{2} < x < 0$

6 $-\frac{11}{3} < x < 2$

7 $-\frac{10}{3} < x < 2$

9 $-\frac{1}{7} < x < 1$

10 $-6 < x < 1$

11 $2 < x < \frac{10}{3}$

13 $\{x \mid x < -\frac{16}{3}\} \cup \{x \mid x > 4\}$

14 $\{x \mid x < -5\} \cup \{x \mid x > 2\}$

15 $\{x \mid x < -8\} \cup \{x \mid x > 3\}$

17 $\{x \mid x < -\frac{3}{2}\} \cup \{x \mid x > 2\}$

18 $\{x \mid x < -\frac{2}{3}\} \cup \{x \mid x > 4\}$

19 $\{x \mid x < -\frac{6}{7}\} \cup \{x \mid x > 2\}$

21 $\{x \mid x < -1\} \cup \{x \mid x > \frac{11}{5}\}$

22 $\{x \mid x < 2\} \cup \{x \mid x > \frac{5}{2}\}$

23 $\{x \mid x < -3\} \cup \{x \mid x > -2\}$

25 $\{x \mid 3 < x < 6\}$

26 $\{x \mid 3 < x < 6\}$, same as **25**

27 $\{x \mid x < -4\} \cup \{x \mid x > \frac{2}{3}\}$

EXERCISE 14.4

21 $(0, 3), (1, -2), (3, 2)$

22 $(1, 3), (4, 1), (2, -2)$

23 $(-3, 2), (3, 1), (2, -1)$

25 $(0, 3), (2, 1), (0, -2), (-3, 0)$

26 $(-2, 0), (3, 2), (3, -3), (0, -4)$

27 $(-1, 0), (0, -1), (2, 0), (3, 1), (0, 2)$

EXERCISE 14.5

1 Maximum of 38 occurs at $(4, 7)$; minimum of 9 occurs at $(1, 3)$

2 Maximum of 18 occurs at $(2, 4)$; minimum of 5 occurs at $(1, 1)$

3 Maximum of 27 occurs at $(5, 2)$; minimum of 5 occurs at $(3, 4)$

5 Maximum of 27 occurs at $(1, 5)$; minimum of 8 occurs at $(2, 1)$

6 Maximum of 101 occurs at $(11, 2)$; minimum of 41 occurs at $(4, 3)$

7 Maximum of 80 occurs at $(5, 7)$; minimum of 21 occurs at $(2, 2)$

9 3 at $(1, 2)$, 18 at $(7, 5)$

10 -14 at $(-3, 0)$ $\frac{45}{7}$ at $(\frac{1}{7}, \frac{11}{7})$

11 14 at $(0, 3)$, 43 at $(5, 4)$

13 -1 at $(1, 0)$, 20 at $(3, 3)$

14 -5 at $(0, 0)$, 41 at $(8, 2)$

15 $\frac{74}{5}$ at $(\frac{4}{5}, 1)$, 46 at $(5, 3)$

17 10 of type A, 35 of type B, \$240

18 10 of type A, 35 of type B, \$260

19 8 of type A, 36 of type B, \$224

21 25 acres of wheat, 70 acres of corn, \$2830

22 $\frac{75}{2}$ acres of wheat, $\frac{415}{7}$ acres of corn, \$2919.29

23 25 acres of wheat, 70 acres of corn, $3120

25 26 acres of lettuce, 4 acres of pepper, 17 acres of tomatoes, $17,550

26 26 acres of lettuce, 4 acres of pepper, 17 acres of tomatoes; $18,825

27 22 acres of lettuce, 8 acres of pepper, 17 acres of tomatoes; $17,150

EXERCISE 14.6

1 $x > 2$

2 $x < 2$

3 $\{x \mid x > 2\} \cup \{x \mid x < -7\}$

5 $x > -2$

6 $-2 < x < \frac{12}{5}$

7 $\{x \mid x > 1\} \cup \{x \mid x < -\frac{3}{2}\}$

9 $\{x \mid x > \frac{5}{2}\} \cup \{x \mid x < -3\}$

10 $-3 < x < \frac{5}{2}$

11 $-\frac{2}{3} < x < 4$

13 $\{x \mid x > 3\} \cup \{x \mid -\frac{3}{2} < x < 1\}$

14 $\{x \mid x < -\frac{9}{4}\} \cup \{x \mid -1 < x < \frac{3}{2}\}$

15 $\{x \mid x < \frac{1}{2}\} \cup \{x \mid 2 < x < 3\}$

23 $(2, 5), (3, 2), (1, 0)$

25 Minimum of 10 occurs at $(3, -1)$; maximum of 53 occurs at $(8, 5)$

26 Minimum of 23 occurs at $(1, 2)$; maximum of 141 occurs at $(9, 4)$

27 Five of type A, 16 of type B

29 Nine of type A, 8 of type B

index